RNA Infrastructure and Networks

ADVANCES IN EXPERIMENTAL MEDICINE AND BIOLOGY

Recent Volumes in this Series

Volume 713
CELL FUSION IN HEALTH AND DISEASE, I: CELL FUSION IN HEALTH
Thomas Dittmar

Volume 714
CELL FUSION IN HEALTH AND DISEASE, II: CELL FUSION IN DISEASE
Thomas Dittmar

Volume 715
BACTERIAL ADHESION: BIOLOGY, CHEMISTRY AND PHYSICS
Dirk Linke and Adrian Goldman

Volume 716
MAST CELL BIOLOGY: CONTEMPORARY AND EMERGING TOPICS
Alasdair M. Gilfillan and Dean D. Metcalfe

Volume 717
KAINATE RECEPTORS: NOVEL SIGNALING INSIGHTS
Antonio Rodriguez-Moreno and Talvinder S. Sihra

Volume 718
BRAIN INSPIRED COGNITIVE SYSTEMS 2010
Ricardo Sanz, Jaime Gomez and Carlos Hernandez

Volume 719
HOT TOPICS IN INFECTION AND IMMUNITY IN CHILDREN VIII
Nigel Curtis

Volume 720
HUMAN CELL TRANSFORMATION
Johng S. Rhim and Richard Kremer

Volume 721
SPHINGOLIPIDS AND METABOLIC DISEASE
L. Ashley Cowart

Volume 722
RNA INFRASTRUCTURE AND NETWORKS
Lesley J. Collins

RNA Infrastructure and Networks

Edited by

Lesley J. Collins
Institute of Fundamental Sciences, Massey University, Palmerston North, New Zealand

Springer Science+Business Media, LLC

Landes Bioscience

Springer Science+Business Media, LLC
Landes Bioscience

Printed in the USA.

Springer Science+Business Media, LLC, 233 Spring Street, New York, New York 10013, USA
http://www.springer.com

Please address all inquiries to the publishers:
Landes Bioscience, 1806 Rio Grande, Austin, Texas 78701, USA
Phone: 512/ 637 6050; FAX: 512/ 637 6079
http://www.landesbioscience.com

The chapters in this book are available in the Madame Curie Bioscience Database.
http://www.landesbioscience.com/curie

RNA Infrastructure and Networks, edited by Lesley J. Collins. Landes Bioscience / Springer Science+Business Media, LLC dual imprint / Springer series: Advances in Experimental Medicine and Biology.

ISBN: 978-1-4614-0331-9

Library of Congress Cataloging-in-Publication Data

RNA infrastructure and networks / edited by Lesley J. Collins.
p. ; cm. -- (Advances in experimental medicine and biology ; v. 722)
Includes bibliographical references and index.
ISBN 978-1-4614-0331-9
1. RNA. I. Collins, Lesley J., 1968- II. Series: Advances in experimental medicine and biology ; v. 722.
0065-2598
[DNLM: 1. RNA. W1 AD559 v.722 2011 / QU 58.7]
QP623.R5727 2011
579.2'5--dc23

2011018033

DEDICATION

This book is dedicated to all my family, especially Maurice and Shannen, the central hub of my RNA world.

PREFACE

Previously, RNA was investigated merely as an intermediate between DNA and proteins. Studies of regulatory pathways led to the discovery that small RNAs are major regulators in animal cells, and other non-coding RNAs responsible for RNA cleavage and modification. Now studies reveal that small RNA regulation exists throughout eukaryotes and prokaryotes, and long non-coding RNAs exert a large influence over epigenetics. We can also see that the RNA components of the cell are not merely 'fixed accidents' of an ancient RNA world, but instead are continuing to evolve and affect the metabolism of all life.

RNAs form complexes with proteins and other RNAs. The RNA-infrastructure represents the spatiotemporal interaction of these proteins and RNAs in a cell-wide network. This volume brings together these ideas to illustrate the scope of RNA-based biology, and how connecting RNA mechanisms is a powerful tool to investigate regulatory pathways. The first chapter is an introduction to the RNA-infrastructure and how RNAs and proteins interact in networks. Following this is a chapter on RNA interactions, explaining how RNA folds upon itself, and then interacts with other RNAs as well as DNAs and proteins.

The second section focuses on RNA interference (RNAi), where siRNA is shown to be important in defending plants against viruses. We then explore how miRNAs act in the regulation of many gene networks, then the spatiotemporal aspects of miRNA-mediated gene regulation. This networking is further investigated with two specific examples; (i) spliceosomal RNA infrastructure and (ii) RNA-binding proteins, expanding on traditional analysis of protein-protein interaction networks to include regulatory RNAs as well as other interacting RNAs.

The third section describes some of the lesser understood RNA mechanisms. First we examine how post-transcriptional control is reinforced by RNA protein complexes (RNPs), responsible for controlling the abundance of gene expression during development. The next chapter summarizes how non-random tRNA fragments can guide mRNA cleavage, inhibit translation and promote morphological changes. A chapter on programmed DNA elimination in the protozoan ciliate *Tetrahymena* then describes a mechanism for RNAi-directed heterochromatin formation. Long non-coding RNAs are next introduced in their context of epigenomics, and a promoter-associated long non-coding RNA that binds to the TLS protein, is examined in more detail.

RNA networks are also found throughout prokaryotes, and the fourth section reviews prokaryotic viral defence (CRISPRs), regulation (riboswitches and small RNAs) and other RNAs surrounding tRNA processing, as well as the spatiotemporal nature of the prokaryotic RNA infrastructure.

The last section reviews some of the latest technology that has greatly increased our knowledge of RNAs including small RNA discovery using high-throughput approaches. We conclude by discussing evolutionary aspects of RNA networking and examining the mechanisms we see today, how such RNA-mechanisms evolved and whether some date back to the ancient RNA world.

This work is but a taste of the wide range of RNA-based mechanisms that connect in the RNA infrastructure. We have only begun to untangle this complex cellular web, and have still an awful lot to learn about the cell.

Lesley J. Collins

ABOUT THE EDITOR...

Photo Credit : Robin Atherton, New Zealand.

LESLEY J. COLLINS is a Senior Research Fellow in the Institute of Fundamental Sciences at Massey University, New Zealand. With an early background in molecular biology she now uses that knowledge in evolutionary genomics and bioinformatics. Her research interests focus on the evolution of RNA networks, especially on how they relate to the evolution of ancient eukaryotes. With the rise of new sequencing technology she was involved in the establishment of next generation sequencing at Massey University. In her spare time she is a bioinformatics consultant, a practitioner of Tai Chi, a knitter and an avid reader.

PARTICIPANTS

Patrick J. Biggs
Institute of Veterinary, Animal and Biomedical Sciences
Massey University
Palmerston North
New Zealand

Věra Čapková
Laboratory of Pollen Biology
Institute of Experimental Botany AS CR
Rozvojová
Czech Republic

X. Sylvia Chen
Department of Biochemistry
University of Otago
Dunedin
New Zealand

Lesley J. Collins
Institute of Fundamental Sciences
Massey University
Palmerston North
New Zealand

Tamas Dalmay
School of Computing Sciences
University of East Anglia
Norwich
UK

Toni Daly
Allan Wilson Centre for Molecular Ecology and Evolution
Massey University
Palmerston North
New Zealand

Said Hafidh
Laboratory of Pollen Biology
Institute of Experimental Botany AS CR
Rozvojová
Czech Republic

David Honys
Laboratory of Pollen Biology
Institute of Experimental Botany AS CR
Rozvojová
and
Department of Plant Experimental Biology
Faculty of Science
Charles University in Prague
Viničná
Czech Republic

Rebecca L. Hurto
Department of Molecular Genetics
The Ohio State University
Columbus, Ohio
USA

Sarath Chandra Janga
MRC Laboratory of Molecular Biology
Cambridge
UK
and
Institute of Genomic Biology
University of Illiniois at Urbana-Champaign
Urbana, Illinois
USA

Chandrasekhar Kanduri
Department of Genetics and Pathology
Uppsala University
Uppsala
Sweden

Kensuke Kataoka
Institute of Molecular Biotechnology of the Austrian Academy of Sciences
Vienna
Austria

Kenneth C. Keiler
Department of Biochemistry and Molecular Biology
Penn State University
Althouse Laboratory
University Park, Pennslyvania
USA

Riki Kurokawa
Division of Gene Structure and Function
Research Center for Genomic Medicine
Saitama Medical University
Yamane, Hidaka-shi, Saitama-Ken
Japan

Manja Marz
Bioinformatics Group
Department of Computer Science and Interdisciplinary Center for Bioinformatics
University of Leipzig
Leipzig
and
RNA-Bioinformatics Group
Institut für Pharmazeutische Chemie
Philipps-Universität Marburg
Marburg
Germany

Nitish Mittal
Department of Biotechnology
National Institute of Pharmaceutical Education and Research
SAS Nagar
Punjab
India

Kazufumi Mochizuki
Institute of Molecular Biotechnology of the Austrian Academy of Sciences
Vienna
Austria

Vincent Moulton
School of Computing Sciences
University of East Anglia
Norwich
UK

Simon Moxon
School of Computing Sciences
University of East Anglia
Norwich
UK

Helio Pais
School of Computing Sciences
University of East Anglia
Norwich
UK

Vitantonio Pantaleo
Istituto di Virologia Vegetale del CNR
Torino
Italy

David Penny
Institute of Molecular BioSciences
Massey University
Palmerston North
New Zealand

Joris Pothof
Department of Genetics
Erasmus MC
University Medical Center
Netherlands

Peter F. Stadler
Bioinformatics Group
Department of Computer Science and Interdisciplinary Center for Bioinformatics
University of Leipzig
Leipzig
Germany;
Institute for Theoretical Chemistry
University of Vienna
Wien
Austria;
Fraunhofer Institute for Cell Therapy and Immunology
Leipzig
Germany;
and
Santa Fe Institute
Santa Fe, New Mexico
USA

Swathi Vallabhaneni
Institute of Genomic Biology
University of Illiniois at Urbana Champaign
Urbana, Illinois
USA

Dik C. van Gent
Department of Genetics
Erasmus MC
University Medical Center
Rotterdam
Netherlands

CONTENTS

ACKNOWLEDGEMENTS

I would like to acknowledge the wonderful assistance from David Penny for mentoring, highly stimulating conversation and funding to delve deep into the Pandora's Box we called the RNA infrastructure. This work was also assisted greatly by the Institute of Fundamental Sciences at Massey University. Not in the very least, my greatest thanks to all the colleagues who contributed, reviewed and revised chapters to help me bring together this book that highlights our modern RNA World.

CHAPTER 1

THE RNA INFRASTRUCTURE: An Introduction to ncRNA Networks

Lesley J. Collins
Institute of Fundamental Sciences, Massey University, Palmerston North, New Zealand
Email: l.j.collins@massey.ac.nz

Abstract: The RNA infrastructure connects RNA-based functions. With transcription-to-translation processing forming the core of the network, we can visualise how RNA-based regulation, cleavage and modification are the backbone of cellular function. The key to interpreting the RNA-infrastructure is in understanding how core RNAs (tRNA, mRNA and rRNA) and other ncRNAs operate in a spatial-temporal manner, moving around the nucleus, cytoplasm and organelles during processing, or in response to environmental cues. This chapter summarises the concept of the RNA-infrastructure, and highlights examples of RNA-based networking within prokaryotes and eukaryotes. It describes how transcription-to-translation processes are tightly connected, and explores some similarities and differences between prokaryotic and eukaryotic RNA networking.

INTRODUCTION

RNA biology in both eukaryotes and prokaryotes exists in a spatiotemporal network of processes we call the RNA-infrastructure. In eukaryotes, there are numerous subtypes of noncoding (nc) RNA genes involved including rRNA, mRNA, tRNA, snRNA, snoRNAs, several classes of regulatory RNAs (RNAi) and many long ncRNAs. In prokaryotes, in addition to tRNAs, mRNAs and rRNAs, we can have small RNAs, CRISPRs and tmRNAs, and even viruses can contain small RNAs. ncRNAs are generally involved in the transcription-to-translation processes surrounding the conversion and regulation of information from DNA to protein, implicated in viral defence mechanisms, or are involved in gene regulation (e.g., RNA interference; RNAi). What we are only beginning

RNA Infrastructure and Networks, edited by Lesley J. Collins.

to understand is how these processes are integrated, and how RNA plays a previously understated role in the overall regulation of the cell.

There are some key differences in cells that are differentiated (i.e., from multicellular eukaryotes), single celled eukaryotes and prokaryotes, but there are also striking similarities in how RNA processing and regulation works in different types of cells, giving us clues to their evolution. Although we are more familiar with RNA networks from eukaryotes, prokaryotic noncoding RNA research is using concepts developed from eukaryotic work to discover new RNA-based systems in bacteria and archaea. Although finding RNA genes is becoming a standard step in genomic investigations, it is now clear that discovering connections between these genes, and their associated proteins is just as important. Once we add in regulatory and epigenetic elements (such as methylation and histone modification) our regulatory networks can become very complex, but these complex networks have the ability to indicate linkages between cellular machineries not previously observed. The examples in this chapter will show how RNA-based processes within both prokaryotic and eukaryotic cells interact in networks in both a spatial and temporal manner.

RNAs PROCESSING OTHER RNAs

A good example in how RNA-processes are connected comes from examining the transcription-to-translation processes which form the core of the RNA-infrastructure (Fig. 1).[1] The processing of the three core-RNAs in eukaryotes (mRNA, tRNA and rRNA) includes the RNA-based mechanisms of RNA cleavage and modification (Fig. 1A). In eukaryotes these are: rRNA by RNase MRP and snoRNAs; tRNA by RNaseP; and mRNAs spliced by snRNAs within the spliceosome. We can then expand this idea to include spatial movement and regulation during RNA-processing (Fig. 1B). In prokaryotes we still have tRNAs and rRNAs being processed either directly by RNAs (e.g., RNase P of tRNA and tmRNA) or indirectly (where rRNAs are released by tRNA processing) (Fig. 1C).

Examining the connections between these processes in more detail we see networking between different mRNA machineries. For example, transcription by RNA Polymerase II (Pol II) and mRNA splicing in mammals are carried out in close proximity,[2] and this coupling may protect the newly synthesised RNA from degradation[3] before the termination of transcription.[4,5] Some splicing may occur cotransciptionally and this significantly improves processing efficiency (reviewed in ref. 6). At the other end of the transcript, 3′-end cleavage and polyadenylation of mRNA can be promoted by splicing proteins (e.g., U2AF65 reviewed in ref. 7). It is clear that splicing (the processing of mRNA with snRNAs) connects to other mRNA-processes including RNA localisation, translational yield and mRNA decay.[8] In another example, the Exon Junction Complex (EJC) is a set of proteins deposited on 5′ end of the exon during the second step of the splicing cycle, and remain bound to the spliced mRNA as it is exported to the cytoplasm.[9] This complex interacts transiently with many factors that connect the mRNA to the downstream RNA processing network,[10,11] as it is a major link between mRNA-splicing and mRNA export, as well as having a potential role in RNA degradation. The EJC appears to relay the previous location of introns,[8] and thus detects incorrect splicing that introduces premature stop-codons. It has been shown that in mammals at least, spliceosomal proteins and especially those involved in exon-definition, remain associated with the pre-mRNA to be available for the splicing of the next introns. This allows for efficient splice site recognition for subsequent introns since splice site recognition only needs to be carried out once for a site.[6] With splicing central to downstream

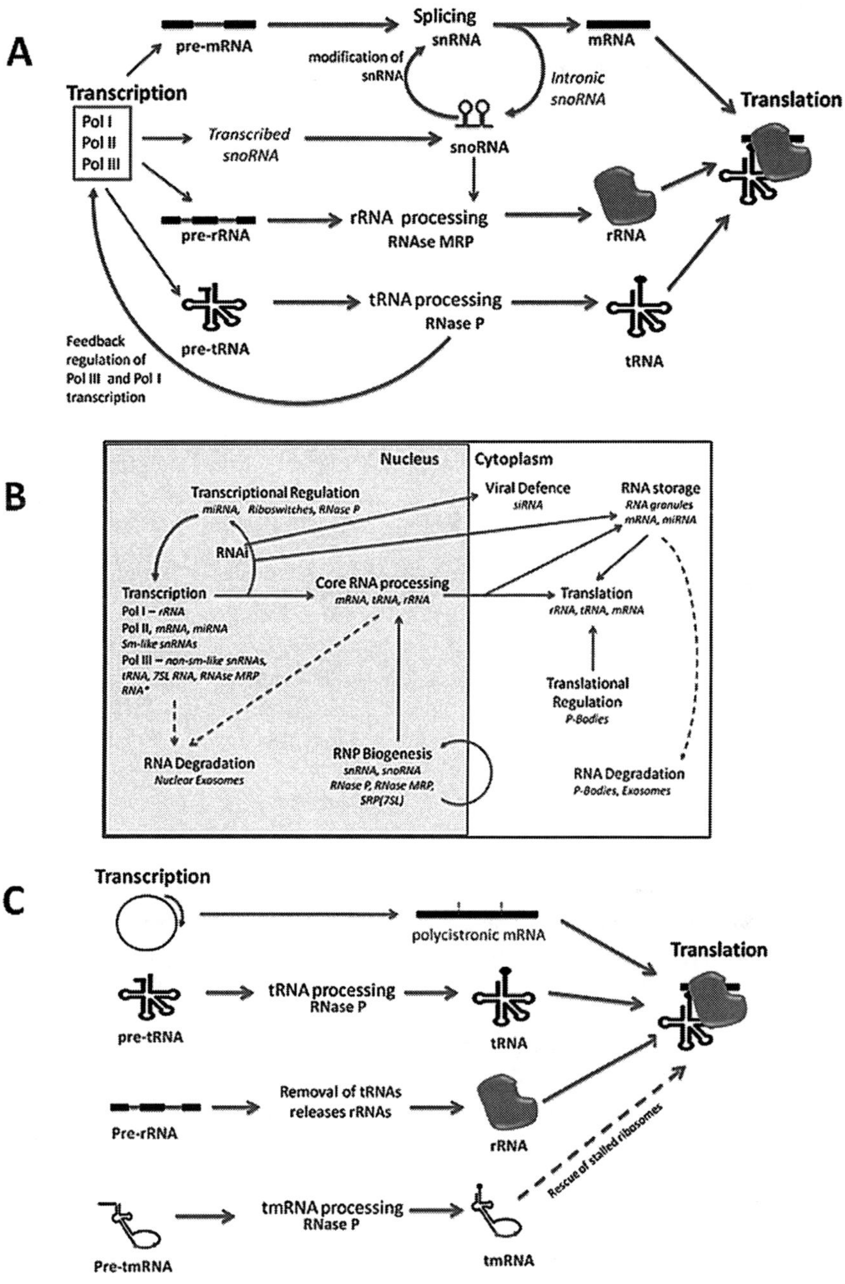

Figure 1. RNAs processing other RNAs are the key feature of the RNA-infrastructure. A) In transcription-to-translation machineries in eukaryotes, ncRNAs are involved in the processing of mRNA, tRNA and rRNA. B) In eukaryotes, compartmentalisation and biogenesis pathways permit regulation of these processes in the RNA infrastructure. *MRP RNA may not be in all eukaryotes. C) In prokaryotes in general, there is still processing of tRNAs and rRNAs but less mRNA processing. A and B adapted from[1] with permission from authors. A color version of this figure is available at www.landesbioscience.com/curie.

RNA processing it is not surprising that many proteins are now seen as having roles in splicing as well as their own function (e.g., transcription or capping). However, it remains to be seen whether these proteins actually influence catalysis in the spliceosome, or are detected due to the close proximity of these RNA processing complexes.

Similarly, transcribed pre-tRNAs require processing before being able to function as amino acid transfer molecules for translation. Leader sequences at the 5′ and 3′ ends of the pre-tRNAs require cleaving, introns within the tRNA may need to be removed and in some cases a 3′ CCA tail needs to be added.[12] In addition, certain nucleotides within the tRNA require modification by aminoacylation. The ribonucleoprotein RNase P is responsible for cleaving the 5′ leader sequence of pre-tRNAs in all cells, although the overall structure of this protein-RNA complex differs in eukaryotes, bacteria and archaea. In bacteria there is one small protein that plays diverse roles such as enhancing substrate binding, altering substrate recognition, stabilising RNA conformation, and aiding catalysis by discriminating between the substrate and product by binding to the 5′ leader sequence of the pre-tRNA.[13,14] Eukaryotes have 9-10 proteins in the complex with a single RNA. Archaeal RNase P also has multiple proteins (five including the ribosomal protein L7Ae) which do show some homology to some of the eukaryotic RNase P proteins. The RNase P RNA from some representatives from each kingdom can be induced to perform weak catalysis without its accompanying proteins, but only with high salt and high cation conditions in vitro (summarized in ref. 15).

RNase P plays key networking roles in both the eukaryote's and prokaryote's RNA infrastructure, resulting in the cleavage of additional substrates and the repression of transcription (Fig. 2).[16,17] In bacteria, as well as cleaving the 5′ leader sequence of tRNAs, it cleaves a similar leader sequence for tmRNA. tmRNA (transfer-messenger RNA) is a specialised tRNA molecule that together with the SmpB protein (small protein B) rescues stalled ribosomes in a process called trans-translation (reviewed in ref. 18).With a structure partly a tRNA molecule and partly an mRNA molecule,[19] the tRNA part binds to the stalled ribosome, allows the translation to proceed along the mRNA part which encodes a distinctive degradation signal and a translation stop signal. When the mistranslated protein is released after the stop signal it is targeted for degradation. This process of trans-translation is conserved throughout bacteria and is also present in some mitochondria and chloroplasts.[20,21] In prokaryotes other cleavage products by RNase P include some riboswitches and some viral RNAs as well as the 4.5S rRNA which is part of the Signal Recognition Particle involved in post-translational transport (reviewed in ref. 22).The eukaryotic counterpart of the 4.5S rRNA (7SL RNA) is also cleaved by RNase P (reviewed in ref. 13). In yeast the HRA1RNA and some C/D box snoRNAs are processed by RNase P although whether cleavage is the exact mechanism is yet to be completely determined. In humans MALAT1, another long ncRNA, is cleaved by RNase P. tRNAs in organelles within eukaryotes (in some species) either encode their own RNase P RNA (e.g., the yeast *S. cerevisiae*), use the nuclear counterpart (e.g., humans) or occasionally do without the RNA component altogether.[23] Additionally, in the archaeans *Nanoarchaeum equitans and Pyrobaculum aerophilum* and the hyperthermophilic bacterium *Aquifex aeolicus*, there does not appear to be any RNase P-like RNA sequence in their genomes.[24] In *N. equitans* the requirement for RNase P has been replaced by a strict placement of the promoter 26 nucleotides upstream of the mature tRNA sequence allowing transcription of leaderless tRNAs.[25]

There is a feedback affect of RNase P on its own polymerase RNA Pol III[16] and the polymerase affecting rRNA transcription, RNA Pol I.[17] Thus, RNase P in eukaryotes has a large effect on other aspects of RNA processing including splicing (U6 snRNA is transcribed by RNA Pol III), and RNA modification (by some yeast C/D box snoRNAs).

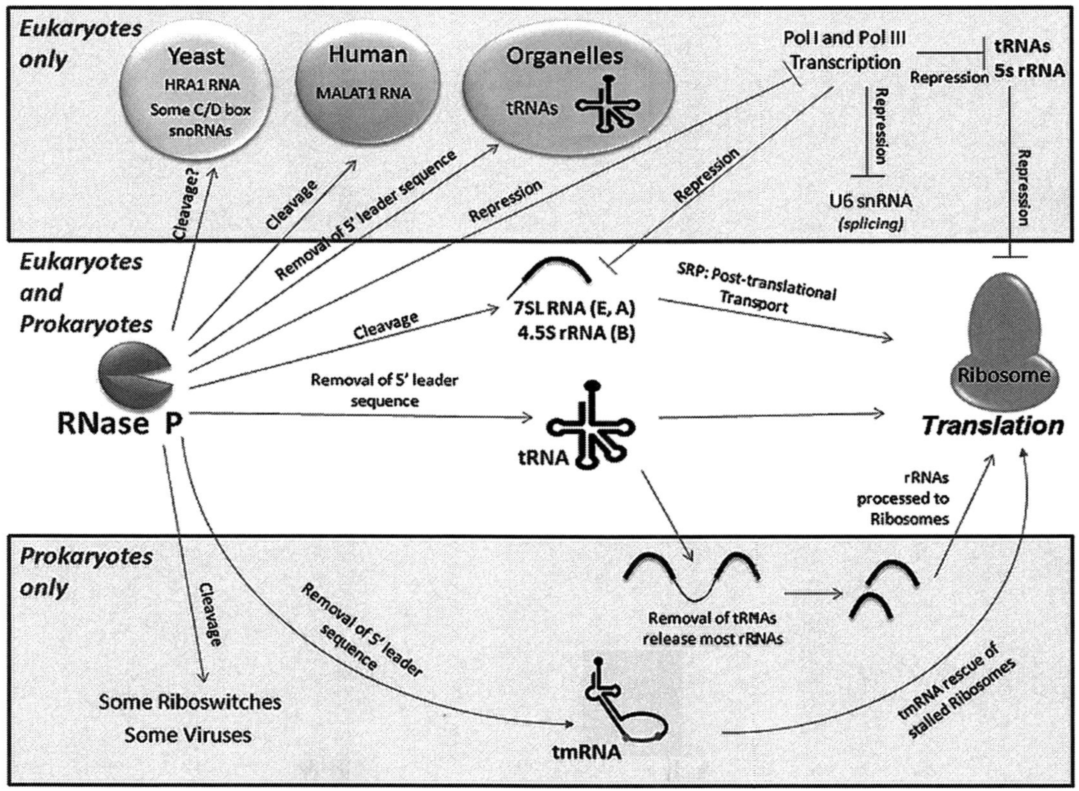

Figure 2. RNase P is central to RNA processing in eukaryotes and prokaryotes. In both eukaryotes and prokaryotes, RNase P cleaves tRNA leader sequences and also the SRP RNA (7SL RNA, called 4.5S rRNA in bacteria). Within eukaryotes RNase P also interacts with other RNAs in repressing Pol I and Pol III transcription which affects splicing and rRNA maturation. Within prokarytoes, RNase P also affects the processing of tmRNA which rescues stalled ribosomes, as well as affecting rRNA processing and cleaving some riboswitches and viruses. A color version of this figure is available at www.landesbioscience.com/curie.

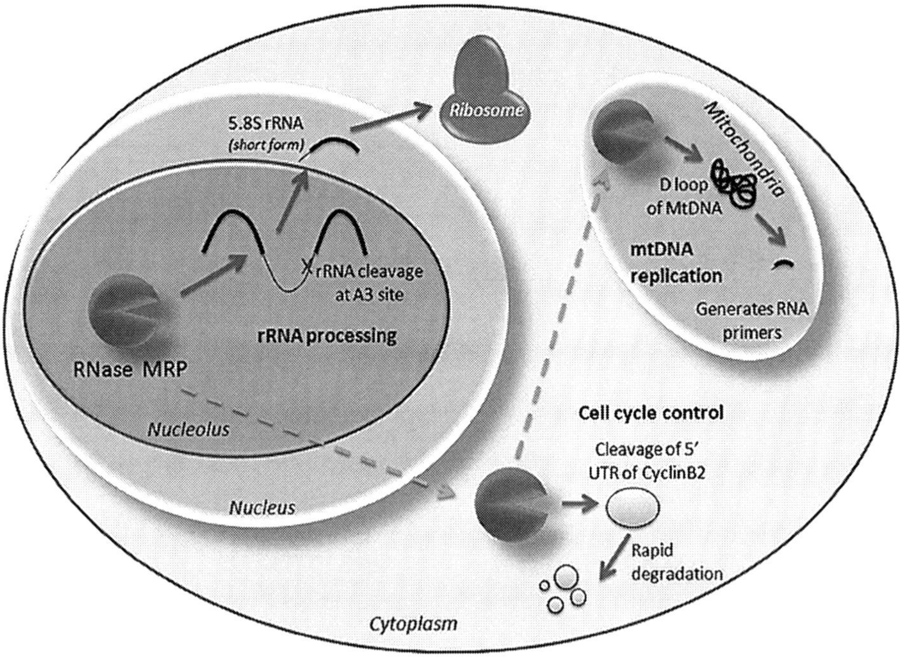

Figure 3. RNase MRP has different functions in different cellular compartments. Within the nucleolus it is involved in rRNA processing. However, in the cytoplasm it cleaves the leader sequence of the cell cycle control protein CyclinB2, and in the mitochondria it is crucial for mtRNA replication where it cleaves the D loop of the mtDNA to generate RNA primers. Whether this macromolecule consisting of one catalytic RNA and ~9 proteins moves as a whole through the cell, or disassembles and re-assembles at the different areas in which it functions, is not yet known. A color version of this figure is available at www.landesbioscience.com/curie.

It clearly plays a central role in the RNA infrastructure of both eukaryotes and prokaryotes and it is likely that other substrates and processing connections, especially in prokaryotes, are still to be uncovered.

RNase MRP is a ribonucleoprotein found only in eukaryotes, but closely related and sharing many of the same proteins with RNase P (for a review see ref. 13). It too has multiple roles (Fig. 3), processing the A3 site of rRNA in the nucleolus, a critical cell cycle control protein (Cyclin B2) in the cytoplasm, and the D-loop of mitochondrial DNA (MtDNA) in the mitochondria to generate RNA primers for Mt DNA replication. This is a good illustration of the spatial nature of RNA-Protein complexes that have different roles in different cellular compartments. RNase MRP is transcribed by RNA Pol III and thus is affected by the RNase P feedback on the polymerase.

Other aspects of rRNA processing in eukaryotes are linked to transcription and downstream rRNA maturation. Extensive modification of the pre-rRNAs includes methylation of 2′ hydroxyl groups of ribose (guided by C/D box snoRNAs [small nucleolar RNAs]) and pseudouridine formation from uracil (guided by H/ACA snoRNAs).[26] In vertebrates, these snoRNAs are mostly found within introns, and are spliced out by snRNAs, illustrating the strong network of RNA biogenesis and splicing machineries. Yeast models (primarily in *S. cerevisiae*) indicate that RNA Pol I, elongation factors and rRNA sequence elements appear to optimize transcription elongation and co-ordinate interactions (including those with snoRNAs) with the pre-rRNA for correct rRNA processing and ribosome assembly.[27] In addition, a protein complex of three transcription factors (the CURI complex comprising of Rap1, Fhl1 and Ifh1) links ribosomal protein production and pre-rRNA processing.[28] Thus, rRNA processing also uses feedback from the later stages of processing to regulate transcription.

SPATIAL REGULATION OF EUKARYOTIC RNA PROCESSING

Spatial placement of both RNA and protein macromolecular components plays an important part in the regulation of RNA-processing. In eukaryotes, this is clearly demonstrated by how RNAs move through nuclear bodies (such as Cajal bodies, Gems and nucleoli) and for some of them, into cytoplasmic bodies such as P-bodies and RNA granules. As an example, Figure 4 illustrates the biogenesis of snRNAs and snoRNAs in

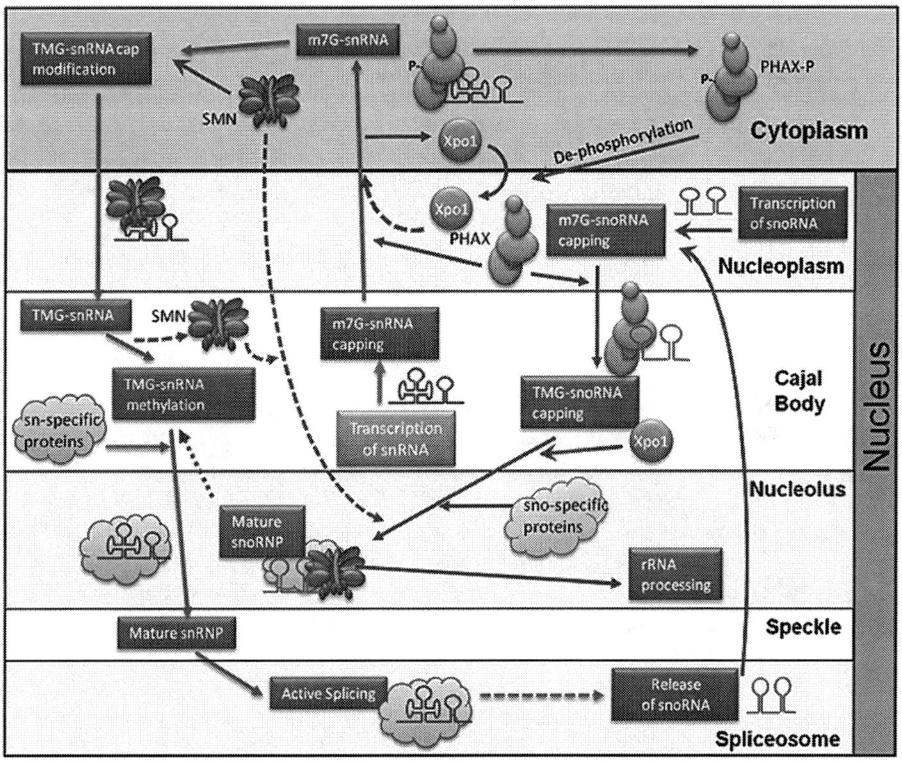

Figure 4. The network of Sm-class snRNA and snoRNA biogenesis pathways connected by the SMN complex and the PHAX complex. Transcribed snRNAs move through nuclear compartments during initial processing then into the cytoplasm using the PHAX complex where they gain the SMN complex. After this, the snRNA/SMN macromolecule moves back into the nucleus for further maturation before being used for active splicing. In contrast snoRNAs do not enter the cytoplasm but instead use the PHAX complex for intranuclear transport and the SMN complex for macromolecule maturation. Figure adapted with permission of authors. A color version of this figure is available at www.landesbioscience.com/curie.

humans. Typically there are different stages of RNA-processing taking place within different nuclear sub-compartments, but for the Sm-class-snRNAs, the processing moves to the cytoplasm, before the re-import of the snRNP-complexes back to the nucleus. In contrast, the Lsm-class snRNAs (U6 and $U6_{atac}$ snRNAs) in humans never leave the nucleus,[29] although in yeast there may be some nuclear export and re-import of U6 snRNA.[29] Cajal Bodies in particular appear to be important sub-nuclear compartments for RNPs since they are not only repositories for the biogenesis of RNPs. Mature snRNPs travel through Cajal Bodies, sometimes moving from one Cajal body to another suggesting that the Cajal Body is being used as a 'recycling center', enabling the re-assembly of the tri-snRNPs.[30] In contrast, the assembly of C/D box snoRNPs appears to occur cotranscriptionally, but much of the intra-nuclear and intra-cellular trafficking of snoRNPs remain to be characterised.[31]

A feature of intra-cellular RNP trafficking is how some proteins assist these different RNPs in different manners. One such group of proteins linking snRNA and snoRNA biogenesis (Fig. 4) is the PHAX complex (consisting of PHAX, Cap Binding Protein (CBC), CRM1 and RanGTP) which in humans at least, transports snRNAs from the nucleus to the cytoplasm as well as transport of some snoRNAs (especially U3, U8, U13) around the nucleus to speckles, Cajal bodies and nucleoli.[32] Although PHAX is a metazoan protein there has been a similar protein characterised in the protist *Cryptosporidium parvum*.[32] Another important RNA-escorting macromolecule is the SMN protein complex, which is found in the nucleoplasm and nuclear bodies called Gems.[33] The SMN complex scrutinizes cellular RNAs to ensure that Sm cores (of highly reactive RNA-binding Sm proteins) are only assembled on proper snRNAs,[34] and the Gemin5 protein of this SMN complex can distinguish snRNAs from other cellular RNAs for snRNP biogenesis.[34] The SMN complex also plays a role in other biogenesis pathways including those for hnRNPs and microRNPs.[33] The above pathways for snRNP and snoRNP biogenesis have been largely characterised for mammalian and yeast systems, and although there is now some plant information,[35] there is little known about how these RNPs complexes form in the many different groups of protists. As with plants we expect some different proteins to be involved and there will likely be different pathways.

After nuclear export some mRNAs are translated immediately, but many mRNAs are recruited to RNA granules (See Chapter 8, Table 1 pg 124 for cellular component definitions) in the cytoplasm until developmental or environmental cues signal their translation.[36] Cytoplasmic RNA granules (reviewed in refs. 36,37)include Processing-bodies and Stress Granules as well as compartments found in germ cells (polar and germinal granules) and neurons (neuronal granules). Processing-bodies (P-bodies or GW bodies) are involved with post-transcriptional processes, including mRNA degradation, nonsense-mediated mRNA decay (NMD), translational repression and RNA-mediated gene silencing (reviewed in ref. 38). mRNA degradation is initiated by the deadenylation (shortening) of the 3′ polyA-tail followed by decapping.[39] Stress Granules are a cytoplasmic RNA granule that typically forms during stress response (whereas P-bodies are present continuously).[40] Stress Granules contain polyadenylated transcripts and are not degraded, making them available for rapid re-initiation after stress recovery,[40] whereas mRNAs recruited to P-bodies are largely deadenylated.[37] mRNAs within stress granules P-bodies and Stress Granules constantly exchange RNAs and proteins with the cytosol[37] and mRNAs can move from one to the other. P-bodies have been investigated in yeasts, plants, trypanosomatids, insects and vertebrates (reviewed in ref. 37) and thus are likely to be important eukaryotic RNA-based cellular features. Evidence is suggesting that Stress Granules are the consequence, not the cause of the shut off of translation during stress, and the formation of critical macromolecules may be linked to the sequestering of key components (reviewed in ref. 37).

Although RNAs such as miRNAs, siRNAs and tRNAs typically act in the cytoplasm, there are some miRNAs that may re-enter the nucleus, possibly playing a role in modification or nuclear component assembly processes (summarized in refs. 41,42). tRNAs in particular show interesting nuclear-cytoplasmic dynamics. tRNA transcription and 5′ processing is typically in the nucleolus, however 3′ processing has been found mostly in the nuceloplasm; tRNA modification is usually in the cytoplasm but in some species, tRNA splicing takes place on the mitochondrial cytoplasmic surface (reviewed in ref. 42). Subsequently however, a retrograde pathway exists where the tRNAs are imported from the cytoplasm back into the nucleus[43] but can then be re-exported to the cytoplasm in response to nutrient availability.[44] Although major studies of tRNA cellular dynamics has been mainly in yeast, the retrograde process (moving cytoplasm to nucleus) at least appears conserved in vertebrates (summarized in ref. 42).

Spatiotemporal movement of RNAs or key components of RNA-based machineries, is not restricted to eukaryotes. The Gram-negative bacteria *Caulobacter crescentus* is dimorphic in that it has a stalked form that adheres to surfaces (with a holdfast and stalk), and a swarmer form that is mobile with a flagellum. Often used as a model for bacterial cell cycle and cell differentiation studies, *C. crescentus* shows substructure localisation and temporal timing of trans-translation.[45-47] tmRNA and its small protein SmpB are colocalised to a helix-like pattern in swarmer and predivisional cells but they are delocalised in stalked cells.[46] However, the protein RNase R which interacts with the tmRNA is localised separately to another helix-like pattern that is out of phase with the tmRNA-SmpB pattern. Trans-translation requires that the individual tmRNA-SmpB molecules would have to disassociate from the helix-like structure in order to pass through the ribosome, and it is feasible that these structures facilitate the regulation of trans-translation.[46] In a possible feedback mechanism, the tmRNA of *C. crescentus* is regulated in the cell cycle by temporally controlled transcription and translation.[45] With trans-translation required for many functions across bacteria, including sporolation in *Bacillus subtilis*,[48] symbiosis in *Bradyrhizobium japonicum*[109], and pathogenecity in *Salmonella enteria* (summarized in ref. 47), the tight regulation of the RNA component is not unexpected. It is even possible that these helix-like structures seen in *C. crescentus* are analogous to the P-bodies we find in eukaryotes.

RNA REGULATION, CONNECTING COMPONENTS OF THE RNA-INFRASTRUCTURE

RNAi Networks

RNA regulation (including RNAi, riboswitches and RNA-editing), storage and degradation are linked to the processes discussed in the earlier sections. RNAi (RNA interference involving miRNAs, siRNAs and piRNAs are reviewed in refs. 49, 50). Although best known for roles in regulating mRNA levels, RNAi is also directly involved in many cellular processes including chromatin-mediated gene silencing and DNA re-arrangements. It is also not a matter of one target to one regulator. It has been shown[51] that a single miRNA can directly or indirectly down-regulate the production of thousands of genes. Although RNAi as a mechanism appears general in eukaryotes, the timing and location of miRNA expression varies even between vertebrates due to changes in miRNA copy number, genomic context (either exclusively intergenic, or intronic and intergenic) or both.[52] There can also be expression differences when there is conservation of the miRNA sequence.[52]

Other forms of RNA-based transcriptional regulation include regulation by RNase P which has a positive effect on Pol III promoter activity.[16,53] RNase P associates with the chromatin of tRNA and 5S rRNA genes which contain the Type-1 Pol III promoter sequences, but not with the U6 snRNA and 7SL-RNA that have Type-3 Pol III promoter sequences. Transcription of these Pol III transcribed ncRNAs declines sharply in extracts depleted of active RNase P.[16] RNase P may also have a role in the splicing-independent maturation of snoRNAs as recently demonstrated in yeast[54] linking the production of tRNAs and rRNA.

A number of longer ncRNAs directly target transcription (reviewed in ref. 55) including SRA (a transcriptional co-activator for several steroid-hormone receptors), NRSE (Neuron-restrictive silencer element dsRNA), HSR1 (heat shock RNA-1) and 7SK RNA. This latter ncRNA is transcribed by Pol III and represses transcript elongation by Pol II (also reviewed in ref. 55). Another instance is a regulatory transcript from a minor promoter interfering with the expression of the main transcript.[56] With the ongoing discovery[57] of new ncRNAs in a wider range of eukaryotes we certainly expect the identification of other direct transcriptional regulators.

An interesting trend is the discovery that small regulatory RNAs can be derived from other ncRNAs. A number of studies have characterised miRNA-like RNAs derived from snoRNA-derived RNAs[58] and RNAs derived from the Vault RNA.[59] tRNA-derived RNAs are thought to be involved in translational repression.[60,61] Studied in mammals, plants, fungi, and the protists *Giardia* and *Tetrahymena*, tRNAs are cleaved by members of the Ribonuclease A or T2 protein families in the anticodon loop forming 5′ and 3′ tRNA halves. Although how these different tRNA halves regulate translation inhibition is still very much under investigation, in mammals it has been shown that 5′ tRNA halves induces Stress Granule formation[62] and that the original cleavage is enhanced by stress. In Trypanosomes, granules are formed that are distinct from Stress Granules.[37,63] Other translational-inhibition small RNAs include qiRNAs (QDE-2 associated RNAs) from the fungus Neurospora which inhibit protein translation during DNA damage response.[64] With mass RNA sequencing still in its early days, these may represent only a fraction of the real amount of derived regulatory RNAs.

Transcription-initiation RNAs* are typically transcribed from a repeat motif called a 'spanion cluster'.[65] These RNAs have a strong preference towards transcription initiation sites. Other small regulatory RNAs of note are the tiny RNAs of mammals, which are 17-18 bp in length and have a connection to splicing in that their 3′ ends map precisely to the splice donor site of internal exons.[66] A subgroup of these splice-site RNAs are seen to be associated with highly expressed genes.[66] How widespread these types of regulatory RNAs are remains to be investigated, but high throughput sequencing technology has enabled researchers to uncover these types and more.

RNAi is seen thus as a typical eukaryotic feature but there are some lineages that have lost their RNAi proteins but some still maintain some form of ncRNA-based regulation. The yeast *S. cerevisiae* does not have the 'standard' RNAi system since it lacks Dicer-like RNases, Argonaute or Piwi-like proteins, but it does have ncRNAs

*In a wave of confusion we are also seeing different types of regulatory RNAs given the same prefix. Two types of small RNA, transcription-initiation RNAs,[65] and tiny RNAs[66] have both been given the name tiRNAs. To avoid confusion the expanded name rather than tiRNA will be used for both cases.

that act in the regulation of its genes.[67] These ncRNAs including the 'cryptic unstable transripts' (CUTS), tend to stem from bidirectional transcription and may be passive by-products of transcriptional noise rather than any specific mechanism.[67] In single celled protists, some species of Trypanosomes have lost their RNAi systems while others have retained them.[68] Additionally, for some Trypanosomes the retention of associated viruses is necessary and it has been suggested that the loss of RNAi has facilitated viral retention. However genome plastidity is also a potential effect of this loss.[68] Whether these RNAi-less Trypanosomes have evolved a different type of RNA-regulation system to compensate, is as yet unknown.

RNA Networks and Epigenetics

ncRNAs play a major role in epigenetics[69-74] and include networks consisting of long ncRNAs (such as XIST and HOTAIR), and short ncRNAs[69,74,75] such as miRNAs, siRNAs and piRNAs. miRNAs in particular have been shown to be important in RNA networks behind stem-cell self-renewal and differentiation (reviewed in ref. 76). In general, there are two types of stem-cell, tissue stem cells (which include somatic and germline cells which develop, maintain and repair tissues in developing and adult organisms), and embryonic stem cells (ES) which develop from an embryo to give rise to the foetus. In one example of a miRNA-epigenetic network, the expression of the miR-290-295 miRNA cluster (a group of miRNAs that share a 5′ proximal AAGUGC motif) increases during pre-implantation development and remains high in undifferentiated ES cells, but then decreases after ES cell differentiation.[77] These miRNAs act as post-transcriptional regulators of retinoblastoma-like 2 (Rbl2) which in turn acts as a transcriptional repressor of DNA methyl transferases (DNMTs), Dnmt3a and Dnmt3b. DNMTs epigenetically silence OCT4, a key transcription factor of ES cell renewal and differentiation.[77,78] Alternatively if Dicer is knocked out, miRNAs are depleted and the methylation of the Oct4 promoter is severely impaired during differentiation. Many other candidate targets of the AAGUGC seed-containing miRNAs have been identified as well as many indirectly regulated targets,[77] but it remains to be seen how other aspects of self-renewal and differentiation are affected by the miR-290 cluster.

Networks involving multiple long ncRNAs (defined generally as having a length > 200 nt) are also known, with a classic example being the long ncRNA control of X-chromosome inactivation (reviewed in refs.74,79). In mammals the potential double dosage of gene expression from the X chromosome in XX females (when compared to XY males) is controlled by inactivating one of the X chromosomes. In mice there are two forms of X Chromosome inactivation (XCI)[79] where XCI is imprinted in extra-embryonic tissues and the paternal X (Xp) is inactivated. Further along in development just before the embryo proper, the Xp is re-activated after which random XCI is initiated during early embryonic development. In humans XCI is randomly activated but it is not clear if the imprinted form is present.[79] During random XCI in humans (Fig. 5), the long ncRNA Xist is repressed on the future active X chromosome Xa by another long ncRNA Tsix,[80] and activated on the future inactive X chromosome by a third long ncRNA Jpx.[81]

This complex network of long ncRNA and methylation processes can be seen if we examine this system in more detail (Fig. 5). In mammals, in pre-XCI embryonic stem cells (ES) Tsix is transcribed at a higher level than Xist, and triggers H3-K4 dimethylation along both the *Xist* and *Tsix* genes. Xist becomes elevated when the major pluripotency factors Nanog, Oct3/4 and Sox2 dissociate from intron 1 within *Xist* initiating XCI. One

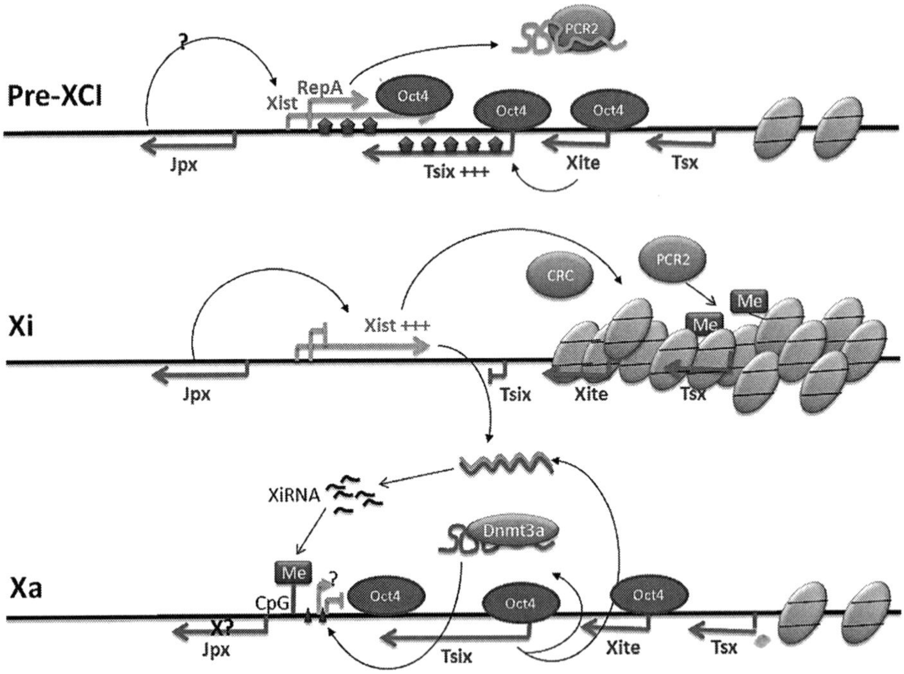

Figure 5. RNA networking during mammalian random X chromosome inactivation (XCI). During Pre-XCI, Tsix and RepA, compete for binding to the polycomb repressive complex PRC2. Tsix is expressed at a high level upon XCI and triggers dimethylation (wide vertical arrows) along both the *Xist* and *Tsix* genes, leading to active transcription of Xist and Tsix. On the future inactive X chromosome Xi, Oct4 binding is lost so Tsix is downregulated and Xist induced and is further enhanced by Jpx. A coating of Xist RNA forms a chromatin compartment and recruits the chromatin repressive complex (CRC) to the Xi. The inactive status of this chromosome is maintained by PCR2. On the future active X chromosome Xa, Oct4 is retained and maintains Tsix expression. Tsix, associated with methyltransferase Dnm3a, directs the methylation on the *Xist* promoter and Xist is repressed. Dicer-dependent XiRNAs are possibly produced from Xist and Tsix ncRNA duplexes and could direct methylation along the future Xi, and also direct methylation of CpG islands of the Xist promoter region in the Xa. Gene distances are not to scale. Figure adapted from[74] with permission from authors. A color version of this figure is available at www.landesbioscience.com/curie.

of these proteins Oct4 is known to active Tsix and another RNA region Xite which is an activator of Tsix. Oct4 also acts as a repressor of Xist aiding in the control of the Xist:Tsix balance in XCI.[82] During XCI different events occur upon the future active X chromosome (Xa) and the future inactive X chromosome (Xi).

On the future inactive X chromosome (Xi), Oct4 binding is lost so Tsix is downregulated and Xist is induced.[82] A coating of Xist RNA forms a silent chromatin compartment where X-linked genes become 'localised' through binding to the Xist RNA.[83] Xist RNA also recruits the chromatin repressive complex (CRC) to Xi. The inactive status is maintained by the polycomb repressive complex PCR2. In pre-XCI Tsix and another RNA, Repeat A (RepA), compete for PRC2 binding[84] but upon XCI, RepA recruits PRC2 to Xist and PRC2 methylates Xist at H3K27 to upregulate Xist[84]. RepA also collaborates with the long ncRNA Jpx in an as yet unknown mechanism to transcriptionally activate Xist.[81] On the future active X chromosome Xa, Oct4 is retained and maintains Tsix expression. Tsix, associated with methyltransferase Dnm3a, directs the methylation on the *Xist* promoter to repress Xist. With Xist repressed, the Xa chromosome is active. Whether Jpx is still expressed or controlled by methylation is as yet unknown.

In addition, small ncRNAs may also be involved in X inactivation. Dicer-dependent XiRNAs can be produced from Xist and Tsix ncRNA duplexes[80] and may direct methylation along the future Xi, and also direct methylation of CpG islands of the Xist promoter region in the Xa.[85] Although XiRNAs are produced with Dicer, RNAi may not be directly involved in X chromosome inactivation, instead it could maintain the steady-state level of the Xist RNA.[86] Recently other reports have contradicted the role of Dicer in XCI (summarized in ref. 87), and questions remain as where Xist:Tsix duplexes form, where Dicer could act, and the possible transport of XiRNAs between the cytoplasm and the nucleus. Although Xist and Tsix are long ncRNAs, they are both spliced and polyadenylated (reviewed in ref. 79), and it will be interesting to understand more in the future about how these mechanisms which have been demonstrated to have a regulatory affect, aid in XCI.

This type of Xist-linked random XCI is specific for eutherian mammals and is not seen in marsupials, flies or nematodes.[88] Marsupial mammals lack Xist and only use imprinted XCI with the paternal X being inactivated.[88] The presence of active and repressive histone modifications does suggest a related mechanism of dosage compensation and the presence of as yet unidentified ncRNAs is not ruled out.[88] Insects possess a mechanism which instead of silencing one X copy in the female, instead promotes 2x transcriptional activity on the male X chromosome involving two long ncRNAs, rox1 and rox2.[89] Rox1 and rox2 transcripts spread along the X chromosome recruiting the histone deacetylation protein complex, which then generates an open chromatin conformation to facilitate active transcription.[90,91] Related mechanisms in other organisms is thus very likely (for more on this subject see Chapter 11, pages 180-184).

RNA Regulation in Prokaryotes

Studies on RNA-based regulation tended in the past to concentrate on eukaryotes because the wealth of nonprotein-coding transcripts offered a large territory in which to find them. Prokaryotes have far less nonprotein coding territory, but in the last decade the importance of their RNA-based regulation has been found. There are some key networks of RNA regulation that have been studies mostly in bacteria but have some characterisation in archaea. This chapter will touch only briefly on these networks and will be expanded on in later chapters (see Chapters 13-15 for more detail on prokaryotic RNA infrastructure). Two areas of interest for prokaryotic RNA regulatory networks are how the CRISPR system interacts with cellular and viral RNAs to produce an effective viral defence mechanism, and secondly how environmental responses can be regulated using small RNA networks.

Clustered, regularly interspaced, short-palindromic repeats (CRISPRs) are an integral part of invader defence in bacteria and archaea (for reviews see refs. 92-94). Each CRISPR loci contains a short invader-derived spacer sequence (~40 nt) flanked by partial repeats (~30 nt), preceded by a leader sequence (150-550 nt). These units form clusters from one to over a hundred depending on species.[93] Generally, upon invasion by viruses (also called in some species bacteriophages) or some plasmids, complexes of Cas proteins target and cleave short recognition motifs in the genome of the invader. These cleaved sequences are then incorporated into the leader sequence of the CRISPR locus. When the CRISPR locus is transcribed the CRISPR RNAs (crRNAs) then serve as templates to target invader sequences upon further infection (summarized in refs. 92,94). This prokaryotic immune mechanism allows for the acquisition of strain-specific immunity, a memory of past infections and has systems to ensure host integrity through self/nonself discrimination.[94,95]

Protein composition of the CRISPR system also varies with species. There are six core protein genes (*cas1-cas6*) having varied distribution throughout eukaryotes, with only *cas1* nearly always being present[92]. These *cas* genes fall in clusters physically linked to the CRISPR loci. Another set of *cmr* genes are also clustered and some of them can be physically linked to the CRISPR loci.[93] These proteins are involved in the maintenance of the CRISPR loci, the crRNA biogenesis and crRNA mediated resistance to invaders.[92] The CRISPR/Cas system is found in most archaea with 70% of these carrying CRISPR/Cmr modules. In bacteria ~40% carry the CRISPR/Cas system and only 30% of these have CRISPR/Cmr modules. An important difference is that the CRISPR/Cas system specifically targets DNA,[93,96] whereas the CRISPR/Cmr system targets RNA.[93] Despite these two systems being mechanistically linked and sometimes physically linked it has been suggested that they have evolved interdependently.[93]

The closest parallel with the eukaryotes are the piRNAs that are processed from pre-piRNA cluster transcripts without the use of Dicer. These piRNAs increase in informational content by the insertion of transposon sequences. Unlike the CRISPR mechanism this increase is passive rather than active.[93] There is no sequence homology between any of the CRISPR-associated proteins and proteins involved in RNAi in eukaryotes.[92,93]

CRISPR systems have an impact on plasmid stability and can affect antibiotic resistance. The CRISPR/Cas system has been demonstrated to target antibiotic resistance genes in *Steptococcus thermophilus* and cause plasmid loss.[97] In another case, *Enterococcus faecalis* and *Enterococcus faecium* stains that had lost their CRISPR-cas systems, also lacked a plasmid mediated hospital-acquired antibiotic resistance[98]. It is possible in this case that modern drug therapy has shifted the evolution of these *Enterococci* with the loss of the CRISPR system, thus permitting the maintenance of an invading plasmid.[98]

Within bacteria there are other small RNAs responsible for negative and positive regulation of chromosomal genes and genes encoded on extra-chromosomal elements.[99] Cis-antisense-RNAs are encoded in cis on the DNA strand opposite to their target share regions of complementarity (often 75 nt or more) to generally repress the target mRNA (reviewed in ref. 100). Many of these cis-RNAs reside in plasmids or other mobile elements, and are responsible for maintaining copy number, whereas others repress the translation of toxic proteins that remove cells that have lost the mobile element. Some chromosomally expressed cis-RNAs have been shown to repress the translation of proteins that are toxic at high levels (reviewed in ref. 100). Trans-encoded small RNAs share only limited complementarity with their targets and regulate expression by the pairing of the small RNA with their mRNA target. This group of small RNAs has been shown to interact in RNA networks that are critical for bacterial responses to stress and environmental cues, and regulate many outer membrane proteins.[101]

Most of these small RNAs repress translation (reviewed in refs. 99,102), but some can activate expression where pairing disrupts inhibitory secondary structure which would otherwise sequester the ribosome binding site (reviewed in ref. 100). The RNA chaperone protein Hfq binds preferentially to single-stranded RNAs interacting with both small RNAs and some target mRNAs by binding to a U-rich motif and an A-rich motif.[103,104] This Sm-like protein is well conserved throughout bacteria,[104] however, the precise mechanism by which it facilitates binding is not yet understood.[103] The Hfq protein has been shown to localise close to the bacterial membrane and in *Salmonella* half of the small RNAs with known targets regulate the expression of outer membrane proteins (OMPs).[104] From expression studies in *Salmonella*,[105] about a fifth of all *Salmonella* genes are under the control of Hfq. The Hfq protein is not found in all bacteria (such as *Bacillus subtilis* and *Staphlococcus aureus*) and is not seen to be widespread in archaea, thus it remains to be seen if these organisms use another RNA chaperone protein to regulate large numbers of genes.[99]

The use of tRNAs in regulation is also not restricted to eukaryotes. The archaean, *Nanoarchaeum equitans* has a normal complement of tRNA genes (34 continuous and four with introns), but it also has six tRNAs genes that are coded in halves which become pieced together during processing (summarized in ref. 106). Other split tRNA genes have been found in *Caldivirga maquilingensis* where examples of tRNAs in three fragments have been discovered.[106] It is possible that these split tRNAs could have resulted from genome re-arrangements at intron-containing tRNAs. In fact one split tRNA gene in *N. equitans* is located directly adjacent to a CRISPR element raising the idea that the insertion and excision of CRISPR elements could have played a role in the spreading of tRNA halves.[106] How the splitting of these tRNAs fits into the regulation of tRNA processing, or are used in other cellular processes, are not yet known.

CONCLUSION

Within the cell we see networks of RNAs involved in cleaving, modifiying or guiding the processing of other RNAs, and over recent time these networks have appeared to have exponentially expanded in complexity. Of course, the genes and their interactions were there for a long time, and it is our knowledge that has grown. The modern RNA world is a tight-knit one where molecular machineries share proteins and sometimes RNAs. With the environment of the cells being so tightly packed (i.e., molecular crowding),[107] this sharing permits connected machineries to respond together to cellular signals. Whether universal RNAs have remained because they are intrinsically better suited for this role remains an interesting evolutionary question.

It is clear that our view of the prokaryotic modern RNA world is changing as the emphasis on the biogenesis and processing of the core RNAs (mRNA, tRNA and rRNA) has shifted now to tackle RNA-based regulation. Although prokaryotic genomes do not carry anywhere near the same degree of noncoding transcripts we see in eukaryotes, they nevertheless carry a wealth of RNA genes with which we can already see networking. As more is uncovered other interesting studies will no doubt tackle the similarities and differences between prokaryotic and eukaryotic RNA networks. From here we can investigate whether these networks evolved separately or have stemmed from an 'RNP-world' ancestor.[108] What is very clear is that in modern cellular networks, the role of RNA genes and how they connect in cellular networks makes them as important as their protein-coding cousins.

ACKNOWLEDGEMENTS

Many thanks go to Prof. David Penny, Prof. Mike Hendy and Prof. Pete Lockhart for stimulating discussions and support.

REFERENCES

1. Collins LJ, Penny D. RNA-infrastructure: dark matter of the eukaryotic cell? Trends in Genetics 2009; 25(3):120-128.
2. Kornblihtt AR, de la Mata M, Fededa JP et al. Multiple links between transcription and splicing. RNA 2004; 10(10):1489-1498.
3. Hicks MJ, Yang CR, Kotlajich MV et al. Linking splicing to Pol II transcription stabilizes pre-mRNAs and influences splicing patterns. PLoS Biol 2006; 4(6):e147.
4. Gornemann J, Kotovic KM, Hujer K et al. Cotranscriptional spliceosome assembly occurs in a stepwise fashion and requires the cap binding complex. Mol Cell 2005; 19(1):53-63.
5. Lacadie SA, Rosbash M. Cotranscriptional spliceosome assembly dynamics and the role of U1 snRNA: 5′ss base pairing in yeast. Mol Cell 2005; 19(1):65-75.
6. Crabb TL, Lam BJ, Hertel KJ. Retention of spliceosomal components along ligated exons ensures efficient removal of multiple introns. RNA 2010; 16(9):1786-1796.
7. Millevoi S, Loulergue C, Dettwiler S et al. An interaction between U2AF 65 and CF I(m) links the splicing and 3′ end processing machineries. The EMBO Journal 2006; 25(20):4854-4864.
8. Chang YF, Imam JS, Wilkinson MF. The nonsense-mediated decay RNA surveillance pathway. Annu Rev Biochem 2007.
9. Nojima T, Hirose T, Kimura H et al. The interaction between cap-binding complex and RNA export factor is required for intronless mRNA export. J Biol Chem 2007.
10. Shibuya T, Tange TO, Stroupe ME et al. Mutational analysis of human eIF4AIII identifies regions necessary for exon junction complex formation and nonsense-mediated mRNA decay. RNA 2006; 12(3):360-374.
11. Tange TO, Nott A, Moore MJ. The ever-increasing complexities of the exon junction complex. Curr Opin Cell Biol 2004; 16(3):279-284.
12. Phizicky EM, Hopper AK. tRNA biology charges to the front. Genes Dev 24(17):1832-1860.
13. Esakova O, Krasilnikov AS. Of proteins and RNA: the RNase P/MRP family. RNA 2010; 16(9):1725-1747.
14. Lai LB, Vioque A, Kirsebom LA et al. Unexpected diversity of RNase P, an ancient tRNA processing enzyme: challenges and prospects. FEBS Lett 2010;584(2):287-296.
15. Pulukkunat DK, Gopalan V. Studies on Methanocaldococcus jannaschii RNase P reveal insights into the roles of RNA and protein cofactors in RNase P catalysis. Nucleic Acids Res 2008;36(12):4172-4180.
16. Reiner R, Ben-Asouli Y, Krilovetzky I et al. A role for the catalytic ribonucleoprotein RNase P in RNA polymerase III transcription. Genes Dev 2006; 20(12):1621-1635.
17. Reiner R, Krasnov-Yoeli N, Dehtiar Y et al. Function and assembly of a chromatin-associated RNase P that is required for efficient transcription by RNA polymerase I. PLoS One 2008; 3(12):e4072.
18. Hayes CS, Keiler KC. Beyond ribosome rescue: tmRNA and co-translational processes. FEBS Lett 584(2):413-419.
19. Wower J, Wower IK, Zwieb C. Making the jump: new insights into the mechanism of trans-translation. J Biol 2008; 7(5):17.
20. Une M, Kurita D, Muto A et al. Trans-translation by tmRNA and SmpB. Nucleic Acids Symp Ser (Oxf) 2009; 53:305-306.
21. Hayes CS, Keiler KC. Beyond ribosome rescue: tmRNA and co-translational processes. FEBS Lett 2010; 584(2):413-419.
22. Marvin MC, Engelke DR. Broadening the mission of an RNA enzyme. J Cell Biochem 2009; 108(6):1244-1251.
23. Holzmann J, Frank P, Loffler E et al. RNase P without RNA: identification and functional reconstitution of the human mitochondrial tRNA processing enzyme. Cell 2008; 135(3):462-474.
24. Li Y, Altman S. In search of RNase P RNA from microbial genomes. RNA 2004; 10(10):1533-1540.
25. Randau L, Schroder I, Soll D. Life without RNase P. Nature 2008; 453(7191):120-123.
26. Moss T, Langlois F, GagnonKugler T et al. A housekeeper with power of attorney: the rRNA genes in ribosome biogenesis. Cell Mol Life Sci 2007; 64(1):29-49.
27. Schneider DA, Michel A, Sikes ML et al. Transcription elongation by RNA polymerase I is linked to efficient rRNA processing and ribosome assembly. Mol Cell 2007; 26(2):217-229.
28. Rudra D, Mallick J, Zhao Y et al. Potential interface between ribosomal protein production and pre-rRNA processing. Mol Cell Biol 2007; 27(13):4815-4824.

29. Hopper AK. Cellular dynamics of small RNAs. Crit Rev Biochem Mol Biol 2006; 41(1):3-19.
30. Stanek D, Pridalova-Hnilicova J, Novotny I et al. spliceosomal small nuclear ribonucleoprotein particles repeatedly cycle through cajal bodies. Mol Biol Cell 2008; 19(6):2534-2543.
31. Matera AG, Terns RM, Terns MP. Non-coding RNAs: lessons from the small nuclear and small nucleolar RNAs. Nat Rev Genet 2007; 8(3):209-220.
32. Mourao A, Varrot A, Mackereth CD et al. Structure and RNA recognition by the snRNA and snoRNA transport factor PHAX. RNA 16(6):1205-1216.
33. Gubitz AK, Feng W, Dreyfuss G. The SMN complex. Exp Cell Res. 2004; 296(1):51-56.
34. Battle DJ, Lau CK, Wan L et al. The Gemin5 protein of the SMN complex identifies snRNAs. Mol Cell 2006; 23(2):273-279.
35. Rodor J, Letelier I, Holuigue L et al. Nucleolar RNPs: from genes to functional snoRNAs in plants. Biochem Soc Trans 2010; 38(2):672-676.
36. Anderson P, Kedersha N. RNA granules. J Cell Biol 2006; 172(6):803-808.
37. Thomas MG, Loschi M, Desbats MA et al. RNA granules: The good, the bad and the ugly. Cell Signal 2010; 23(2):324-334.
38. Jakymiw A, Pauley KM, Li S et al. The role of GW/P-bodies in RNA processing and silencing. J Cell Sci 2007; 120(Pt 8):1317-1323.
39. Brengues M, Parker R. Accumulation of Polyadenylated mRNA, Pab1p, eIF4E, and eIF4G with P-Bodies in Saccharomyces cerevisiae. Mol Biol Cell 2007.
40. Kedersha N, Stoecklin G, Ayodele M et al. Stress granules and processing bodies are dynamically linked sites of mRNP remodeling. J Cell Biol 2005; 169(6):871-884.
41. Politz JC, Hogan EM, Pederson T. MicroRNAs with a nucleolar location. RNA 2009; 15(9):1705-1715.
42. Phizicky EM, Hopper AK. tRNA biology charges to the front. Genes Dev 2010; 24(17): 1832-1860.
43. Shaheen HH, Hopper AK. Retrograde movement of tRNAs from the cytoplasm to the nucleus in Saccharomyces cerevisiae. Proc Natl Acad Sci U S A 2005;102(32):11290-11295.
44. Whitney ML, Hurto RL, Shaheen HH et al. Rapid and reversible nuclear accumulation of cytoplasmic tRNA in response to nutrient availability. Mol Biol Cell 2007; 18(7):2678-2686.
45. Keiler KC, Shapiro L. tmRNA is required for correct timing of DNA replication in Caulobacter crescentus. J Bacteriol 2003; 185(2):573-580.
46. Russell JH, Keiler KC. Subcellular localization of a bacterial regulatory RNA. Proc Natl Acad Sci U S A 2009; 106(38):16405-16409.
47. Cheng L, Keiler KC. Correct timing of dnaA transcription and initiation of DNA replication requires trans translation. J Bacteriol 2009; 191(13):4268-4275.
48. Abe T, Sakaki K, Fujihara A et al. tmRNA-dependent trans-translation is required for sporulation in Bacillus subtilis. Mol Microbiol 2008; 69(6):1491-1498.
49. Munroe SH, Zhu J. Overlapping transcripts, double-stranded RNA and antisense regulation: a genomic perspective. Cell Mol Life Sci 2006; 63(18):2102-2118.
50. Hartig JV, Tomari Y, Forstemann K. piRNAs—the ancient hunters of genome invaders. Genes Dev 2007; 21(14):1707-1713.
51. Selbach M, Schwanhausser B, Thierfelder N et al. Widespread changes in protein synthesis induced by microRNAs. Nature 2008; 455(7209):58-63.
52. Ason B, Darnell DK, Wittbrodt B et al. Differences in vertebrate microRNA expression. Proc Natl Acad Sci U S A 2006; 103(39):14385-14389.
53. Jarrous N, Reiner R. Human RNase P: a tRNA-processing enzyme and transcription factor. Nucleic Acids Res 2007; 35(11):3519-3524.
54. Coughlin DJ, Pleiss JA, Walker SC et al. Genome-wide search for yeast RNase P substrates reveals role in maturation of intron-encoded box C/D small nucleolar RNAs. Proc Natl Acad Sci U S A 2008; 105(34):12218-12223.
55. Goodrich JA, Kugel JF. Non-coding-RNA regulators of RNA polymerase II transcription. Nat Rev Mol Cell Biol 2006; 7(8):612-616.
56. Martianov I, Ramadass A, Serra Barros A et al. Repression of the human dihydrofolate reductase gene by a non-coding interfering transcript. Nature 2007; 445(7128):666-670.
57. Washietl S, Pedersen JS, Korbel JO et al. Structured RNAs in the ENCODE selected regions of the human genome. Genome Res 2007; 17(6):852-864.
58. Taft RJ, Glazov EA, Lassmann T et al. Small RNAs derived from snoRNAs. RNA 2009; 15(7):1233-1240.
59. Persson H, Kvist A, Vallon-Christersson J et al. The non-coding RNA of the multidrug resistance-linked vault particle encodes multiple regulatory small RNAs. Nat Cell Biol 2009; 11(10):1268-1271.
60. Cole C, Sobala A, Lu C et al. Filtering of deep sequencing data reveals the existence of abundant Dicer-dependent small RNAs derived from tRNAs. RNA 2009; 15(12):2147-2160.
61. Pederson T. Regulatory RNAs derived from transfer RNA? RNA 2010; 16(10):1865-1869.

62. Emara MM, Ivanov P, Hickman T et al. Angiogenin-induced tRNA-derived stress-induced RNAs promote stress-induced stress granule assembly. J Biol Chem 2010; 285(14):10959-10968.
63. Garcia-Silva MR, Frugier M, Tosar JP et al. A population of tRNA-derived small RNAs is actively produced in Trypanosoma cruzi and recruited to specific cytoplasmic granules. Mol Biochem Parasitol 2010; 171(2):64-73.
64. Lee HC, Chang SS, Choudhary S et al. qiRNA is a new type of small interfering RNA induced by DNA damage. Nature 2009; 459(7244):274-277.
65. Cserzo M, Turu G, Varnai P et al. Relating underrepresented genomic DNA patterns and tiRNAs: the rule behind the observation and beyond. Biol Direct 2010; 5:56.
66. Taft RJ, Simons C, Nahkuri S et al. Nuclear-localized tiny RNAs are associated with transcription initiation and splice sites in metazoans. Nat Struct Mol Biol 2010; 17(8):1030-1034.
67. Harrison BR, Yazgan O, Krebs JE. Life without RNAi: noncoding RNAs and their functions in Saccharomyces cerevisiae. Biochem Cell Biol 2009; 87(5):767-779.
68. Lye LF, Owens K, Shi H et al. Retention and loss of RNA interference pathways in trypanosomatid protozoans. PLoS Pathog 2010; 6(10):e1001161.
69. Royo H, Cavaille J. Non-coding RNAs in imprinted gene clusters. Biol Cell 2008; 100(3):149-166.
70. Whitehead J, Pandey GK, Kanduri C. Regulation of the mammalian epigenome by long noncoding RNAs. Biochim Biophys Acta 2008.
71. Kurokawa R, Rosenfeld MG, Glass CK. Transcriptional regulation through noncoding RNAs and epigenetic modifications. RNA Biol 2009; 6(3).
72. Verdel A, Vavasseur A, Le Gorrec M et al. Common themes in siRNA-mediated epigenetic silencing pathways. Int J Dev Biol 2009; 53(2-3):245-257.
73. Costanzo M, Baryshnikova A, Bellay J et al. The genetic landscape of a cell. Science 2010; 327(5964):425-431.
74. Collins LJ, Chen XS, Schonfeld B. The Epigenetics of Non-coding RNA. In: Tollefsbol T, ed. Handbook of Epigenetics. Oxford: Academic Press; 2010:49-61.
75. Collins LJ, Chen XS. Ancestral RNA: The RNA biology of the eukaryotic ancestor. RNA Biology 2009; Invited review, 2009;6(5):495-502.
76. Gangaraju VK, Lin H. MicroRNAs: key regulators of stem cells. Nat Rev Mol Cell Biol 2009; 10(2):116-125.
77. Sinkkonen L, Hugenschmidt T, Berninger P et al. MicroRNAs control de novo DNA methylation through regulation of transcriptional repressors in mouse embryonic stem cells. Nat Struct Mol Biol 2008; 15(3):259-267.
78. Benetti R, Gonzalo S, Jaco I et al. A mammalian microRNA cluster controls DNA methylation and telomere recombination via Rbl2-dependent regulation of DNA methyltransferases. Nat Struct Mol Biol 2008; 15(3):268-279.
79. Barakat TS, Jonkers I, Monkhorst K et al. X-changing information on X inactivation. Exp Cell Res 2010; 316(5):679-687.
80. Navarro P, Pichard S, Ciaudo C et al. Tsix transcription across the Xist gene alters chromatin conformation without affecting Xist transcription: implications for X-chromosome inactivation. Genes Dev 2005; 19(12):1474-1484.
81. Tian D, Sun S, Lee JT. The long noncoding RNA, Jpx, is a molecular switch for X chromosome inactivation. Cell 2010; 143(3):390-403.
82. Donohoe ME, Silva SS, Pinter SF et al. The pluripotency factor Oct4 interacts with Ctcf and also controls X-chromosome pairing and counting. Nature 2009; 460(7251):128-132.
83. Chaumeil J, Le Baccon P, Wutz A et al. A novel role for Xist RNA in the formation of a repressive nuclear compartment into which genes are recruited when silenced. Genes Dev 2006; 20(16):2223-2237.
84. Zhao J, Sun BK, Erwin JA et al. Polycomb proteins targeted by a short repeat RNA to the mouse X chromosome. Science 2008; 322(5902):750-756.
85. Ogawa Y, Sun BK, Lee JT. Intersection of the RNA interference and X-inactivation pathways. Science 2008; 320(5881):1336-1341.
86. Kanellopoulou C, Muljo SA, Dimitrov SD et al. X chromosome inactivation in the absence of Dicer. Proc Natl Acad Sci U S A 2009; 106(4):1122-1127.
87. Kota SK. RNAi in X inactivation: contrasting findings on the role of interference. Bioessays 2009; 31(12):1280-1283.
88. Rens W, Wallduck MS, Lovell FL et al. Epigenetic modifications on X chromosomes in marsupial and monotreme mammals and implications for evolution of dosage compensation. Proc Natl Acad Sci U S A 2010; 107(41):17657-17662.
89. Franke A, Baker BS. The rox1 and rox2 RNAs are essential components of the compensasome, which mediates dosage compensation in Drosophila. Mol Cell 1999; 4(1):117-122.
90. Brown JW, Simpson CG, Thow G et al. Splicing signals and factors in plant intron removal. Biochem Soc Trans 2002; 30(2):146-149.

91. Oh H, Park Y, Kuroda MI. Local spreading of MSL complexes from roX genes on the Drosophila X chromosome. Genes Dev 2003; 17(11):1334-1339.
92. Hale CR, Zhao P, Olson S et al. RNA-guided RNA cleavage by a CRISPR RNA-Cas protein complex. Cell 2009; 139(5):945-956.
93. Shah SA, Garrett RA. CRISPR/Cas and Cmr modules, mobility and evolution of adaptive immune systems. Res Microbiol 2010.
94. Vale PF, Little TJ. CRISPR-mediated phage resistance and the ghost of coevolution past. Proc Biol Sci 2010; 277(1691):2097-2103.
95. Marraffini LA, Sontheimer EJ. Self versus nonself discrimination during CRISPR RNA-directed immunity. Nature 2010; 463(7280):568-571.
96. Marraffini LA, Sontheimer EJ. Invasive DNA, chopped and in the CRISPR. Structure 2009; 17(6):786-788.
97. Garneau JE, Dupuis ME, Villion M et al. The CRISPR/Cas bacterial immune system cleaves bacteriophage and plasmid DNA. Nature 2010; 468(7320):67-71.
98. Palmer KL, Gilmore MS. multidrug-resistant enterococci lack CRISPR-cas. MBio 2010; 1(4).
99. Gottesman S, Storz G. bacterial small RNA Regulators: versatile roles and rapidly evolving variations. Cold Spring Harb Perspect Biol 2010.
100. Waters LS, Storz G. Regulatory RNAs in bacteria. Cell 2009; 136(4):615-628.
101. Vogel J, Papenfort K. Small noncoding RNAs and the bacterial outer membrane. Curr Opin Microbiol 2006; 9(6):605-611.
102. Gottesman S. Small RNAs shed some light. Cell 2004; 118(1):1-2.
103. Soper T, Mandin P, Majdalani N et al. Positive regulation by small RNAs and the role of Hfq. Proc Natl Acad Sci U S A 2010; 107(21):9602-9607.
104. Diestra E, Cayrol B, Arluison V et al. Cellular electron microscopy imaging reveals the localization of the Hfq protein close to the bacterial membrane. PLoS One 2009; 4(12):e8301.
105. Sittka A, Lucchini S, Papenfort K et al. Deep sequencing analysis of small noncoding RNA and mRNA targets of the global posttranscriptional regulator, Hfq. PLoS Genet 2008; 4(8):e1000163.
106. Heinemann IU, Soll D, Randau L. Transfer RNA processing in archaea: unusual pathways and enzymes. FEBS Lett 2010; 584(2):303-309.
107. Ellis RJ. Macromolecular crowding: obvious but underappreciated. Trends Biochem Sci 2001; 26(10): 597-604.
108. Collins LJ, Kurland CG, Biggs P et al. The modern RNP world of eukaryotes. J Hered 2009; 100(5): 597-604.
109. Ebeling S, Kundig C and Hennecke H. Discovery of a rhizobial RNA that is essential for symbiotic root nodule development. J. Bacteriol. 1991; 173(20): 6373-6382.

CHAPTER 2

RNA INTERACTIONS

Manja Marz*,[1,2] and Peter F. Stadler[1,3-5]

[1]*Bioinformatics Group, Department of Computer Science and Interdisciplinary Center for Bioinformatics, University of Leipzig, Leipzig, Germany;* [2]*RNA-Bioinformatics Group, Institut für Pharmazeutische Chemie, Philipps-Universität Marburg, Marburg, Germany;* [3]*Institute for Theoretical Chemistry, University of Vienna, Wien, Austria;* [4]*Fraunhofer Institute for Cell Therapy und Immunology, Leipzig, Germany;* [5]*Santa Fe Institute, Santa Fe, New Mexico, USA*

**Corresponding Author: Manja Marz—Email: manja@staff.uni-marburg.de*

Abstract: Noncoding RNAs form an indispensible component of the cellular information processing networks, a role that crucially depends on the specificity of their interactions among each other as well as with DNA and protein. Patterns of intramolecular and intermolecular base pairs govern most RNA interactions. Specific base pairs dominate the structure formation of nucleic acids. Only little details distinguish intramolecular secondary structures from those cofolding molecules. RNA-protein interactions, on the other hand, are strongly dependent on the RNA structure as well since the sequence content of helical regions is largely unreadable, so that sequence specificity is mostly restricted to unpaired loop regions. Conservation of both sequence and structure thus this can give indications of the functioning of the diversity of ncRNAs.

INTRODUCTION

The discovery of a universe of noncoding RNAs (ncRNAs) of unexpected diversity during the last decade has profoundly changed our understanding of the role RNA in particular in eukaryotic cells. Merely a decade ago, RNA was mostly thought of as a (rather boring) intermediate in protein expression (mRNAs). A few classes of ancient RNAs, some with catalytic functions (RNase P RNA, ribosomal RNAs and self-splicing introns), were seen as remnants of an ancient RNA World, an evolutionary dead end in protein-dominated modern cells. In striking contrast, we now see ncRNAs as an integral part of the regulatory machinery in modern cells. In fact, there is mounting evidence that

RNA Infrastructure and Networks, edited by Lesley J. Collins.

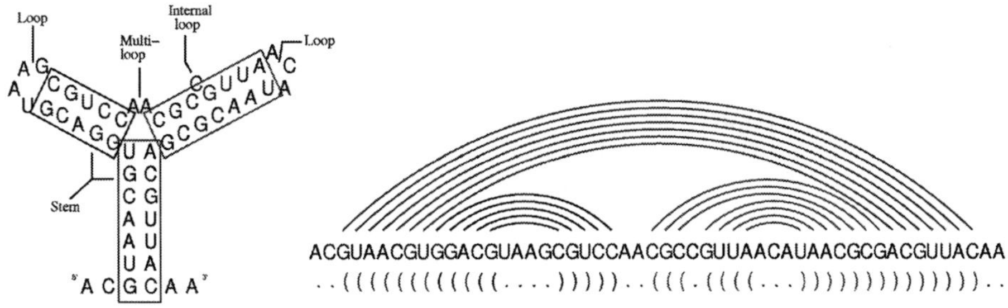

Figure 1. Example of an RNA secondary structure in three different representations: secondary structure graph (left), in bracket notation (right, bottom) and as arc plot (right, top). A pseudoknot is most easily defined by crossing arcs in its arc plot.

RNAs form a complex network of regulatory interactions among each other, with the genome and with a large variety of proteins.[1,2]

The spatio-temporal structure of the cellular "RNA infrastructure"[2] is shaped by the diversity of interactions of RNA molecules. Formation of base pairs between complementary nucleic acids is the key for both the structure formation of individual RNAs and for interactions between RNA and/or DNA molecules. The patterns of base pairing constitute the secondary structures (Fig. 1). They characterize functional classes of ncRNAs and are often well-conserved over large evolutionary time-scales (reviewed in see refs.3,4). Specific base pairing also contributes to the tertiary structure of the RNA.[5] The importance of specific base pairing is not limited to the structure of a single RNA, however. It has a crucial impact on the tertiary (3D) structure of the RNA and thus on way in which RNA is recognized by protein partners. Secondary structures are also formed by interacting RNAs, thus determining both strength and exact position of hybridization between two or more partners.

In this chapter we focus on how ncRNAs interact with their partners in a large number of different molecular and functional contexts. Base-pairing patterns play a central role because of their large impact on the structure of binding sites for proteins, or more directly, because the interaction consists of intermolecular base pairs. We organize our contribution by the composition of the players: RNA with DNA, RNA with RNA and RNA with proteins. Our presentation does not strive to be exhaustive and attempts to give a broad overview rather than an in-depth discussion of specific examples. Wherever possible, we thus cite recent reviews rather than the original literature.

ncRNA-DNA INTERACTIONS

RNA:DNA hybrids play a crucial role in transcription termination in bacteria: the formation of the terminator hairpin in the nascent RNA transcripts shortens the RNA:DNA duplex associated with the polymerase complex and facilitates dissociation.[6] Another important case are the RNA primers of the Okazaki fragment on the lagging strand during DNA replication.[7] Despite their use in biotechnology,[8,9] very little is known about the potential of triple-helices as form of direct RNA:DNA interaction.

Comprehensive surveys such as the ENCODE and FANTOM projects[10,11] demonstrated that the genomes of higher eukaryotes are pervasively transcribed. In mammals, many of these large intergenic noncoding RNAs (lincRNAs) associate with chromatin-modifying complexes and affect gene expression.[12-15] Apparently, there are diverse roles for lincRNAs in processes from embryonic stem cell pluripotency to cell proliferation, based on the observation that these transcripts are differentially regulated by key transcription factors such as p53, NFkappaB, or Nanog. Similar mechanisms are at work also in the yeast *Saccharomyces cerevisiae*[16] and in plants.[17,18] A significant fraction of long noncoding RNAs are subject to at least moderate stabilizing selection (where mutants of well adapted functional molecules are associated with a detrimental effect on fitness and hence are selected against) on the exon,[19] corroborating the functionality of these transcripts. Even transcripts with little or no sequence conservation may be functional as shown by the deep conservation of the gene structure itself.[20]

At present, multiple models for the mode of action of these long ncRNA transcripts have been proposed, reviewed e.g., in[21] and there is at least circumstantial evidence that different transcripts may be governed by different mechanisms (Fig. 2). Epigenetic regulator proteins directly or indirectly mediate chromatin regulation. By the first mechanism they are recruited from transcribed ncRNA or released from the acting site during transcription. By the latter mechanism regulative epigenetic markers may interact directly or via antisense RNA mediator with ncRNAs in order to bind to the target DNA. Therewith expression states might be switched and epigenetic memory is established.

Y RNAs were originally discovered as an RNA component of Ro RNPs, binding to Ro60 which at first is not related to DNA interaction. This RNP is active in RNA quality control[22] requiring RNA-RNA interactions between Y RNA and a variety of target RNAs. However, more recently Y RNAs were shown to be essential for chromosomal DNA replication.[23] Here, a cluster of Y RNAs is associated with the protein RPA and the sliding clamp PCNA and other proteins.[24-26] Y RNAs are essential for DNA replication. Not surprisingly, these ncRNAs are over expressed in many human cancer cells. So far, Y RNAs have been described in vertebrates and nematodes.[27-29]

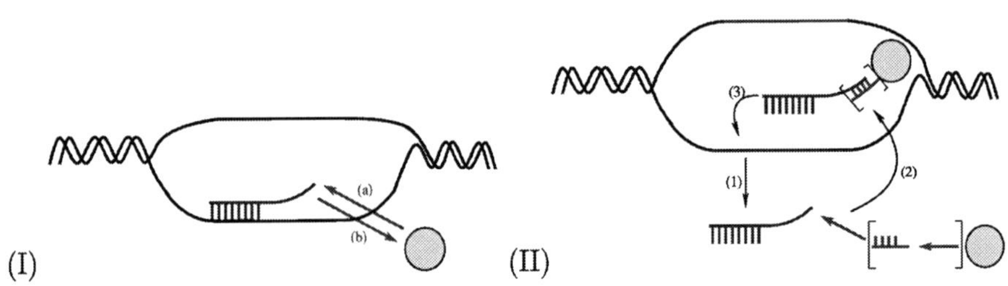

Figure 2. Putative mechanisms of chromatin-associated RNAs (modified from ref. 21: (I) Direct RNA mediated chromatin regulation during transcription. Epigenetic regulator proteins (circles) are (a) recruited directly from transcribed ncRNA or released from acting site during transcription (b). (II) Indirect association to DNA regulatory elements after transcription. After transcription of long ncRNAs from DNA (1), regulative epigenetic regulators (circles) may interact directly (without brackets) or via antisense RNA mediator (brackets) with ncRNAs (2) in order to bind to DNA (3). With this mechanism the recruitment of epigenetic regulatory proteins, switches off gene expression states and maintenance of epigenetic memory is performed.

RNA-RNA INTERACTIONS

General Properties

RNA-RNA interactions provide one of the fundamental mechanisms of cellular regulation. Single-stranded nucleic acids readily form complex interaction structures (cofolds) stabilized by complementary base pairing, thereby achieving a high sequence specificity. This recognition principle is utilized for wide variety of biological functions, including the decoding of the genetic code. Each codon is recognized by a complementary anti-coding presented by the corresponding tRNA. In this case, the interaction covers only three nucleotides, which is stabilized by the surrounding ribosomal machinery.[30] The direct binding of small RNAs to each other also plays a crucial role for the catalytic activity of the spliceosome, where snRNAs and the pre-mRNA cofold.[31]

At present, we lack high-throughput methods to assess RNA-RNA interactions both in vitro and in vivo. The direct proof of RNA-RNA binding in vivo is at least very difficult so that most if not all experimental reports on two interacting RNA-RNA molecules come from in vitro experiments. To this end chemical probing[32] is used to determine the secondary structure of both the isolated binding partners and their interaction structures. In practice, chemical probing, which determines by means of context-specific degradation whether a nucleotide is paired or unpaired only provides constraints which in some cases imply an unique RNA secondary structure model, while in other cases ambiguities remain.

In contrast to the experimental difficulties, RNA secondary structures and RNA-RNA interactions can be predicted efficiently by computational methods. The folding problem is posed as the following combinatorial matching problem with certain constraints. A matching in a graph consists of edges (here representing the base pairs) that have no vertices (here the nucleotides) in common. The task of computing a matching with the maximal number of edges is a classical problem in combinatorial optimization that can be solved efficiently. The nucleotides of the two sequences are represented as vertices of a graph, whose edges encode the logically allowed Watson-Crick and GU base pairs. We then search for a matching in this graph (i.e., a subset of edges such that every vertex/nucleotides take part in only one edge) that satisfies further structural constraints and maximizes an energy function. For RNA secondary structure prediction, for instance, one requires that base-pairs do not cross each other. The energy function accounts for base-pair stacking and loop entropies, i.e., it favours parallel base pairs and discounts long unpaired regions.[33] Although the RNA-RNA interaction problem (RIP) and the closely related RNA folding problem with arbitrary pseudoknots (see Fig. 1) are NP-complete (i.e., not be solved efficiently as the required computation scales exponentially with the length of the RNA sequence) in their most general forms,[34] efficient polynomial-time dynamic programming algorithms (a solution is recursively composed of solutions of smaller sub-programs) can be derived by restricting the space of allowed configurations.

The simplest approaches concatenate the two interacting sequences and subsequently employ a slightly modified standard secondary structure folding algorithm,[35] possibly allowing some pseudoknots.[36] The resulting model, however, still does not generate all relevant interaction structures. Alternatively, internal base-pairs in the interaction partners are neglected.[37] RNAup[38] and intaRNA[39] restrict interactions to a single contact interval; this model class has proved particularly useful for bacterial sRNA/mRNA interactions. To-date only a handful of interaction structures are known that are even more complex, some of which we will encounter later in this section.

MicroRNAs and RNA Interference

MicroRNAs were the first small regulatory RNAs found in animals. During the past few years, a wide variety of additional classes of small RNAs associated with the RNA interference (RNAi) pathways[40] were discovered throughout Eukarya. They share many functional properties and utilize at least in part the same processing machinery (see refs. 41, 42 for recent reviews). Their size of about 20-30nt and their final destination define these RNAs as a reasonably homogeneous group. Without going into detail about their biochemical processing, they guide large protein complexes to their targets, thus comprising the "RNA sensor" allowing sequence specific binding of the proteins. Both miRNAs and siRNAs form subclasses of this large class of small ncRNAs involved predominantly in gene silencing. MiRNAs stand out from the other small RNAs in several ways: they are processed from an extremely stable hairpin-structured precursor and they are typically highly conserved over long evolutionary timescales,[43] indicating crucial regulatory functions.

Although microRNAs clearly interact by basepairing with their targets, our understanding of the rules that govern the recognition of microRNA target sites is limited. Surprisingly, even exact complementarity is insufficient in some cases, while rather poor interactions are still functional in some cases. For animal microRNAs, the "seed rule", for instance, postulates that exact complementarity between nucleotides 2-7 with the mRNAs is necessary. There are several well-documented exceptions, however[44] (Fig. 3). In case of *C.elegans* insertions in the seed region are known as well as for plants mismatches beside several GU pairs. On the other hand, no mRNA cleavage was shown for perfect helix formations (mir-834/CIP4.1).

MicroRNAs may act in two distinct ways to downregulate protein expression either by translational repression or by *slicing*, i.e., endonuclease cleavage of the target mRNA. The latter pathway is the principle mode of action of siRNAs and possibly other sub-classes of short RNAs. MicroRNA target complementarity (Fig. 3) does not necessarily predict the regulatory output of the interaction. In particular complementarity does not imply slicing, while central mismatches prevent slicing, reviewed in reference 44. This observation is

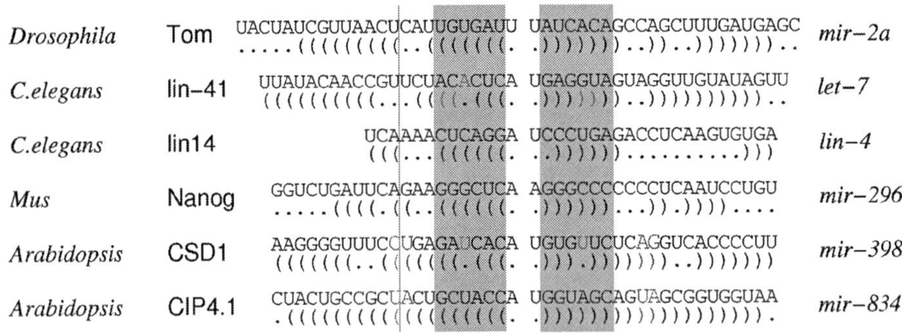

Figure 3. Structures of miRNA/target interactions adapted from reference 44. Let-7/lin-41, mir-398a/CSD1 and mir-298/Nanog violate the seed hypothesis by having mismatches or GU pairs in the (shaded) seed region. The mir-2a/Tom and lin-4/lin-14 interactions are functional but show little complementarity outside the seed region. Although mir-834/CIP4.1 forms a perfect helix, it does not lead to mRNA cleavage, mir-398a/CSD1, on the other hand, can act according to both mechanisms. The vertical line indicates the putative mRNA cleavage position.

consistent with structural models that suggest that the mRNA faces the Ago RNase H active site about 10nt from the beginning of the miRNA–mRNA duplex.[45] Imperfectly paired target sites in animal transcripts often occur multiple times in the same mRNA. Cooperative action of multiple sites has also been observed.

Small Bacterial RNAs and Translational Control

Small noncoding RNAs (sRNAs) form an important class of regulators of gene expression. In bacteria, their role in the cell was not fully appreciated until the discovery of hundreds of sRNAs in *E. coli* and many other species.[46,47] While the chaperone protein Hfq plays an important role in facilitating sRNA/mRNA interactions, sRNAs are also abundant in species such as *Helicobacter pylori* that lack Hfq[48] [(Fig. 4)(see Chapter 14, pages 224-227 for more detail on Hfq related regulation)].

The exact location and structure of the sRNA/target duplex has a decisive impact on the function of the sRNA. While most sRNAs act to downregulate protein expression exceptions are well known. Typically, the sRNAs regulate translation by influencing the accessibility of the Shine-Dalgarno (SD) sequence. If the SD is contained in a stable secondary structure, translation is inhibited. Most sRNAs achieve this by binding directly to the SD, thus covering the SD by a sRNA/mRNA interaction.[49] Several sRNAs, however, (among them DsrA, GlmZ, RNAIII, RprA, RyhB and Qrr) activate translation by binding the 5′ mRNA region in such a way that the mRNA refolds to liberate a sequestered ribosome binding site.[50] The pairing of GadY sRNA to the 3′-end alters processing and increases mRNA stability.

The binding of a small RNA to an mRNA can have long-range effects through a refolding of the mRNA. The binding affinity of *HuR* protein to human mRNAs, for example, can be tuned by artificial "opener" and "closer" RNAs that interact far away from the *HuR* binding motif.[51,52] So far, it remains open to what extent such induced RNA refolding plays a role in eukaryotes. Changes in RNA secondary structure can also be induced by temperature changes or metabolite binding. A wide variety of procaryotic riboswitches, reviewed in,[53] utilizes this mechanism to control either translation (by hiding or exposing the SD) or transcription (by forming a premature terminator signal upstream of the coding sequence). Complex sensors consisting of elaborate RNA structures also play a role in recently-evolved eukaryotic mRNAs.[54,55] A recent study of the primary transcriptome of *Helicobacter pylori*[48] demonstrated an unexpectedly high number of anti-sense transcripts and alternative transcription start sites, indicating that there may be an additional layer of regulatory complexity based on these novel transcripts.

Small Nucleolar RNAs and Chemical Modifications

Chemical modifications of RNAs are an evolutionarily old phenomenon. In particular, tRNAs are heavily edited. In total, over 100 structurally distinguishable modified nucleosides have been encountered so far in different types of RNAs from many diverse organisms of all domains of life.[56,57] These modifications increase the chemical diversity of RNA and are beneficial or even critical for proper folding and function of the RNA molecule. Most types of modifications are carried out by specialized, site-specific enzymes. A large number of pseudouridine and 2′-O-methylated nucleotides however, are synthesized by generic RNP complexes. They recognize their target sites in rRNAs, snRNAs and tRNAs through their RNA components, the H/ACA and C/D snoRNAs,

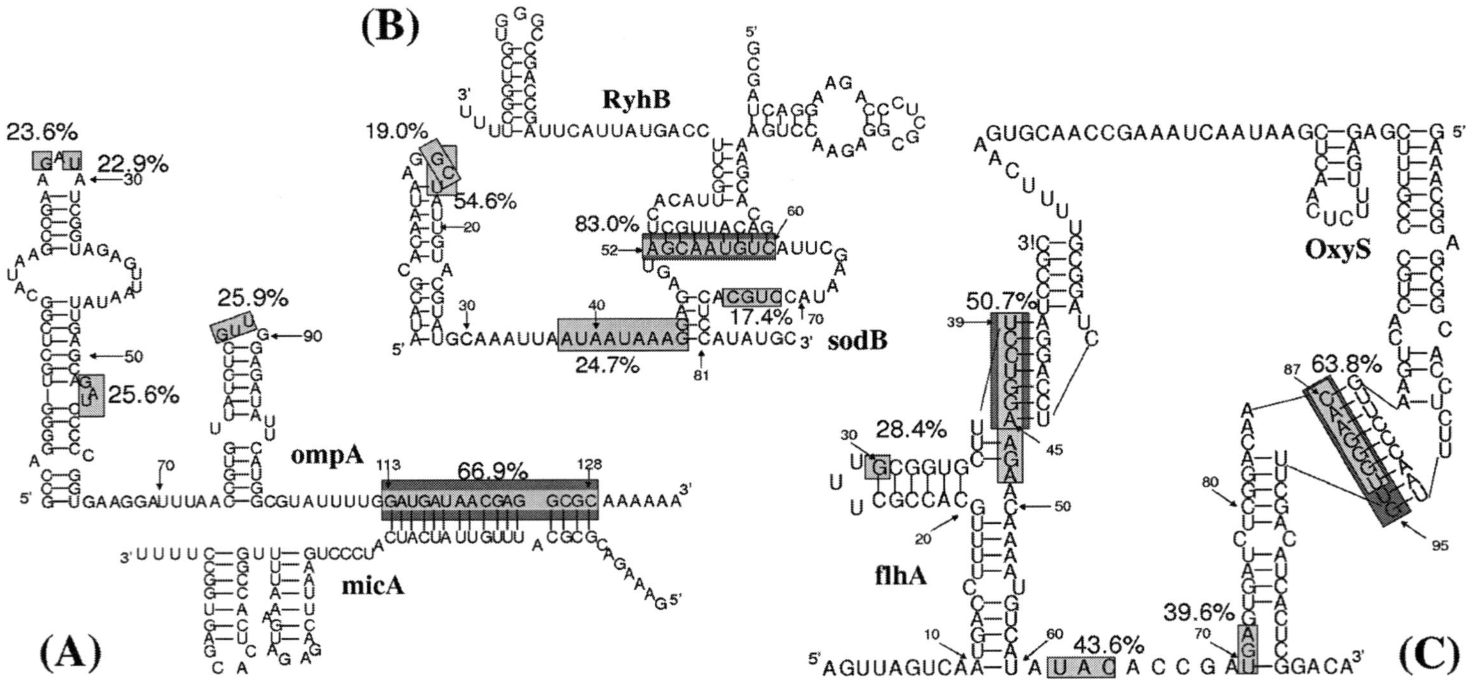

Figure 4. Three examples of interaction between bacterial sRNA and mRNAs. The primary interaction region(s), corresponding to the published structures are highlighted with double boxes, hybridization probabilities computed by rip2[135] are annotated by shaded boxes, identifying additional hybridization regions that may further stabilize the interaction. A) The small RNA micA accumulates in *E. coli* upon entry into stationary phase and down-regulates expression of ompA (outer membrane protein A) by binding across the Shine-Dalgarno sequence of this mRNA. B) RyhB represses translation of the superoxide dismutase SodB in response to low iron concentration in several bacteria. C) OxyS is induced in response to oxidative stress in *E. coli* and represses the translation of flhA, a transcriptional activator for formate metabolism.

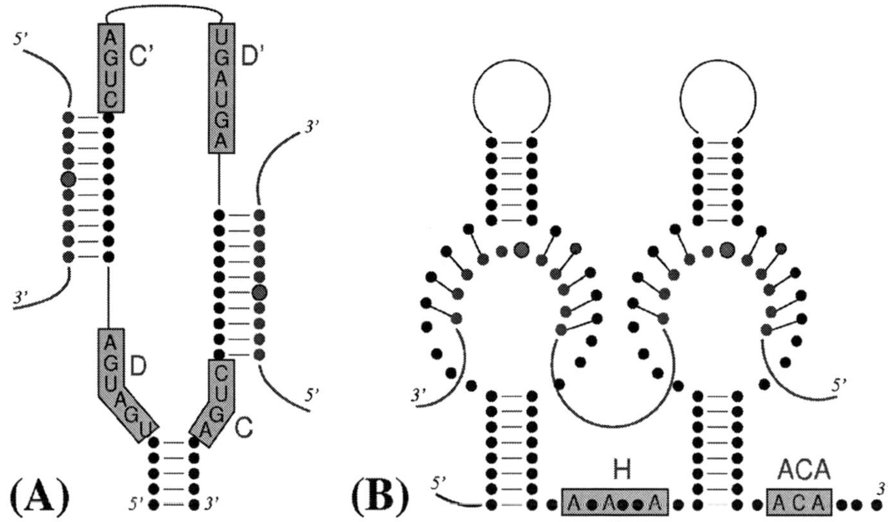

Figure 5. The two classes of snoRNA have different structures and different interactions with their targets. Box C/D snoRNAs (A) guide 2′ O-methylation to the fifth nucleotide from the D and D′ boxes. Some C/D snoRNAs lack a discernible D′ box and modify only a single target. In contrast, the target site of box H/ACA snoRNAs (B) is determined by the structure. The target uracil (shaded) is positioned by two specific interactions of the flanking target RNA sequence with the complementary sequence of the recognition loop of the snoRNA.[136]

respectively.[58] These two classes of enzymatically active RNPs are ubiquitous in eukaryotic and archaeal organisms but are not found in bacteria. H/ACA and C/D snoRNAs are evolutionarily and structurally unrelated (Fig. 5).

A small class of hybrid snoRNAs contain an H/ACA and a box C/D motif. An example is U85, guiding both the pseudouridylation of base U46 and the 2′O-ribose methylation of base C45 of the U5 snRNA.[59] Several snoRNAs, in particular those of hybrid structure but also several otherwise canonical ones, contain a CAB box (AGAG, typically located in a hairpin loop). The signal directs the snoRNAs to the Cajal body, where they guide modifications of the RNA polymerase II-transcribed snRNAs (U1, U2, U4 and U5). In Archaea, the snoRNAs also target tRNA precursors.[60]

An increasing number of orphan snoRNAs, i.e., snoRNAs lacking targets in rRNAs or snRNAs, has been described in different eukaryotes. In particular, a subgroup of snoRNAs expressed in the mammalian brain does not appear to be involved in modifications of rRNAs or snRNAs. Instead some of these snoRNAs make use of the same type of interaction to target specific mRNAs and appear to interfere with A-to-I editing, which in turn can alter the functional properties of proteins, silence constitutive activity and modulate RNA translation, localization and stability.[61,62]

RNAs in RNA End-Processing

The overwhelming majority of protein-coding pol-II transcripts ends in a poly(A) tail that is generated by endonucleolytic cleavage followed by polyadenylation.[63] This is not an absolute rule, however. The mRNAs of the replication-dependent histones lack

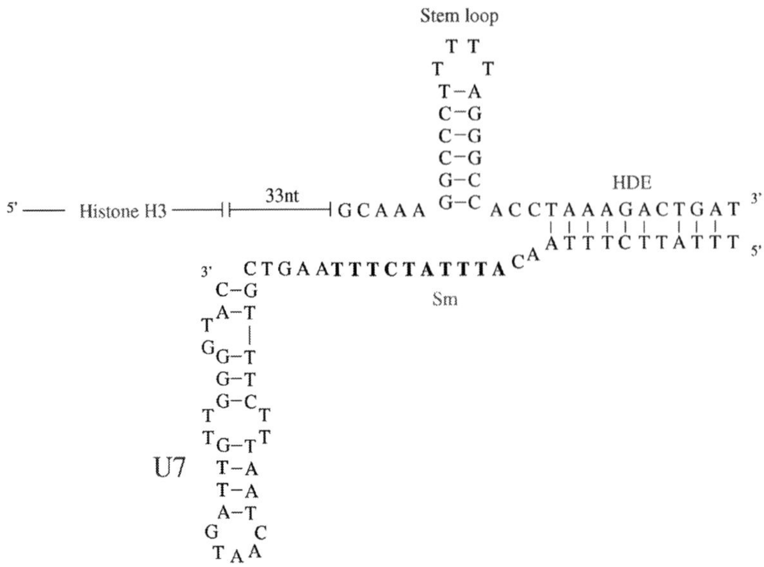

Figure 6. U7-directed processing of histone 3′ ends. U7 snRNA interacts upstream of the Sm protein binding site with histone downstream element (HDE), about 30nt downstream of histone-pre-mRNA.

poly(A) and instead feature a highly conserved stem–loop structure in their 3′ UTRs. It binds the stem–loop binding protein that ensures RNA stability and enhances translational efficiency. The 3′ end is determined by base-pairing of the histone-down-stream element (HDE) with the U7 snRNA (Fig. 6) which directs endonucleolytic cleavage.[64] In contrast, the 3′ ends of the pol-II transcribed snRNAs are produced by the *Integrator*, a specialized protein complex.[65] The 3′ end of the telomerase RNA is processed by the spliceosome in *Schizosaccharomyces pombe*.[66] Finally, RNase P RNA processes the 3′ ends of the two long ncRNAs MALAT-1 and MENε, utilizing a tRNA-like element to attract the tRNA processing machinery. Besides tRNAs, RNase P RNA in addition cleaves the yeast HRA1 ncRNAs, bacterial riboswitches and possibly also some box C/D snoRNAs [(see ref. 67 for a review)(see Chapter 1, Figure 2, page 5 for a summary of RNase P activity, and Chapter 14, pages 222-223 for more detail on RNase P in prokaryotes)].

The rRNA operon is transcribed by pol-I. The maturation of the ribosomal RNAs involves a complex cascade of processing steps. In Eukarya and possibly Archaea, this involves also the use of the atypical box C/D U3 snoRNA[68] that acts as an RNA-chaperone mediating the correct structure conformations of the pre-rRNA for endonuclease cleavage.[69] Unlike other small nucleolar RNAs, U3 snoRNA acts not by a direct 2′-O-methylation, instead the 5′ domain (consisting of A and A′ box) and the single stranded hinge region (connecting the the 5′ and 3′ domain of U3 snoRNA) bind to the prerRNA. The 3′ domain of U3 snoRNA hypermethylates itself, arranges RNA stability and has essential conserved protein binding sites.

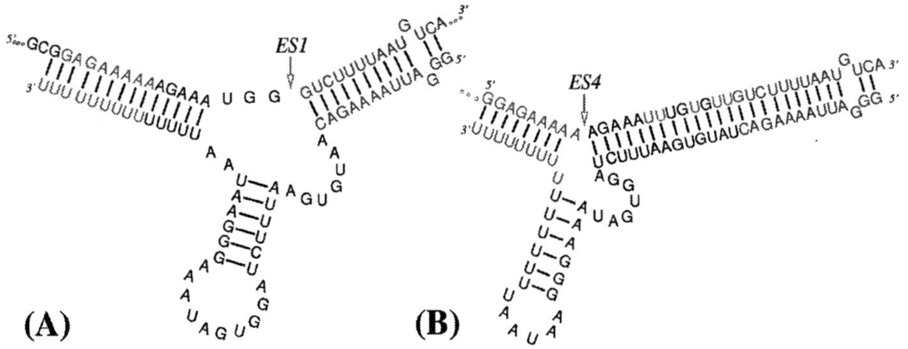

Figure 7. U insertion editing of the CYb mRNA of *Trypanosoma*.[137] A single guide RNA (bottom strand) encodes multiple editing sites. The gRNA and its mRNA target (upper strand) form a 3-stem structure that is refolded as the processing moves from one editing site to the next.

Guide RNAs and RNA Editing

Many of the mRNAs of the kinetoplasts (mitochondria) of Trypanosomes and Leishmanias are edited by inserting and/or deletion of uridines, a process that depends on a specialized class on ncRNAs, the guide RNAs (gRNAs). In contrast to the ncRNA-guided editing system in Kinetoplastida, the C->U editing in plant organelles is directed by *cis*-acting elements and a plethora of PPR enzymes (reviewed in see ref. 70).

The gRNAs of kinetoplastids are typically 50-70nt long and contain three functional elements. A 5-21 nt region on their 5′ side acts as anchor specifically recognizing the target mRNA. The "guide region" in the middle of the molecule serves a template for editing. It is complementary (allowing GU pairs, however) to the mature, edited mRNA. The 3′-tail consists of posttranscriptionally added poly-U tail. The editing process takes place in the editosome, a complex structure comprising of more than 20 proteins arranged around the mRNA/gRNA pair.[71] The editosome typically performs several successive rounds of enzymatic reactions templated by a gRNA (Fig. 7). Most of the gRNAs are not encoded in the major maxicircle DNA, which also contains the mitochondrial rRNA and protein-coding genes. Instead, large numbers of gRNA genes are located on the minicircles.[72,73]

The U insertions and deletions can be extensive. The 1246 nts ND7 mRNA of *Trypanosoma brucei*, for instance, is processed by inserting 551 and deleting 86 uridines. It is no surprise, therefore, that the U insertions and deletions typically change the open reading frames of the mRNAs. Intriguingly, the premRNAs of the *Trypanosoma brucei* mitochondrion are edited in sometimes alternative ways to yield distinctive protein sequences. Multiple gRNAs that target the same primary transcript[72] expand the diversity of mitochondrial proteins since they can produce significantly different alternative ORFs. These editing variants are of functional importance as the case of AEP-1 (alternatively edited protein 1) shows, which is involved in the maintenance of kinetoplastid DNA.

CRISPRs

Immunity against viruses and plasmids in 40% of Eubacteria and 90% of Archaea is connected to clustered regularly interspaced short palindromic repeat sequences

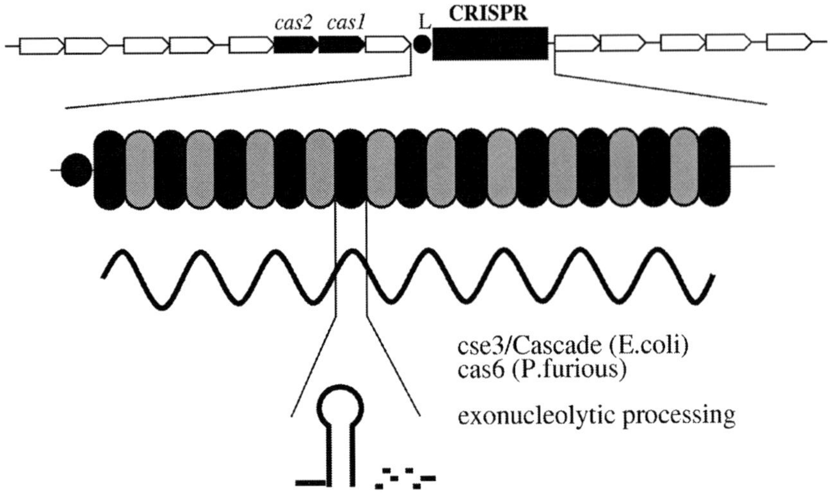

Figure 8. Processing of CRISPR content into crRNAs (redrawn from refs. 74,79). Cas genes (black constrained, white optionally) upstream and downstream of CRISPR cluster. The Leader sequence (L) is located directly upstream of CRISPR. Transcribed CRISPRs are cleaved within the repeat by cse3 or cas6. Additionally, precrRNA is exonucleolytic processed into mature crRNA consisting of a repeat tag (folding into a single hairpin) and the spacer sequence.

(CRISPRs). Transcripts from these loci target foreign nucleic acids in by specific-sequence detection. In addition to other well-described defense strategies, such as prevention of adsorption, blocking of injection and abortive infection,[74] the CRISPR clusters store invading genetic information in preparation for an "immune response" during a second infection. Immunization is established in three phases:[75,76] (1) incorporation of new spacers into CRISPR arrays, (2) expression and processing of CRISPR RNAs (crRNA) and (3) CRISPR interference.

Various *cas* proteins are located upstream and optionally downstream of CRISPR cluster (Fig. 8). Six core genes *cas1* to *cas6*, located <1kb around CRISPR have been identified to interact beside repeat-associated mysterious proteins (RAMP) with mature processed CRISPR (crRNA). *cas1* acts as endonuclease within immunization process and *cas2 and cas6* are a sequence specific endoribonuclease. The CRISPR leader located upstream of repeat tags is defined by a low-complexity A-T-rich region. This area acts as promoter for the CRISPR transcript. The spacer addition provides novel phage resistance, whereas spacer deletion was shown to result in a loss of phage resistance.[77] With the CRISPR/Cas system, horizontal gene transfer between distant organisms was shown by resistance of bacteria that were never invaded by that phage before.[78]

Recently, a possible relationship between CRISPR interference and eukaryotic RNA interference (RNAi) by siRNA, miRNA and piRNA has been discussed,[76] since there are many obvious similarities: (1) RNA guiding effector apparatus to the target, (2) the gene function is affected in a programmable and sequence directed manner, (3) adaptive and heritable components used to establish recoverable genomic records of past invasions (e.g., piRNA of eukaryotic silencing). On the other hand, the two systems clearly are not homologous: both the protein machinery is completely different and crRNAs are not amplified posttranscriptionally from single stranded precursors. For more details, we refer to the reviews in references 74, 76 and 79 (see Chapter 13, pages 213-218 for more detail on the CRISPR system).

Ribozymes

A ribonucleic acid enzyme (ribozyme) is an RNA molecule that, like the much more familiar protein-based enzymes, catalyzes a chemical reaction. Although most of them function together with protein components in vivo, experiments in in vitro prove that the catalytic activity resides in the RNA component. Natural ribozymes include, for example, the peptidyl transferase activity of the 23S rRNA, the hammerhead and hairpin ribozyme, and RNase P RNA. Similar to the ribosome, the spliceosome is in essence also a ribozyme.[80] The efficiency of the catalysis, however, depends to a certain extent on secondary structure features of the mRNA to be catalyzed.[81]

Ribonuclease P (RNase P) and mitochondrial RNA processing (RNase MRP) are ribozymes acting in processing tRNA and rRNA, respectively. The RNA subunits are evolutionarily related. RNase MRP is eukaryotic specific whereas RNase P is present in all kingdoms of life.[82] The high similarity of P and MRP RNA secondary structures[83] and similarity of the protein contents and interactions of RNase P and MRP[84,85] suggest that P and MRP RNAs are paralogues. Both enzymes consist of an RNA molecule and 9-12 protein subunits, most of which are part of both enzymes. An *in silico* relationship was shown for Pop8 and Rpp14/Pop5 as well as Rpp25 and Pop6.[86] The RNA subunit and its interacting proteins build a coevoluting network.

Beyond the relatively small collection of natural ribozymes, it is quite easy to "breed" catalytically active RNAs (and even single-stranded DNAs) in in vitro selection (SELEX) experiments, see for example a recent review.[87] In fact, most of the chemical transformations that are catalyzed by protein enzymes in extant organism can be implemented in moderate-size RNAs. This lends further credibility not only to the RNA World Hypothesis but opens the possibility that at least some ncRNAs might also be catalytically active. There is, however, no indication that RNA catalysis is wide-spread phenomenon beyond some classes of "selfish elements" such as viroids.

ncRNA-PROTEIN INTERACTIONS

Ribonucleoparticles

Ribonucleoparticles (RNPs) are composed of both RNA and protein components. In some cases, such as RNase P, where the RNA is catalytically active, one speaks of ribonucleoproteins. In the following we briefly explore the huge variety of RNA-protein interactions using 7SK RNA, SRP RNA, Telomerase RNA, spliceosomal RNAs and vault RNA as examples.

The 7SK RNA has been described in a diverse set of animals.[88-92] It negatively controls transcription by regulating the activity of the Positive Transcription Elongation Factor b (P-TEFb) for Polymerase II.[93,94] In this process HEXIM1/2 proteins dissolve from P-TEFb and bind to the polymerase III transcript 7SK RNA.[95-97] Furthermore the La-related protein 7 (LARP7) was proven to regulate the stability of 7SK RNA.[88,98,99] The network of 7SK RNA and its specific protein partners is a metazoan invention.[90-92,100] The well characterized LARP7 homologs[101] are found in most metazoan clades, HEXIM was found in all clades of metazoans excluding Platyhelminthes. Two copies are known for eutheria.[92,102] In contrast, another protein known to interact with 7SK RNA, MePCE/BCDIN3 has a much broader phylogenetic distribution, indicating that it has other important functions beyond its interaction with 7SK RNA.

The signal recognition particle (SRP) is present in all living organisms. This ribonucleoparticle targets ribosomes to the endoplasmatic reticulum (ER) in order to translate proteins at the ER. The SRP complex consists of a highly conserved 300 nt SRP RNA (or 7S RNA) and six proteins (9, 14, 19, 54, 68 and 72 SRPs) in eukaryotes. In archaeas four protein particles are absent and SRP19 and SRP54 exists compared to one Ffh protein (homolog to SRP54) and a RNA molecule (also known as 4.5S RNA or 6S RNA in *Bacillus*).[103] Although the network of SRP RNA and corresponding proteins is highly conserved, substantial variations between the three main kingdoms of life are clearly visible.[104]

Eukaryotic cells have been shown to decrease protein synthesis and increasing the expression of protein quality control mechanisms, such as chaperones and proteases. The bacterial SRP receptor, FtsY, inhibits the translation of both SRP-dependent and SRP-independent proteins.[105] In higher plants two different SRP-dependent mechanisms are known: once proteins are posttranslational transferred to chloroplasts and on the other hand proteins are cotranslational encoded by the plastid genome.[106] Although for the first system in general no RNA component seems to be necessary. However in single plastids from red algae and chlorophyta a reminiscent eubacterial SRP was identified. The cotranslational protein requires the SRP RNA, which accelerates the interaction between the SRP and SRP receptor 200-fold.[107] This SRP RNA is missing in the chloroplast SRP (cpSRP) pathway. Instead, the cpSRP and cpSRP receptor (cpFtsY) by themselves can interact 200-fold faster than their bacterial homologues.[107]

Linear chromosomal ends of most eukaryotes are replicated by a telomerase enzyme which consists of a catalytic protein component, the telomerase reverse transcriptase (TERT) and the core functional unit telomerase RNA.[66] Telomerase dates back to the origin of eukaryotes.[108] Notable exceptions are diptera including *Anopheles* and *Drosophila*, which use retrotransposons or unequal recombination instead of a telomerase enzyme. Although TERT proteins are experimentally validated for a wide range of eukaryotes, the knowledge of RNA:protein interacting networks implies that telomerase RNA exists for more organisms than the experimentally validated small phylogenetic groups within vertebrates, yeasts, ciliates and plasmodia.[109,110] Telomerase is normally active only early in development and in stem cells and the germline. It extends the ends of linear chromosomes and thus counteracts the gradual shortening of the ends caused by intrisic limitations of the replication machinery. In cancer research, telomerase has received a lot of interest because of this crucial role in cell proliferation and the observation that telomerase is also active in many cancer cells.[111]

Several small ncRNAs are involved in splicing processes. Small nuclear RNAs (snRNAs) act with up to 200 proteins as a large RNP (spliceosome) in eukaryotes to splice exons of protein-coding genes together.[112] We find two homologous systems for splicing in eukaryotes: The major spliceosome involves U1, U2, U4, U5 and U6 snRNA and splices more than 90% of (human) protein-coding genes. In contrast, the minor spliceosome utilizes the alternative U11, U12, U4atac, U5 and U6atac RNAs and plays an ancillary role in the nucleus. Although the splicing reactions seem to date back until the last unknown common ancestor (LUCA), its extensive use in protein expression seems to be an eukaryotic innovation.[113] Contrary to *cis*-splicing, in eight eukaryotic phyla (kinetoplastids, nematodes, platyhelminthes, cnidaria, arthropods, tunicates, rotifera, dinoflagellates) a short leader sequence derived from small SL RNA is added to the 5′ end of a mRNA by *trans*-splicing.[114,115] The Sm protein binds to SL RNAs in order to act in this spliceosome-catalyzed process. In nematoda SmY RNAs are hypothesized to recycle the spliceosome proteins after SL RNAs are consumed in the *trans*-splicing reaction.[116] Direct interactions are not known.

While the spliceosome and its RNAs are well understood, there are several much more enigmatic ncRNA species. One example is the vault ribonucleoprotein complex, one of the largest RNPs. It consists of 8-16 noncoding vault RNAs, a major vault protein (MVP) and two minor vault proteins (VPARP and TEP1).[117] It is involved in transport processes to and from the nuclear membrane. Vault RNAs are polymerase III transcripts with length of about 100nt which are known throughout the deuterostomes[118] and in a few protostomes.[119] The RNA components are not necessary for the structural assembly of vault particles.[120] They strongly bind to a variety of organic molecules including chemotherapy drugs, which might explain the relation of the vault complex with chemoresistance.

Long ncRNA as Coat-Hangers?

A handful of long noncoding RNAs appear to play a crucial role in the organization of complex RNPs, acting like a "coat-hanger" to bring together different proteins. An example is the huge telomerase RNA of yeasts.[121] Another example is the heat-shock ncRNAs hsrω of *Drosophila*. Different isoforms of hsrω are expressed nearly ubiquitously in a developmentally regulated pattern. Upon heat shock, the nuclear-retained longest isoform sequesters a variety of nuclear RNA processing factors.[122] Similarly, NEAT1 (also known as MENε) may act as organizing factor of the paraspeckles,[123,124] which are ribonucleoprotein bodies found in the interchromatin space of mammalian cell nuclei. The hsrω-n RNA of *Drosophila* probably plays a similar function in the organization of the ω speckles.[122] It is a crucial component of the stress response.

At present, it is unknown to what extent other long ncRNAs function as stabilizing backbones of large RNPs, bringing together specific combinations of proteins and what role such complexes might play.

CONCLUSION

There is compelling evidence, that ncRNAs and especially long ncRNAs do not constitute a homogeneous class of transcripts but rather differ dramatically in structure, function, mechanisms and processing. While some are retained in the nucleus, others are predominantly cytoplasmic. As in the case of *Drosophila* hsrω, different splice variants and different poly-adenylation sites can be associated with differences in cellular localization.[122] Some mRNA-like noncoding RNs (mlncRNAs) are primarily involved in shaping spatial structures. Others, such as the noncoding host genes of snoRNAs and microRNAs[125] are expressed to deliver their pay-load of small structured RNAs to the appropriate processing machinery. Yet another group is processed into a large collection of short RNAs such as the precursors of piRNAs,[126] and nuclear ncRNAs interact directly with the transcription machinery.[127]

It may not come as a surprise that functionalities are not strictly separated. Many of the evolutionary ancient and well-conserved "housekeeping RNAs", including tRNAs, snoRNAs and snRNAs, give rise to smaller processing products[118,128-130] that at least in some cases have their own function in gene regulation.[131-133] Similarly, MALAT1 and MENβ are prolific sources of small RNAs.[134] We may conclude therefore, that the network of a cell's RNA infrastructure is not only shaped by complex patterns of interactions among RNAs and between RNA and protein and DNA, but also by an intricate tangle of processing pathways that extract multiple functions from different "developmental stages" of the same primary transcript.

ACKNOWLEDGEMENTS

This work was supported in part by the State of Saxony through a "Landesstipendium".

REFERENCES

1. Amaral PP, Dinger ME, Mercer TR et al. The eukaryotic genome as an RNA machine. Science 2008; 319:1787-1789.
2. Collins LC, Penny D. The RNA infrastructure: dark matter of the eukaryotic cell? Trends Genet 2009; 25:120-128.
3. The Athanasius F. Bompfunewerer RNA Consortium: RNA's everywhere: genome-wide annotation of structured RNA's. J Exp Zool B: Mol Dev Evol 2007; 308B:1-25.
4. Bompfünewerer AF, Flamm C, Fried C et al. Evolutionary patterns of noncoding RNAs. Theory Biosci 2005; 123:301-369.
5. Leontis NB, Westhof E. Analysis of RNA motifs. Curr Opin Struct Biol 2003; 13:300-308.
6. Komissarova N, Becker J, Solter S et al. Shortening of RNA:DNA hybrid in the elongation complex of RNA polymerase is a prerequisite for transcription termination. Molecular Cell 2002; 10:1151-1162.
7. MacNeill SA. DNA replication: partners in the Okazaki two-step. Curr Biol 2001; 11(20):R842-R844.
8. Nadal A, Eritja R, Esteve T et al. "Parallel" and "antiparallel tail-clamps" increase the efficiency of triplex formation with structured DNA and RNA targets. Chembiochem 2005; 6:1034-1042.
9. Suzuki T. Targeted gene modification by oligonucleotides and small DNA fragments in eukaryotes. Front Biosci 2008; 13:737-744.
10. The ENCODE Project Consortium. Identification and analysis of functional elements in 1% of the human genome by the ENCODE pilot project. Nature 2007; 447:799-816.
11. The FANTOM Consortium. The Transcriptional Landscape of the Mammalian Genome. Science 2005; 309:1159-1563.
12. Khalil AM, Guttman M, Huarte M et al. Many human large intergenic noncoding RNAs associate with chromatin-modifying complexes and affect gene expression. Proc Natl Acad Sci USA 2009; 106:11675-11680.
13. Sheik MJ, Gaughwin PM, Lim B et al. Conserved long noncoding RNAs transcriptionally regulated by Oct4 and Nanog modulate pluripotency in mouse embryonic stem cells. RNA 2010; 16:324-337.
14. Mattick JS, Amaral PP, Dinger ME et al. RNA regulation of epigenetic processes. Bioessays 2009; 31:51-59.
15. Mondal T, Rasmussen M, Pandey GK et al. Characterization of the RNA content of chromatin. Nat Methods. 2010;7(8):582
16. Harrison BR, Yazgan O, Krebs JE. Life without RNAi: noncoding RNAs and their functions in Saccharomyces cerevisiae. Biochem Cell Biol 2009; 87:767-779.
17. Matzke M, Kanno T, Daxinger L et al. RNA-mediated chromatin-based silencing in plants. Curr Opin Cell Biol 2009; 21:367-376.
18. Swiezewski S, Liu F, Magusin A et al. Cold-induced silencing by long antisense transcripts of an Arabidopsis Polycomb target. Nature 2009; 462:799-802.
19. Marques AC, Ponting CP. Catalogues of mammalian long noncoding RNAs: modest conservation and incompleteness. Genome Biol 2009; 10:R124.
20. Hiller M, Findeiß S, Lein S et al. Conserved introns reveal novel transcripts in Drosophila melanogaster. Genome Res 2009; 19:1289-1300.
21. Hekimoglu B, Ringrose L. Noncoding RNAs in polycomb/trithorax regulation. RNA Biol 2009; 6:129-137.
22. Stein AJ, Fuchs G, Fu C et al. Structural insights into RNA quality control: the Ro autoantigen binds misfolded RNAs via its central cavity. Cell 2005; 121:529-539.

23. Christov CP, Gardiner TJ, Szüts D et al. Functional requirement of noncoding Y RNAs for human chromosomal DNA replication. Mol Cell Biol 2006; 26:6993-7004.
24. Krude T. Non-coding RNAs: New players in the field of eukaryotic DNA replication. In: Nasheuer HP, ed. Genome Stability and Human Diseases. New York: Springer, 2010:105-118.
25. Szüts D, Christov C, Kitching L et al. Distinct populations of human PCNA are required for initiation of chromosomal DNA replication and concurrent DNA repair. Exp Cell Res 2005; 311:240-250.
26. Szüts D, Kitching L, Christov C et al. RPA is an initiation factor for human chromosomal DNA replication. Nucleic Acids Res 2003; 31:1725-1734.
27. Mosig A, Guofeng M, Stadler BMR et al. Evolution of the vertebrate Y RNA cluster. Theory Biosci 2007; 126:9-14.
28. Perreault J, Perreault JP, Boire G. Ro-associated Y RNAs in metazoans: evolution and diversification. Mol Biol Evol 2007; 24:1678-1689.
29. Boria I, Gruber AR, Tanzer A et al. Nematode sbRNAs: homologs of vertebrate Y RNAs. J Mol Evol 2010.
30. Aitken CE, Petrov A, Puglisi JD. Single ribosome dynamics and the mechanism of translation. Annu Rev Biophys 2010; 39.
31. Valadkhan S. snRNAs as the catalysts of pre-mRNA splicing. Currl Op1 Chem1 Biol 2005; 9:603-608.
32. Brunel C, Romby P. Probing RNA structure and RNA-ligand complexes with chemical probes. Methods Enzymol 2000; 318:3-21.
33. Mathews D, Sabina J, Zuker M et al. Expanded sequence dependence of thermodynamic parameters improves prediction of scRNA secondary structure. J Mol Biol 1999; 288:911-940.
34. Alkan C, Karakoc E, Nadeau JH et al. RNA-RNA interaction prediction and antisense RNA target search. J Comput Biol 2006; 13:267-282.
35. Hofacker IL, Fontana W, Stadler PF et al. Fast folding and comparison of RNA secondary structures. Monatsh Chem 1994; 125:167-188.
36. Rivas E, Eddy SR. A Dynamic programming algorithms for RNA structure prediction including pseudoknots. J Mol Biol 1999; 285:2053-2068.
37. Rehmsmeier M, Steffen P, Höchsmann M et al. Fast and effective prediction of microRNA/target duplexes. Gene 2004; 10:1507-1517.
38. Mückstein U, Tafer H, Hackermüller J et al. Thermodynamics of RNA-RNA binding. Bioinformatics 2006; 22:1177-1182.
39. Busch A, Richter AS, Backofen R. IntaRNA: efficient prediction of bacterial sRNA targets incorporating target site accessibility and seed regions. Bioinformatics 2008; 24:2849-2856.
40. Kim D, Rossi J. RNAi mechanisms and applications. Biotechniques 2008; 44:613-616.
41. Carthew RW, Sontheimer EJ. Origins and mechanisms of miRNAs and siRNAs. Cell 2009; 136:642-655.
42. Verdel A, Vavasseur A, Le Gorrec M et al. Common themes in siRNA-mediated epigenetic silencing pathways. Int J Dev Biol 2009; 53:245-257.
43. Hertel J, Lindemeyer M, Missal K et al. The expansion of the metazoan MicroRNA repertoire. BMC Genomics 2006; 7:15.
44. Brodersen P, Voinnet O. Revisiting the principles of microRNA target recognition and mode of action. Nat Rev Mol Cell Biol 2009; 10:141-148.
45. Song JJ, Smith SK, Hannon GJ et al. Crystal structure of Argonaute and its implications for RISC slicer activity. Science 2004; 305:1434-1437.
46. Vogel J, Sharma CM. How to find small noncoding RNAs in bacteria. Biol Chem 2005; 386:1219-1238.
47. Waters LS, Storz G. Regulatory RNAs in bacteria. Cell 2009; 136:615-628.
48. Sharma CM, Hoffmann S, Darfeuille F et al. The primary transcriptome of the major human pathogen Helicobacter pylori. Nature 2010; 464:250-255.
49. Sharma CM, Darfeuille F, Plantinga TH et al. A small RNA regulates multiple ABC transporter mRNAs by targeting C/A-rich elements inside and upstream of ribosome-binding sites. Genes Dev 2007; 21:2804-2817.
50. Fröhlich KS, Vogel J. Activation of gene expression by small RNA. Curr Opin Microbiol 2009; 12:674-682.
51. Meisner NC, Hackermüller J, Uhl V et al. mRNA openers and closers: modulating AU-rich element-controlled mRNA stability by a molecular switch in mRNA secondary structure. Chembiochem 2004; 5:1432-1447.
52. Hackermüller J, Meisner NC, Auer M et al. The effect of RNA secondary structures on RNA-ligand binding and the modifier RNA mechanism: a quantitative model. Gene 2005; 345:3-12.
53. Henkin TM. Riboswitch RNAs: using RNA to sense cellular metabolism. Genes Dev 2008; 22:3383-3390.
54. Kaempfer R. RNA sensors: novel regulators of gene expression. EMBO Rep 2003; 4:1043-1047.
55. Ray PS, Jia J, Yao P et al. A stress-responsive RNA switch regulates VEGFA expression. Nature 2009; 457:915-919.
56. Limbach PA, Crain PF, McCloskey JA. Summary: the modified nucleosides of RNA. Nucleic Acids Res 1994; 22:2183-2196.
57. Dunin-Horkawicz S, Czerwoniec A, Gajda MJ et al. MODOMICS: a database of RNA modification pathways. Nucleic Acids Res 2006; 34:D145-D149.

58. Bachellerie JP, Cavaillé J, Hüttenhofer A. The expanding snoRNA world. Biochimie 2002; 84:775-790.
59. Henras AK, Capeyrou R, Henry Y et al. Cbf5p, the putative pseudouridine synthase of H/ACA-type snoRNPs, can form a complex with Gar1p and Nop10p in absence of Nhp2p and box H/ACA snoRNAs. RNA 2004; 10:1704-1712.
60. Singh SK, Gurha P, Gupta R. Dynamic guide-target interactions contribute to sequential 2′-O-methylation by a unique archaeal dual guide box C/D sRNP. RNA 2008; 14:1411-1423.
61. Vitali P, Basyuk E, Le Meur E et al. ADAR2-mediated editing of RNA substrates in the nucleolus is inhibited by C/D small nucleolar RNAs. J Cell Biol 2005; 169:745-753.
62. Rogelj B. Brain-specific small nucleolar RNAs. J Mol Neurosci 2006; 28:103-109.
63. Proudfoot N. New perspectives on connecting messenger RNA 3′ end formation to transcription. Curr Opin Cell Biol 2004; 16:272-278.
64. Marzluff WF. Metazoan replication-dependent histone mRNAs: a distinct set of RNA polymerase II transcripts. Curr Opin Cell Biol 2005; 17:274-280.
65. Baillat D, Hakimi MA, Näär AM et al. Integrator, a multiprotein mediator of small nuclear RNA processing, associates with the C-terminal repeat of RNA polymerase II. Cell 2005; 123:265-276.
66. Box JA, Bunch JT, Tang W et al. Spliceosomal cleavage generates the 3′ end of telomerase RNA. Nature 2008; 456:910-914.
67. Wilusz JE, Spector DL. An unexpected ending: noncanonical 3′ end processing mechanisms. RNA 2010; 16:259-266.
68. Marz M, Stadler PF. Comparative analysis of eukaryotic U3 snoRNA. RNA Biol 2009; 6:503-507.
69. Atzorn V, Fragapane P, Kiss T. U17/snR30 Is a Ubiquitous snoRNA with Two Conserved Sequence Motifs Essential for 18S rRNA Production. Mol Cell Biol 2004; 24:1769-1778.
70. Shikanai T. RNA editing in plant organelles: machinery, physiological function and evolution. Cell Mol Life Sci 2006; 63:698-708.
71. Simpson L, Aphasizhev R, Gao G et al. Mitochondrial proteins and complexes in Leishmania and Trypanosoma involved in U-insertion/deletion RNA editing. RNA 2004; 10:159-170.
72. Ochsenreiter T, Cipriano M, Hajduk SL. Alternative mRNA editing in Trypanosomes is extensive and may contribute to mitochondrial protein diversity. PLoS One 2008; 3:e1566.
73. Madej MJ, Niemann M, Hüttenhofer A et al. Identification of novel guide RNAs from the mitochondria of Trypanosoma brucei. RNA Biology 2008; 5:84-91.
74. Horvath P, Barrangou R. CRISPR/Cas, the immune system of bacteria and archaea. Science 2010; 327:167-170.
75. van der Oost J, Jore MM, Westra ER et al. CRISPR-based adaptive and heritable immunity in prokaryotes. Trends Biochem Sci 2009; 34:401-407.
76. Marraffini LA, Sontheimer EJ. CRISPR interference: RNA-directed adaptive immunity in bacteria and archaea. Nat Rev Genet 2010; 11:181-190.
77. Barrangou R, Fremaux C, Deveau H et al. CRISPR provides acquired resistance against viruses in prokaryotes. Science 2007; 315:1709-1712.
78. Godde JS, Bickerton A. The repetitive DNA elements called CRISPRs and their associated genes: evidence of horizontal transfer among prokaryotes. J Mol Evol 2006; 62:718-729.
79. Karginov FV, Hannon GJ. The CRISPR system: small RNA-guided defense in bacteria and archaea. Mol Cell 2010; 37:7-19.
80. Valadkhan S. The spliceosome: a ribozyme at heart? Biol Chem 2007; 388:693-697.
81. Warf MB, Berglund JA. Role of RNA structure in regulating pre-mRNA splicing. Trends Biochem Sci 2010; 35:169-178.
82. Piccinelli P, Rosenblad MA, Samuelsson T. Identification and analysis of ribonuclease P and MRP RNA in a broad range of eukaryotes. Nucleic Acids Res 2005; 33:4485-4495.
83. Collins LJ, Moulton V, Penny D. Use of RNA secondary structure for studying the evolution of RNase P and RNase MRP. J Mol Evol 2000; 51:194-204.
84. Aspinall TV, Gordon JMB, Bennet HJ et al. Interactions between subunits of Saccharomyces cerevisiae RNase MRP support a conserved eukaryotic RNase P/MRP architecture. Nucleic Acids Res 2007; 35:6439-6450.
85. Walker SC, Engelke DR. Ribonuclease P: the evolution of an ancient RNA enzyme. Crit Rev Biochem Mol Biol 2006; 41:77-102.
86. Rosenblad MA, López MD, Piccinelli P et al. Inventory and analysis of the protein subunits of the ribonucleases P and MRP provides further evidence of homology between the yeast and human enzymes. Nucleic Acids Res 2006; 34:5145-5156.
87. Talini G, Gallori E, Maurel MC. Natural and unnatural ribozymes: back to the primordial RNA world. Res Microbiol 2009; 160:457-465.
88. Krueger BJ, Jeronimo C, Roy BB et al. LARP7 is a stable component of the 7SK snRNP while P-TEFb, HEXIM1 and hnRNP A1 are reversibly associated. Nucleic Acids Res 2008; 36:2219-2229.
89. Murphy S, Di Liegro C, Melli M. The in vitro transcription of the 7SK RNA gene by RNA polymerase III is dependent only on the presence of an upstream promoter. Cell 1987; 51:81-87.

90. Gruber AR, Kilgus C, Mosig A et al. Arthropod 7SK RNA. Mol Biol Evol 2008; 25:1923-1930.
91. Gruber AR, Koper-Emde D, Marz M et al. Invertebrate 7SK snRNAs. J Mol Evol 2008; 66:107-115.
92. Marz M, Donath A, Verstraete N et al. Evolution of 7SK RNA and its protein partners in metazoa. Mol Biol Evol 2009; 26:2821-2830.
93. Peterlin BM, Price DH. Controlling the elongation phase of transcription with P-TEFb. Moll Cell 2006; 23:297-305.
94. Egloff S, Van Herreweghe E, Kiss T. Regulation of polymerase II transcription by 7SK snRNA: two distinct RNA elements direct P-TEFb and HEXIM1 binding. Moll Celll Biol 2006; 26:630-642.
95. Yik JH, Chen R, Nishimura R et al. Inhibition of P-TEFb (CDK9/Cyclin T) kinase and RNA polymerase II transcription by the coordinated actions of HEXIM1 and 7SK snRNA. Mol Cell 2003; 12:971-982.
96. Michels AA, Nguyen VT, Fraldi A et al. MAQ1 and 7SK RNA interact with CDK9/cyclin T complexes in a transcription-dependent manner. Mol Cell Biol 2003; 23:4859-4869.
97. Blazek D, Barboric M, Kohoutek J et al. Oligomerization of HEXIM1 via 7SK snRNA and coiled-coil region directs the inhibition of P-TEFb. Nucleic Acids Res 2005; 33:7000-7010.
98. Markert A, Grimm M, Martinez J et al. The La-related protein LARP7 is a component of the 7SK ribonucleoprotein and affects transcription of cellular and viral polymerase II genes. EMBO Rep 2008; 9:569-575.
99. He S, Su H, Liu C et al. MicroRNA-encoding long noncoding RNAs. BMC Genomics 2008; 21:236.
100. Gürsoy HC, Koper D, Benecke BJ. The vertebrate 7S K RNA separates hagfish (Myxine glutinosa) and lamprey (Lampetra fluviatilis). J1 Moll Evol 2000; 50:456-464.
101. Bousquet-Antonelli C, Deragon JM. A comprehensive analysis of the La-motif protein superfamily. RNA 2009; 15:750-764.
102. Byers SA, Price JP, Cooper JJ et al. HEXIM2:a HEXIM1-related protein, regulates positive transcription elongation factor b through association with 7SK. J Biol Chem 2005; 280:16360-16377.
103. Rosenblad MA, Gorodkin J, Knudsen B et al. SRPDB: signal recognition particle database. Nucleic Acids Res 2003; 31:363-364.
104. Andersen ES, Rosenblad MA, Larsen N et al. The tmRDB and SRPDB resources. Nucleic Acids Res 2006; 34:163-168.
105. Bürk J, Weiche B, Wenk M et al. Depletion of the signal recognition particle receptor inactivates ribosomes in Escherichia coli. J Bacteriol 2009; 191:7017-7026.
106. Rosenblad MA, Samuelsson T. Identification of chloroplast signal recognition particle RNA genes. Plant Cell Physiol 2004; 45:1633-1639.
107. Jaru-Ampornpan P, Nguyen TX, Shan SO. A distinct mechanism to achieve efficient signal recognition particle (SRP)-SRP receptor interaction by the chloroplast srp pathway. Mol Biol Cell 2009; 20:3965-3973.
108. Chen JLWGC. An emerging consensus for telomerase RNA structure. Proc Natl Acad Sci USA 2004; 101:14683-14684.
109. Podlevsky JD, Bley CJ, Omana RV et al. The telomerase database. Nucleic Acids Res 2008; 36:D339-D343.
110. Xie M, Mosig A, Qi X et al. Size variation and structural conservation of vertebrate telomerase RNA. J Biol Chem 2008; 283:2049-2059.
111. Artandi SE, DePinho RA. Telomeres and telomerase in cancer. Carcinogenesis 2010; 31:9-18.
112. Nilsen TW. The spliceosome: the most complex macromolecular machine in the cell? Bioessays 2003; 25:1147-1149.
113. Dávila López M, Rosenblad MA, Samuelsson T. Computational screen for spliceosomal RNA genes aids in defining the phylogenetic distribution of major and minor spliceosomal components. Nucleic Acids Res 2008; 36:3001-10.
114. Hastings KE. SL trans-splicing: easy come or easy go? Trends Genet 2005; 21:240-247.
115. Nilsen TW. Evolutionary origin of SL-addition trans-splicing: still an enigma. Trends Genet 2001; 17:678-680.
116. MacMorris M, Kumar M, Lasda E et al. A novel family of C. elegans snRNPs contains proteins associated with trans-splicing. RNA 2007; 13:511-520.
117. Smith S. The world according to PARP. Trends Biochem Sci 2001; 26:174-179.
118. Stadler PF, Chen JJL, Hackermüller J et al. Evolution of vault RNAs. Mol Biol Evol 2009; 26:1975-1991.
119. Mosig A, Zhu L, Stadler PF. Customized strategies for discovering distant ncRNA homologs. Brief Funct Genomics Proteomics 2009; 8:451-460.
120. van Zon A, Mossink MH, Scheper RJ et al. The vault complex. Cell Mol Life Sci 2003; 60:1828-1837.
121. Zappulla DC, Cech TR. RNA as a flexible scaffold for proteins: yeast telomerase and beyond. Cold Spring Harb Symp Quant Biol 2006; 71:217-224.
122. Jolly C, Lakhotia SC. Human sat III and Drosophila hsrω transcripts: a common paradigm for regulation of nuclear RNA processing in stressed cells. Nucleic Acids Res 2006; 34:5508-5514.
123. Bond A, VanGompel MJW, Sametsky E et al. Balanced gene regulation by an embryonic brain ncRNA is critical for adult hippocampal GABA circuitry. Nature Neuroscience 2009; 12:1020-1027.

124. Sasaki YTF, Ideue T, Sano M et al. MENεβ noncoding RNAs are essential for structural integrity of nuclear paraspeckles. Proc Natl Acad Sci USA 2009; 106:2525-2530.
125. Mourtada-Maarabouni M, Pickard MR, Hedge VL et al. GAS5:a nonprotein-coding RNA, controls apoptosis and is downregulated in breast cancer. Oncogene 2009; 28:195-208.
126. Betel D, Sheridan R, Marks DS et al. Computational analysis of mouse piRNA sequence and biogenesis. PLoS Comput Biol 2007; 3:e222.
127. Bond CS, Fox AH. Paraspeckles: nuclear bodies built on long noncoding RNA. J Cell Biol 2009; 186:637-644.
128. Cole C, Sobala A, Lu C et al. Filtering of deep sequencing data reveals the existence of abundant Dicer-dependent small RNAs derived from tRNAs. RNA 2009; 15:2147-2160.
129. Langenberger D, Bermudez-Santana C, Stadler PF et al. Identification and classification of small RNAs in transcriptome sequence data. Pac1 Symp1 Biocomput 2010; 15:80-87.
130. Taft RJ, Glazov EA, Lassmann T et al. Small RNAs derived from snoRNAs. RNA 2009; 15:1233-1240.
131. Ender C, Krek A, Friedländer MR et al. A human snoRNA with microRNA-like functions. Mol Cell 2008; 32:519-528.
132. Lee YS, Shibata Y, Malhotra A et al. A novel class of small RNAs: tRNA-derived RNA fragments (tRFs). Genes Dev 2009; 23:2639-2649.
133. Persson H, Kvist A, Vallon-Christersson J et al. The noncoding RNA of the multidrug resistance-linked vault particle encodes multiple regulatory small RNAs. Nat Cell Biol 2009; 11:1268-1271.
134. Stadler P. Evolution of the long noncoding RNAs MALAT1 and MENβε. In: Ferreira CE, Satoru M, Stadler P eds. Proceedings of the Brazilian Symposium on Bioinformatics. Lecture Notes in Computer Science. New York:Springer, 2010.
135. Huang FWD, Qin J, Reidys CM et al. Target prediction and a statistical sampling algorithm for RNA-RNA interaction. Bioinformatics 2010; 26:175-181.
136. Ni J, Tien AL, Fournier MJ. Small nucleolar RNAs direct site-specific synthesis of pseudouridine in ribosomal RNA. Cell 1997; 89:565-573.
137. Yu LE, Koslowsky DJ. Interactions of mRNAs and gRNAs involved in trypanosome mitochondrial RNA editing: Structure probing of a gRNA bound to its cognate mRNA. RNA 2006; 12:1050-1060.

CHAPTER 3

PLANT RNA SILENCING IN VIRAL DEFENCE

Vitantonio Pantaleo

Istituto di Virologia Vegetale del CNR, Torino, Italy
Email: v.pantaleo@ivv.cnr.it

Abstract: RNA silencing is described in plants and insects as a defence mechanism against foreign nucleic acids, such as invading viruses. The RNA silencing-based antiviral defence involves the production of virus–derived small interfering RNAs and their association to effector proteins, which together drive the sequence specific inactivation of viruses. The entire process of antiviral defence 'borrows' several plant factors involved in other specialized RNA silencing endogenous pathways. Different viruses use variable strategies to infect different host plants, which render the antiviral RNA silencing a complex phenomenon far to be completely clarified. This chapter reports current advances in understanding the main steps of the plant's RNA-silencing response to viral invasion and discusses some of the key questions still to be answered.

INTRODUCTION

The term RNA interference (RNAi) comes from evidence that experimental introduction of RNA into cells can interfere with the function of endogenous genes (first provided by Fire et al[1] in the nematode *Caenorhabditis elegans* and then extended to other organisms[2,3]). The main effect observed on gene expression and control is generally inhibitory and the corresponding regulatory mechanisms are therefore collectively indicated under the term of RNA silencing, even though some activation mechanisms have been described.[4]

Biochemical and genetic analyses in the last few years have shown that RNA silencing-based regulation of genes and genomes is shared among almost all eukaryotes and can occur at different levels including post-transcriptional regulation, RNA-directed

RNA Infrastructure and Networks, edited by Lesley J. Collins.

DNA methylation and chromatin remodelling.[5-8] The main actors of RNA silencing are small RNAs (sRNAs) and numerous aspects of their origins, structures, associated effector proteins and biological roles have suggested their classification in three main categories: short interfering RNAs (siRNAs), microRNAs (miRNAs) and Piwi-interacting RNAs (piRNAs). siRNAs and miRNAs are widely distributed in both phylogenetic and physiological terms in animals, plants and insects and are characterized by the double-stranded nature of their precursors. In contrast, piRNAs are mainly found in animals (although there is a piRNA-like class in ciliates), exert their regulatory roles in the germline and somatic cells in drosophila and derive from precursors appearing to be single-stranded.[9,10] An additional term of distinction among the three classes of sRNAs refers to the subset of effector proteins with which they are associated: siRNAs and miRNAs bind to members of the Ago clade of Argonaute proteins, whereas piRNAs bind to members of the Piwi clade.[6,7,11] Based on biogenesis and their biological role, in plants siRNAs can be further distinguished into several sub-groups such as *trans*-acting siRNAs (ta-siRNAs), natural *cis*-acting siRNAs (nat-siRNAs) and heterochromatic siRNAs (hc-siRNAs), the latter ones involved in transcriptional gene silencing.[12]

Eukaryotic RNAs form dynamic networks of molecules and processes integrated over space and time, termed RNA infrastructures.[13] Viruses are obligate infectious agents, with most of their life cycle (expression of viral proteins, viral replication and virion assembly) integrated into the host cell RNA infrastructure. The degree of cross talk between viral molecular processes and the host RNA processes regulates the coexistence of both virus and host. RNA silencing is a key part of the antiviral response in plants and invertebrates, whereas analogous responses to viral infection are not induced in mammalian somatic cells and are replaced mainly by an interferon-regulated gene response.[14] It is plausible that in plants and invertebrates, RNA silencing first evolved as a response to viral infection. Indeed, the triggers for RNA silencing in these species are the long double-stranded RNAs (dsRNAs) that form intermediates in the life cycle of all viruses. By analogy with endogenous sRNAs, the main key components of RNA silencing-based antiviral response are the viral small interfering RNAs (vsiRNAs); the first siRNAs to be identified were antiviral siRNAs produced in tobacco cells infected by *Potato virus X* (PVX).[15] Virus-plant systems are very useful to investigate aspects of plant-pathogen interactions and to identify specific endogenous mechanisms of gene regulation and response to stress, which may find parallels in other eukaryotic organisms. This chapter discusses some of the latest research concerning antiviral silencing in plant defence, with particular reference to the generation, amplification and function of small RNAs of viral origin.

GENERATION OF VIRAL SMALL INTERFERING RNAs: VIRUSES, vsiRNAs AND DICERS

In plants, the initial step of virus-induced gene silencing (VIGS) is the generation of vsiRNAs from virus-derived RNAs. At the heart of this pathway are plant Dicer enzymes. Dicers are Type III endonucleases, which specifically cleave dsRNAs or dsRNA-like hairpin regions of single-stranded (ss) RNAs.[16] There are several types of virus-derived RNA that might stimulate the action of Dicers, including long dsRNA replication intermediates, imperfect RNA hairpins of ssRNA viral RNAs, overlapping sense/antisense transcripts produced by circular ssDNA genomes (e.g., geminiviruses)

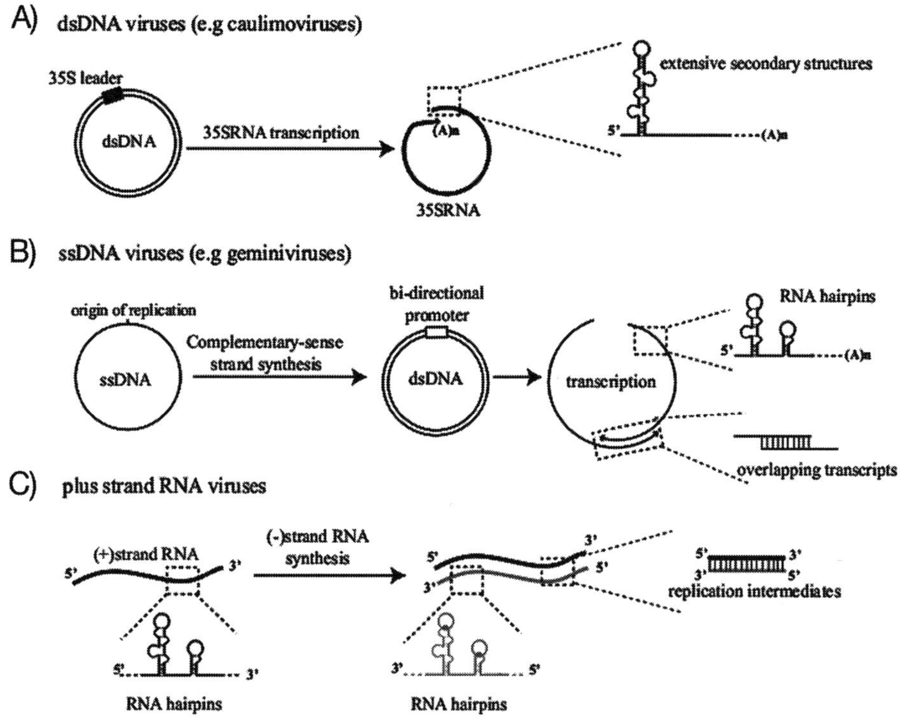

Figure 1. Diagram of primary viral double-stranded (ds) RNA substrates for Dicers. A) The 35S leader of caulimoviruses (e.g., CaMV) promotes the transcription of a long polycistronic viral transcript (35SRNA) carrying an extensive secondary structure at its 5′ end (35S translational leader). This region is the major source of CaMV vsiRNAs. B) Geminiviruses are single-stranded (ss) DNA plant viruses having an intermediate dsDNA step. Their bidirectional promoter drives the generation of long viral RNA transcripts of opposite polarities, which overlap at their 3′ ends, thus forming a perfectly base-paired dsRNA structure. In addition, secondary structures along the viral transcripts may form imperfect RNA hairpins. C) The replication machinery of RNA viruses implies the accumulation of (+) and (–) stranded viral RNAs. Secondary structures on both positive and negative viral RNAs may form imperfect RNA hairpins. In addition, the complementarity of the two viral strands may promote the formation of perfectly base-paired viral dsRNA replication intermediates. The figures are not to scale.

and extensive secondary structures of the polycistronic 35S RNA of dsDNA viruses (e.g., caulimoviruses)[17,18] (Fig. 1).

The model plant *Arabidopsis thaliana* encodes four Dicer-like proteins (DCLs) (Table 2), involved in both endogenous processes and antiviral RNA silencing. While DCL1 is primarily involved in the genesis of miRNAs 21 nt long, DCL4, DCL2 and DCL3 produce siRNAs of 21, 22 and 24 nt, respectively, mostly involved in post-transcriptional and transcriptional gene regulation and having specific dsRNA substrates.[12,19-21] To date, much effort has gone into identifying plant antiviral Dicers, but the functional redundancy and hierarchical action of the four DCLs has disclosed a more complex situation than expected.

Analysis of (+) ssRNA viruses such as *Turnip crinkle virus* (TCV) and *Cucumber mosaic virus* (CMV) (Table 1) infecting single or multiple *dcl* mutants of *A. thaliana* has shown that 21 nt-long vsiRNAs are primarily produced by DCL4 and not by DCL1.

Table 1. Overview of plant viruses and their experimental hosts as reported in the text

Virus Name (Genus)	Acronym	Nucleic Acid	Experimental Host as Reported in the Text	References
Beet yellows virus (Closterovirus)	BYV	(+)ssRNA	*Nicotiana benthamiana*	126
Cabbage leaf curl virus (Begomovirus)	CaLCuV	ssDNA	*Arabidopsis thaliana*	33, 34
Carnation Italian ringspot virus (Tombusvirus)	CIRV	(+)ssRNA	*Nicotiana benthamiana*	59, 126
Cauliflower mosaic virus (Caulimovirus)	CaMV	dsDNA	*Arabidopsis thaliana*	35, 39-42
Citrus tristeza virus (Closterovirus)	CTV	(+)ssDNA	*Nicotiana benthamiana*	131
Cucumber mosaic virus (Cucumovirus)	CMV	(+)ssRNA	*Arabidopsis thaliana* *Nicotiana benthamiana*	22-26, 28, 66, 70, 79, 96, 98
Cucumber yellows virus syn. *Beet pseudoyellows virus (Closterovirus)*	CuYV (BPYV)	(+)ssRNA	*Cucurbita maxima* (pumpkin)	114
Cymbidium ringspot virus (Tombusvirus)	CymRSV	(+)ssRNA	*Nicotiana benthamiana*	29, 73, 100, 105
Oilseed rape mosaic virus (Tobamovirus)	ORMV	(+)ssRNA	*Arabidopsis thaliana* *Nicotiana benthamiana*	33, 62
Pepper golden mosaic virus (Begomovirus)	PepGMV	ssDNA	*Capsicum annuum* (pepper)	36
Plum pox virus (Potyvirus)	PPV	(+)ssRNA	*Nicotiana benthamiana*	99
Potato virus X (Potexvirus)	PVX	(+)ssRNA	*Arabidopsis thaliana*	15, 83, 98, 99, 102, 122
Potato virus Y (Potyvirus)	PVY	(+)ssRNA	*Nicotiana benthamiana*	98
Tabacco mosaic virus (Tobamovirus)	TMV	(+)ssRNA	*Nicotiana benthamiana* *Nicotiana tabacum* *Arabidopsis thaliana* (TMV-Cg)	32, 62, 98, 101, 113
Tobacco etch virus (Potyvirus)	TEV	(+)ssRNA	*Nicotiana benthamiana*	59
Tobacco rattle virus (Tobravirus)	TRV	(+)ssRNA	*Arabidopsis thaliana*	98, 101
Tomato aspermy virus (Cucumovirus)	TAV	(+)ssRNA	*Nicotiana benthamiana*	126, 127
Tomato yellow leaf curl virus (Begomovirus)	TYLCV	ssDNA	*Nicotiana benthamiana*	130
Turnip crinkle virus (Carmovirus)	TCV	(+)ssRNA	*Arabidopsis thaliana*	22, 23, 28, 43, 62, 96, 98
Turnip mosaic virus (Potyvirus)	TuMV	(+)ssRNA	*Arabidopsis thaliana*	28, 96, 103
Turnip yellow mosaic virus (Tymovirus)	TYMV	(+)ssRNA	*Arabidopsis thaliana*	70

(+): positive; ss: single-stranded; ds: double-stranded; syn.: synonym.

DCL2 derived-22 nt-long vsiRNAs also accumulate in infected tissues, but they are hardly detected when DCL4 is fully functional. DCL3 produces 24 nt-long vsiRNAs only in the case of *dcl2-dcl4* double mutants.[22-25] The same situation has been found for viral satellite RNAs such as CMV satellite RNA.[26] Satellite RNAs are subviral RNAs that require a helper virus to supply *trans*-acting factors for replication.[27]

Different Dicers contribute differently to vsiRNA generation; while DCL2 is the major contributor to generation of TCV siRNAs, DCL4 mainly supplies CMV-derived siRNAs.[22,28] This different contribution of Dicers is likely to be due to specific virus-host interactions rather than to a simple question of affinity for dsRNAs. Indeed, TCV seems to inhibit DCL4 through the action of its own viral suppressor, p38 (Table 2).[22,23]

In the case of the *Cymbidium ringspot virus* (CymRSV) (Table 1) in *Nicotiana benthaminana*, analysis of siRNA cDNA libraries has confirmed that vsiRNAs are predominantly 21 nt and 22 nt in length (likely produced by DCL4 and DCL2, respectively, although the loss-of-function *dcl* mutants genetic approach is not available in this plant system since the genome is not available to date) and that 24 nt long vsiRNAs are only represented at trace level.[29,30]

Cloning and sequencing data suggest that secondary structures within the ssRNA viral genome predominantly serve as substrates for DCL-mediated cleavage (Fig. 1C). Hence, most CymRSV vsiRNAs are not perfect duplexes and it is likely that they are generated from imperfect secondary structures on the (+)ssRNA.[29,30] The preferential (+) strand vsiRNA enrichment in response to most (+)ssRNA plant viruses and their satellite RNAs, can probably be explained by the generalized prevalent accumulation of the (+) strand rather than (–) strand RNA during viral replication (Fig. 2).[26,29-33] Moreover, the existence of long dsRNA viral replicative forms has not yet been proved in vivo, and their contribution to vsiRNA creation is still debated.

The involvement of plant Dicers is quite different in the case of DNA virus infections. The viruses most studied in regard to the antiviral action of plant dicers are the nuclear replicating dsDNA *Cauliflower mosaic virus* (CaMV) and ssDNA *Cabbage leaf curl virus* (CaLCuV) (Table 1).[34-36] *A. thaliana* plants infected by either CaMV or CaLCuV generate similar patterns of 21, 22 and 24 nt-long vsiRNAs of both (–) and (+) sense, from both coding and noncoding regions. In contrast to the case of RNA viruses, DCL3 always appears to be the active antiviral Dicer. However, genetic approaches have shown that DCLs compensate for deficiencies in a specific Dicer during the production of vsiRNAs (absence of DCL3 in *dcl3* mutants correlates with the increase of 21 and 22 nt species and vice versa).[34,35] For both classes of DNA viruses, DCL1 seems to cooperate with other DCLs and generates 21 nt long vsiRNAs. In this regard, DCL1 seems to efficiently generate 21 nt-long vsiRNAs when DCL4 is not functional, but this activity seems to interfere with the accumulation of DCL4- and DCL3-dependent vsiRNAs from CaMV 35S polycistronic RNA.[36] Recently Rodriguez-Negrete et al.[37] have provided evidence that 21 and 22 nt-long siRNAs are mainly derived from coding regions, whereas the 24 nt-long siRNAs come from the intergenic regions of *Pepper golden mosaic geminivirus* (Table 1). These data further stimulate the open debate on the affinity of DCLs for different dsRNAs.

The major role of DCL4, the minor role of DCL2 and the small or absent activity of DCL3 and DCL1 in antiviral dicing, at least for RNA viruses, still deserve deeper study. First, while DCLs contain all the required RNase III domains to exercise their activity, they function as components of larger protein complexes, in which cofactors carrying dsRNA binding domains (dRBDs) may be important modifiers of their dsRNA affinity.[38]

Table 2. Overview of plant and viral proteins mentioned in the text and their roles in sRNA pathways

Acronym of Protein	Protein Name	Activity	Pathway	References
Plant proteins				
AGO1	Argonaute1	miRNA slicer siRNA slicer vsiRNA slicer	miRNAs sense transgene PTGS ta-siRNAs antiviral defence	63-68, 70-72
AGO2	Argonaute2	unclear	antiviral defence?	63, 79
AGO4	Argonaute4	unclear	RdRM miRNAs (miR172, miR390) antiviral defence?	63, 84, 90
AGO5	Argonaute5	Unclear	miRNAs (miR169)? antiviral defence?	63, 79
AGO7	Argonaute7	unclear	ta-siRNA (TAS3) miRNAs (miR390) antiviral defence?	43, 63
DCL1	Dicer-like1	RNaseIII endonuclease Generation of 21 nt long sRNAs	miRNAs nat-siRNAs	12, 16, 18, 19, 20, 21, 23, 28, 133, 134
DCL2	Dicer-like2	RNaseIII endonuclease Generation of 22 nt long sRNAs	nat-siRNAs vsiR-NAs	16, 18, 19, 20, 21, 22, 23, 25, 28, 133, 134
DCL3	Dicer-like3	RNaseIII endonuclease Generation of 24 nt long sRNAs	RdDM vsiRNAs	12, 16, 21, 33, 34, 35, 133, 134
DCL4	Dicer-like4	RNaseIII endonuclease Generation of 21 nt long sRNAs	ta-siRNA vsiRNAs	26, 28, 133, 134
DRB1	dsRNA-binding protein1	dsRNA binding in association with DCL1	miRNAs	37, 45, 135, 136
DRB4	dsRNA-binding protein4	dsRNA binding in association with DCL4	ta-siRNAs vsiRNA	37, 38, 42, 43, 45, 135
HEN1	Hua enhancer1	sRNA methyltransferase	All sRNA patways including vsiRNAs	33, 53, 54, 55, 59
RDR1	RNA-directed RNA polymerase1	RdRp	secondary vsiRNAs	28, 86, 87, 101, 102, 103
RDR2	RNA-directed RNA polymerase2	RdRp	RdRM secondary vsiRNAs?	28, 86, 87, 103
RDR6 (SGS2)	RNA-directed RNA polymerase6	RdRp	sense transgene PTGS ta-siRNA secondary vsiRNAs	86, 87, 91-93, 96, 98, 99, 103, 106, 133, 137

continued on next page

Table 2. Continued

Acronym of Protein	Protein Name	Activity	Pathway	References
SGS3	Suppressor of gene silencing3	RNA stabilization in RDR6 pathway	sense transgene PTGS secondary vsiRNAs	92, 96-98
Viral proteins				
2b		Viral movement VSR	Agol-interaction (CMV 2b) siRNA binding (TAV 2b)	70, 127
HcPro		Viral movement Polyprotein processing VSR	siRNA binding	59, 125, 126
P0		Pathogenic determinant VSR	Ago destabilization	72, 129
P19		Viral movement VSR	siRNA binding	126, 128, 59
P21		Replication enhancer VSR	siRNA binding	126
P25		Viral movement VSR	Systemic silencing	122
P38		Coat protein VSR	dsRNA binding	23, 135
P6		Translational transactivator VSR	RDB4 interaction	39, 40, 41, 42
V2		VSR	SGS3 activity	130

sRNA: small RNA; miRNA: micro RNA; nat-siRNA: natural small interfering RNA; vsiRNA: viral small interfering RNA; ta-siRNA: trans-acting siRNAs; ds: double-stranded; RdRM: RNA-dependent DNA methylation; RdRp: RNA-dependent RNA polymerase; VSR: Viral suppressor of RNA silencing.

In plants, the cofactors DRB1 and DRB4 (Table 2) have been described as assisting DCL1 and DCL4-mediated processing of fold-back precursors and long dsRNAs, respectively.[12] The DRB4 protein is localized in the nucleus[39] and seems to be involved in the plant's antiviral strategies against dsDNA viruses. The CaMV protein P6 is multifunctional, being involved in different steps in the virus cycle. It accumulates in infected cells mainly in amorphous cytoplasmic bodies considered to be essential for virus assembly, replication and translational transactivation of 35S RNA and has also been described as a silencing suppressor.[40-42] In addition, a small fraction of P6 enters the importin-α pathway and localizes in the nucleus. Here P6 targets DRB4, so it impairs the activity of DCL4 and it leads to a reduced level of DCL4-dependent 21 nt vsiRNAs, whereas DCL3-dependent 24 nt vsiRNAs are enhanced.[43]

Involvement of DRB4 in antiviral defence against RNA viruses, or rather in stabilizing 21 nt long vsiRNAs, has also been suggested in the case of a TCV mutant in *A. thaliana*. Infected *drb4* mutant plants showed a consistent increase in viral RNA accumulation, although a direct involvement of DRB4 in antiviral strategies should lead to evident change in vsiRNA profiles, which was not reported.[44]

Preferential affinities of the different DCLs for different dsRNA substrates of viral origin may also be dictated by their mutual subcellular location. For instance the remarkable

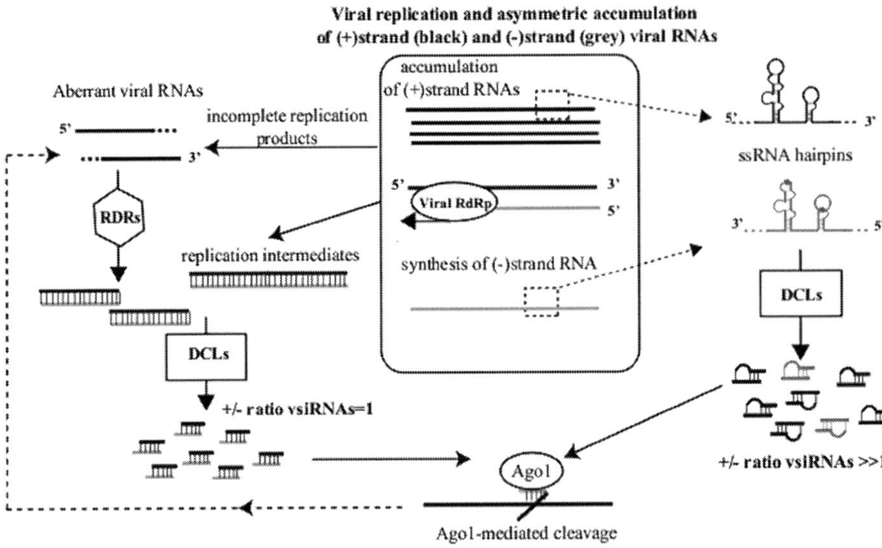

Figure 2. Diagram of the generation of viral (v) siRNAs in the case of positive-strand RNA viruses. These viruses replicate through negative-stranded RNA intermediates via the viral RNA-dependent RNA polymerase (RdRp). It often results in over-accumulation of (+) rather than (−) viral RNAs. Imperfect RNA hairpins within both viral singe-strand RNAs are substrates for Dicers (DCL2, DCL3, DCL4), thus producing an asymmetric accumulation of vsiRNAs (bias for (+) strand-derived vsiRNAs). Instead, when perfect dsRNAs of viral origin are prevalent (i.e. from replication intermediates), symmetric accumulation of vsiRNAs from (+) and (−) strands is expected. Thus, the final (+)/(−) ratio of vsiRNAs likely depends on the viral life cycle and Dicer substrates. In addition, aberrant viral RNAs (i.e. from incomplete replication or from Ago1-mediated slicing) can enter plant RNA-dependent RNA polymerase (RDR1, RDR2, RDR6) pathways, thus amplifying perfect dsRNA of viral origin and determining the final (+)/(−) ratio of vsiRNAs.

antiviral role of DCL3 in nuclear-replicating DNA viruses and its minor role for cytoplasmic replicating viruses should not be considered an accident. Notably, similar observations can be done in the case of viroids, the smallest plant infectious agents composed of non coding RNAs; in contrast to chloroplastic replicating viroids, the nuclear replicating ones generate 24 nt long siRNAs in addition to the 21 nt and 22 nt long species.[136-138]

Localization studies using GFP fusion proteins indicate that all four *Arabidopsis* Dicers localize to the nucleus,[28,45,46] where they most likely generate endogenous sRNAs. On the other hand, most RNA plant viruses replicate entirely in the cytoplasm; how these nuclear enzymes use cytoplasmic substrates to produce vsiRNAs is still a puzzle.

Viruses replicate in association with specific cell membranes, thus modifying different subcellular compartments (such as the endoplasmic reticulum, mitochondria, peroxisomes, chloroplasts and nuclei) and leading to the formation of vesicles either free in the cytoplasm or associated with the organelles of origin. These structures are dedicated to housing the viral replication complex and to overcoming competition from host mRNA translation.[47-52] Since membrane trafficking within cellular compartments is crucial for transport of proteins from translational to functional loci,[53] any pathological alteration of membrane organization may affect protein localization. It is not unlikely that in these circumstances, DCLs can be displaced from the nucleus and this might explain the apparent discrepancy between cytoplasmic viral substrates and nuclear Dicers.

POST-DICING MODIFICATION OF vsiRNAs

The products of Dicer-mediated cleavage are duplexes of small RNAs (sRNAs) with 3′- 2 nt overhangs in each strand.[16] Upon dicing, all plant sRNAs undergo methylation of their 3′-terminal nucleotide at the 2′-hydroxyl group, mediated by HUA ENHANCER1 (HEN1) (Table 2).[54,55] The specific recognition of sRNA substrates and the 3′-end methylation pathway have recently been elucidated by determining the crystal structure of *A. thaliana* HEN1.[56] In *A. thaliana*, the methylation of miRNAs and siRNAs protects them from 3′-end uridylation and subsequent 3′-to-5′ exonuclease degradation.[57,58] The methylation status of sRNAs can be evaluated by the treatment of the RNAs with sodium periodate followed by β-elimination where in RNAs with a free 2′ and 3′ OH (nonmethylated), the last nucleotide is eliminated resulting in faster gels migration.[57,59]

Akbergenov and colleagues[34] have provided the first evidence that HEN1 mediates the methylation of vsiRNAs. In *hen*1-mutant *A. thaliana* plants infected with CaLCuV (Table 1), accumulation of 21, 22 and 24 nt vsiRNAs was significantly reduced and was sensitive to β-elimination (indicating a lack of methylation). Reduced accumulation of vsiRNAs is consistent with the hypothesis that methylation prevents uridylation, thus helping to stabilize the vsiRNA duplexes in the cell. The same authors found a more complicated situation when analysing the *Oilseed rape mosaic virus* (ORMV) (Table 1) which is an RNA virus replicating in the cytoplasm. Unexpectedly, the sensitivity of ORMV-derived vsiRNAs to β-elimination, likely due to the unmodified 3′-temini, was similar both in wild type plants and *hen1* mutants. Later studies have explained this apparent incongruence and have provided further evidence that methylation is a key step in silencing pathways mediated by vsiRNAs.

Immunoprecipitation analysis of small RNAs bound to the *Tobacco etch virus* HcPro and *Carnation Italian ringspot virus* (CIRV) p19 (Tables 1 and 2), has shown that the physical interaction (likely due to spatial and temporal co-expression) between the suppressor protein and the sRNA duplexes (either vsiRNAs or endogenous siRNAs), prevents their 3′ modification.[60] In this regard, the prereadthrough replicase protein of tobamoviruses is also a strong silencing suppressor whose activity is based on binding siRNAs[61,62] and this could explain the observations on ORMV-derived vsiRNAs. Similarly, the methylation prevention mediated by viral suppressors may sometimes engage the sRNA endogenous pathways, since both *Tobacco mosaic virus* (TMV) and ORMV infections lead to accumulation of nonmethylated miRNAs, whereas TCV infection does not.[63] Thus, intracellular location of virus-derived and host factors may play a critical role in RNA silencing.

ANTIVIRAL EFFECTOR COMPLEXES

Proteins of the ARGONAUTE (Ago) family are large molecules (ca 90-100 kDa) conserved across eukaryotes, consisting of a variable N-terminal domain and conserved C-terminal PAZ, MID and PIWI domains. While the MID and the PAZ domains bind the 5′ phosphate and the 3′ end of sRNAs respectively, the PIWI domain has RNaseH-like endonuclease activity that cleaves ssRNAs in the region complementary to the sRNA guide (Table 2).[64,65] In *Arabidopsis*, the Ago family consists of about 10 members,[66] among which Ago1, the first described, is the best characterized. Ago1 plays a role in maintaining the functionality of the miRNA pathway and indeed *Arabidopsis ago1* mutants show both a reduced accumulation of miRNA and an increased corresponding mRNA target accumulation.[67,68] Biochemical studies have shown that Ago1 cleaves mRNAs and is physically associated with several classes of DCL-derived sRNAs including miRNAs, transgene siRNAs and ta-siRNAs.[69,70]

The first evidence that Ago1 is involved in plant antiviral defence came from the observation that *ago1 A. thaliana* hypomorphs (partial loss of function mutants) are much more susceptible to CMV infection than wild type plants.[67] Later, Flag-Ago1 immunoprecipitates recovered from plants infected with CMV and *Turnip yellow mosaic virus* (TYMV) (Table 1) were shown to be enriched in vsiRNAs.[71] Moreover, the polerovirus silencing suppressor promotes the degradation of targeted Ago1.[72,73]

An in vivo CymRSV-*Nicotiana benthamiana* system, has given additional support to the idea that Ago1-loaded vsiRNAs cleave viral RNAs. CymRSV-derived vsiRNAs cofractionate in protein complexes likely corresponding to Ago1 based on their size. In addition, cleavage fragments of either CymRSV viral RNAs or CymRSV-based reporter sequences were characterised and carried nontemplate uracil residues at predicted cut sites, which are marks of Ago1-mediated cleavage.[74,75] Fractionation of protein extracts derived from CymRSV-infected *N. benthamiana* has shown that vsiRNAs are loaded into an even bigger complex of about 670 kDa, which is believed to be the functional effector complex (denoted as the holo RNA-induced silencing complex (holo-RISC)).[74] Alternatively the larger complex may correspond to a Dicer complex.[19] One model derived from Drosophila studies supposes the assembly of a RISC Loading Complex (RLC), which includes Dicer, its DRB cofactor and other still uncharacterized accessory proteins.[76,77] In plants, Jones-Rhoades and colleagues have proposed a similar mechanism for the Ago1/miRNA pathway.[78] Recent experiments with *A. thaliana* show that the sorting of small RNAs into Argonaute complexes is mainly determined by the 5′ terminal nucleotide. Ago1 preferentially incorporates RNAs bearing a 5′ terminal U, while Ago2 and Ago4 select those having a terminal A[79] and Ago5 those with a terminal C.[80] However, 5′-end dependent incorporation is not exclusive; despite a general preference, different small RNAs with the same 5′ terminal nucleotide can be sorted into different Ago proteins.[79] Different Dicers can act in the close vicinity of different Ago proteins and the nascent DCL-generated small RNA might enter into the nearest Ago protein with a 5′-end-independent mechanism. Indeed, miRNAs derive from the action of DCL1 and are mainly incorporated into Ago1, although they do not necessarily possess a 5′ terminal U. Recent studies on mammalian cells suggest that depletion of Dicers reduces the effectiveness of siRNA silencing since hairpin–derived siRNAs are more efficiently incorporated into effector complexes than simple siRNAs.[81-83] This further suggests colocalization of DCLs and AGO proteins.

Finally, although the previous considerations indicate that Ago1 may play a predominant role in antiviral defence, the redundant action of the different Ago proteins in plants should not be completely excluded. In this regard, Northern analysis of CMV-derived vsiRNAs in immuno-precipitated RNA fractions shows that both Ago2 and Ago5 can bind vsiRNAs in *A. thaliana.*[80] In addition, in *A. thaliana, ago7* mutants support the accumulation of TCV mutants at higher level than in wild type plants,[44] suggesting that Ago7 loaded with vsiRNAs may play some role in viral inactivation.

Ago proteins may also be involved in other plant antiviral pathways. Indeed, silencing Ago4 in *N. benthamiana* plants through PVX based-viral vectors compromises the R-mediated gene-for-gene antiviral resistance. In this study Bhattacharjee and colleagues[84] suggested an intriguing engagement of Ago4 in translational control of viral transcripts.

To date, RNA silencing-based viral inactivation has been ascribed to RISC-mediated cleavage of the viral target. However, translational inhibition of viral transcripts may be an alternative to viral cleavage, since Ago proteins are implicated in some examples of endogenous degradation-independent translational control in plants.[85,86] Another intriguing question is whether vsiRNAs can enter into RISC complex and can target plant endogenous mRNAs. To date, a large set of host genes have been identified to be post-transcriptionally regulated by vsiRNAs[33,36] however, their identification has been mostly based on computational analysis and still offer indirect evidences of the phenomena.

To summarize, how vsiRNAs are selected and incorporated into the RISC complex is still an open question. In plants, these aspects of antiviral defence are still obscure and deserve further careful study.

AMPLIFICATION AND SPREAD OF ANTIVIRAL DEFENCE

A. thaliana possesses six putative RNA-directed RNA polymerase proteins (RDR1-6), although only RDR1, RDR2 and RDR6 are characterized in terms of function and have direct orthologs in many plant species (Table 2).[87] In vitro and in vivo studies show that RDRs drive the synthesis of complementary RNAs (cRNAs) from the 3′-terminal nucleotides of ssRNAs or ssDNAs.[88] In vitro, the cRNA remains annealed to the RNA template, thus forming a long dsRNA molecule.[89] RDR2 shows efficient polymerase activity both upon annealing of complementary siRNAs to the 3′end of the RNA template and in its absence.[90] RDR1 and RDR6 are primer-independent polymerases, but their homology with RDR2 suggests that primer-dependence and primer-independence are both possible mechanisms used in plant RDR-mediated 3′-extension.[88]

A further complexity is that, in all cases, plant RDR activities are linked to DCL(s) and AGO(s) functions and act either upstream or downstream of them. RDR2 contributes to the production of DCL3-dependent secondary 24 nt siRNAs, which are involved in RNA-directed DNA methylation (RdDM) and chromatin remodelling, likely the basic processes of transcriptional gene silencing.[12] In this regard, RDR2, DCL3, Ago4 and other proteins involved in RdDM colocalize in distinct nuclear compartments.[91] RDR6 is involved in the production of ta-siRNAs and requires upstream activity of DCL1 and Ago1, until the TAS primary RNAs have been cleaved by the miRNA. Next, RDR6 extends a cRNA on cleaved TAS RNA, generating a dsRNA molecule. RDR6-derived TAS dsRNAs are sequentially cut into phased 21 nt ta-siRNAs by DCL4.[92-94] In the nat-siRNAs pathway, RDR6 seems to act both upstream, in concert with DCL2 and downstream, with Ago-mediated cleavage of RNA precursors.[95]

Besides their roles in endogenous sRNA pathways, plant RDRs are fundamental to the RNA silencing-based antiviral response. *A. thaliana* suppressors of gene silencing 2 and 3 (SGS2 and SGS3) (Table 2) were later described respectively as RDR6 and as an RNA-binding factor involved in stabilization of ssRNA precursors.[92,96,97] By microscopic analyses and complementation experiments Kumakura and colleagues[98] have recently suggested that SGS3 stabilizes RNAs and recruits them to RDR6 to initiate the dsRNA synthesis, a process which seems to take place in SGS3/RDR6 cytoplasmic bodies. SGS2 and SGS3 mutants are very susceptible to CMV infection and this susceptibility is associated with five-fold over-accumulation in viral titre.[97] Notably, *A. thaliana* mutants lacking factors of the Ago1-RDR6-SGS3 pathway behave like *dcl2-dcl4* mutant plants in terms of CMV accumulation.[22,23] Conversely, neither *sgs2* nor *sgs3* mutants affected symptoms or viral accumulation for the potyvirus *Turnip mosaic virus* (TuMV) or the *Tobacco crinkle virus* (TCV) (Table 1).[97] Infection of RDR6-silenced *N. benthamiana* with *Potato virus Y*, PVX and CMV, gave more severe symptoms than on wild type plants; no change of symptomatology was observed in the case of TMV, *Tobacco rattle virus* (TRV) (Table 1) and TCV infections.[99] Likewise, in *N. benthamiana* primary vsiRNAs from both PVX and the potyvirus *Plum pox virus* (Table 1) did not evoke an antiviral response unless RDR6-dependent secondary vsiRNAs were produced.[100] Conversely, high-throughput sequence analysis of CymRSV-infected *N. benthamiana* (both wild type and *rdr6* mutant), showed an unchanged profile in terms of quantity and size of CymRSV-derived vsiRNAs.[30] This suggests the specific involvement of RDR6 in the generation of vsiRNAs for some plant viruses and not for all.

RDR1 is another plant RDR playing a major antiviral role but, unlike RDR6, its expression is induced by salicylic acid (SA). SA is a defence signal that accumulates during aggression by pathogens, including many viruses.[28] In *A. thaliana* RDR1 mutants, TMV and TRV viral RNAs accumulate 4 to 15 times more than in wild type plants both in inoculated and upper, systemically infected leaves.[101] Notably, the relative level of vsiRNAs is lower in the mutant than in the wild type. Moreover, recombinant PVX constructs containing *N. benthamiana* (Nb) RDR1 fragments inoculated on *N. benthamiana*, induce severe symptoms of local necrosis associated with decreased accumulation of endogenous NbRDR1 mRNA transcripts.[102]

The use of viral suppressor-deficient CMV recombinants (termed as CMVΔ2b) has clarified the mechanism of RDR involvement in antiviral plant defence. Indeed, the production of CMV vsiRNAs is largely RDR1-dependent when *A. thaliana* plants are challenged with CMVΔ2b (but not with CMV wild type).[25] Given that CMV 2b impairs the action of Ago1,[71] in wild type plants infected with wild type CMV, 2b-mediated suppression of RDR1 may be explained by possible Ago1 action upstream of RDR1.

The absence of different symptoms or significant change in viral accumulation in early viral infections in *rdr* mutants, does not negate the importance of RDR1 and RDR6 in antiviral defence. Indeed, in experiments with *A. thaliana* infected with TMV-Cg (a strain infecting members of the family Cruciferae, to which *A. thaliana* belongs) small RNA libraries from either wild type or *rdr1* or *rdr6* mutants, show that both *rdr1* and *rdr6* mutants contain substantially lower amounts of vsiRNAs than wild type plants, without any change in viral accumulation. The same analysis reveals a significantly reduced strand bias in the viral small RNA population, with the portion of "antisense/sense" vsiRNA reads increased in both libraries from *rdr* mutants. Importantly, all phased sense but not antisense vsiRNAs were substantially less frequent in the *rdr6* mutant.[33] More recently, Garcia-Ruiz and collaborators have provided evidences that Arabidopsis RDR1 RDR2 and RDR6 are together involved in limiting infections in TuMV-inoculated leaves. In addition, both RDR1 and RDR6 were necessary, but individually insufficient, to prevent viral infections in non-inoculated leaves.[103]

Together these data suggest that most vsiRNAs originate from the dicing of dsRNA viral substrates generated by plant RDRs, rather than of replication intermediates by the viral RdRp (Fig. 2). This is evident for many plant viruses, albeit not all. The apparent

RDR-independency of vsiRNAs in the case of some viruses suggests the presence of still uncharacterized RDRs; it may also suggest the importance of dsRNAs of viral origin in generating vsiRNAs (i.e., RdRp-dependent dsRNAs or secondary structures of ssRNAs[103]) (Fig. 2).

MOVEMENT OF ANTIVIRAL SILENCING

In plants the effects of RNA silencing can extend beyond the initiation site. Virus cell-to-cell movement occurs exclusively through plasmodesmata (the channels traversing cell walls and enabling the transport between cells) and requires viral-encoded proteins, which help to increase the pore size, thus facilitating the movement of RNA-protein complexes.[104]

Virus-induced silencing signals are also believed to move ahead of the infection front through plasmodesmata and these signals may prepare silencing in cells yet to be infected. For example, the lack of the tombusvirus silencing suppressor p19, prevents repeated cell-to-cell invasion of the leaf lamina and uninfected cells show evidence of nucleotide sequence-specific resistance to CymRSV.[105] The same was noted for viral suppressor-deficient TCV.[23]

In the case of transgenes, Himber and colleagues have shown that DCL4-dependent 21 nt sRNAs, rather than DCL3-dependent 24 nt sRNAs, are needed for cell-to-cell movement of the silencing signal.[106] In addition, this movement can be amplified by plant RDRs upon conversion of cleaved target RNAs into new dsRNA molecules. DCL4 then produces secondary sRNAs both 3′ and 5′ of the region initially targeted by primary sRNAs. This phenomenon has been termed "transitivity" and though its involvement in antiviral defence has never been demonstrated, it is a plausible model for amplification and cell-to-cell trafficking of antiviral silencing.[107,108]

Despite these observations, cell-to-cell movement of the antiviral RNA silencing signal is still far from clear in terms of which form of vsiRNA is moving-duplexes, free vsiRNAs, 21 nt or 24 nt siRNAs, or Ago-bound vsiRNAs.[109] Even the involvement of RDRs in movement of the silencing signal is unclear, since vsiRNA profiles, viral accumulation and viral movement seem to be unaffected by the lack of plant RDRs for many virus/plant systems.

Another intriguing and still unclear question is whether enlarged plasmodesmata are preferential or unique channels of RNA silencing signal movement. Given their small size, it has been assumed that siRNAs can move from cell to cell through plasmodesmata by simple diffusion.[110] More precisely, the degree of movement of the silencing signal has been shown to be governed by diffusion and regulated by plasmodesmata aperture.[111]

TMV movement protein (TMV MP) is known to modify the pore size of plasmodesmata, thus facilitating the spread of viral RNA.[112] Interestingly, TMV MP expressed in plants enhances cell-to-cell spreading of silencing signals; this mechanism rely on the ability of MP to enlarge the plasmodesmata, to spread cell to cell and on its ability to interact with nucleic acid in a sequence nonspecific manner.[113] Yoo and colleagues[114] raised the possibility that specific RNA binding proteins could function in cell-to-cell trafficking of sRNAs. Indeed, they have shown that single-stranded sRNAs, rather than double-stranded forms, may move cell-to-cell as stable ribonucleoprotein complexes through dilated plasmodesmata channels.[114]

Long-distance or systemic movement explains the spread of silencing between distant organs and displays kinetics similar to photo-assimilates moving through the plant via the phloem. Grafting experiments on transgenic plants provide evidence that a signal can be transmitted from rootstocks to scions; the silence trigger in the rootstock silences the corresponding transgene in the scion.[115-117] The recovery phenotype, in the case of viral infections, is a sign of movement of viral silencing signals. Initially, the plant shows severe symptoms associated with a high titre of viral RNA, in inoculated tissues or those initially systemically invaded. Later, when the plant recovers, symptoms on new leaves are attenuated and the virus accumulates at lower levels in upper non-inoculated tissues. Moreover, the recovered tissues are resistant to re-inoculation.[74,118-121] In these infected plants, the primary infections are believed to be a source of signals that move at long distance, thus conferring resistance in recovered leaves. Similarly, movement-defective mutants of PVX containing fragments of the plant ribulose bisphosphate carboxylase small subunit, elicit systemic silencing of the corresponding gene, despite confinement of virus replication to a single leaf.[122] Micrograft experiments on transgenic *Arabidopsis*[123] suggested that the long distance silencing signal might not be a product of Dicer, but instead the signal is a long dsRNA precursor. On the other hand, small RNAs generated from transgenes or from *Cucumber yellows virus* (CuYV) (Table 1) can be detected in the phloem and these sRNAs may be carrying the long distance signals.[114] With DCL3-derived 24 nt and the DCL4-derived 21 nt sRNAs, the former are thought to be more directly involved in systemic signalling,[106] however, in the case of viral infection Yoo and colleagues have found that 21nt long CuYV-derived siRNAs are prevalent in the phloem sap.[114]

As with antiviral cell-to-cell signalling, systemic trafficking of the silencing signal appears to be highly complex. How the systemic signal or signals are produced, what is their nature and which are the exact channels of movement, are all questions deserving further study.

VIRAL SUPPRESSORS OF RNA SILENCING

A large number of viral suppressors of RNA silencing (VSRs) have been identified among most plant virus types, including those with (+) and (−) ssRNA, ssDNA and dsDNA genomes (reviewed by Csorba et al[124]). The data shows that viruses have evolved many protein-based counterstrategies against the plant's antiviral silencing defence. Most VSRs are viral structural or nonstructural proteins with already recognized important functions at different levels in the viral life cycle (e.g., coat protein, movement, aphid trasmission, replication).[124] Thus, VSRs are generally multifunctional proteins, which rarely show motifs common to the silencing suppression activity (Table 2).

Theoretically, VSRs might impair any single step of the RNA silencing-based antiviral pathway. For example, the potyviral silencing suppressor HcPro most likely inhibits unwinding of the siRNA duplex and RISC assembly[125] by binding them[126], P25 of PVX suppresses the production or accumulation of mobile silencing signals[122] and P19 of tombusviruses (i.e., CIRV), P21 of *Beet yellows virus* (BYV) and 2b of *Tomato aspermy virus* (TAV) inhibit RNA silencing by binding siRNAs and preventing their processing or incorporation into the RISC (Table 2).[126] The crystal structure analysis for some of the VSRs (i.e., CIRV P19, TAV 2b, has provided further indications of their intimate size selectivity for sRNAs).[127,128]

Among numerous known viral RNA-silencing suppressors, the 2b protein of CMV, the P0 of poleroviruses, the p6 of CaMV and the V2 of *Tomato yellow leaf curl virus* have been shown to interact directly with a protein component of the host's silencing pathway. Specifically, CMV 2b binds AGO1 and inhibits its slicing activity,[71] P0 targets AGO1 for degradation,[73,129] p6 physically interacts with DRB4[43] and V2 interacts with the *A. thaliana* SGS3 or its homolog in tomato (Table 2).[130]

Only CMV 2b and CaMV p6 have been studied in the context of viral infection and there is evidence that their expression during the viral cycle is necessary for altering the antiviral response to the virus. Some plant viruses produce several VSRs, such as the p20, p23 and the CP of *Citrus tristeza virus* and the AC2 and AC4 of geminiviruses.[131,132] Given that viruses modulate their protein expression depending on their life cycle in the cell and on their strategy of replication and expression, it is not unlikely that these spatio-temporal restrictions determine the VSRs' specific functions.

CONCLUSION

In the last decade something has been learned of the RNA silencing mechanisms involved in both endogenous and viral regulatory pathways, but they still require careful study. For example, how is the cytoplasmic localization of a virus compatible with the nuclear activity of DCLs? How can a RISC reach and inactivate a viral RNA sequestered in replication complexes or in membrane-bound vesicles? Where exactly in the cell and in the viral life cycle are the VSRs expressed and how are these elements compatible with the side-effects described in artificial systems? There are no conclusive answers. Even the cell-to-cell and long-distance movement of the antiviral silencing signals are still unclear, probably because the same cell-to-cell and long-distance viral transport is far from being understood for many viral systems.

The study of RNA silencing mechanisms in permissive plant systems in which genetic dissection is possible, may help to clarify many of these points. However, these models generally involve strong symptom expression and high viral multiplication. RNA silencing, besides its important regulatory role in endogenous pathways, is also an antiviral defence mechanism in crop plants. However, in virus-host couples that have cohabited over long times, it should more correctly be considered as a mechanism of surveillance, ensuring the coexistence of both partners and entailing the absence of symptoms. The time is ripe to focus on these plant-virus systems allowing us to contrast them with the situation in crop plants, with the support of ongoing sequencing of plant genomes and the use of high-throughput technology.

ACKNOWLEDGEMENTS

Thank to Robert G. Milne and Zoltán Havelda for critical reading of the manuscript and for their valuable comments.

REFERENCES

1. Fire A, Xu S, Montgomery MK et al. Potent and specific genetic interference by double-stranded RNA in Caenorhabditis elegans. Nature 1998; 391(6669):806-811.
2. Fire A. RNA-triggered gene silencing. Trends Genet 1999; 15(9):358-363.
3. Tuschl T, Zamore PD, Lehmann R et al. Targeted mRNA degradation by double-stranded RNA in vitro. Genes Dev 1999; 13(24):3191-3197.
4. Breving K, Esquela-Kerscher A. The complexities of microRNA regulation: mirandering around the rules. Int J Biochem Cell Biol 2010; 42(8):1316-1329.
5. Hammond SM, Bernstein E, Beach D et al. An RNA-directed nuclease mediates post-transcriptional gene silencing in Drosophila cells. Nature 2000; 404(6775):293-296.
6. Bartel DP. MicroRNAs: genomics, biogenesis, mechanism and function. Cell 2004; 116(2):281-297.
7. Baulcombe D. RNA silencing in plants. Nature 2004; 431(7006):356-363.
8. Verdel A, Jia S, Gerber S et al. RNAi-mediated targeting of heterochromatin by the RITS complex. Science 2004; 303(5658):672-676.
9. Aravin A, Gaidatzis D, Pfeffer S et al. A novel class of small RNAs bind to MILI protein in mouse testes. Nature 2006; 442(7099):203-207.
10. Malone CD, Brennecke J, Dus M et al. Specialized piRNA pathways act in germline and somatic tissues of the Drosophila ovary. Cell 2009; 137(3):522-535.
11. Hammond SM, Boettcher S, Caudy AA et al. Argonaute2, a link between genetic and biochemical analyses of RNAi. Science 2001; 293(5532):1146-1150.
12. Vaucheret H. Post-transcriptional small RNA pathways in plants: mechanisms and regulations. Genes Dev 2006; 20(7):759-771.
13. Collins LJ, Penny D. The RNA infrastructure: dark matter of the eukaryotic cell? Trends Genet 2009; 25(3):120-128.
14. Cullen BR. Is RNA interference involved in intrinsic antiviral immunity in mammals? Nat Immunol 2006; 7(6):563-567.
15. Hamilton AJ, Baulcombe DC. A species of small antisense RNA in post-transcriptional gene silencing in plants. Science 1999; 286(5441):950-952.
16. Hammond SM. Dicing and slicing: the core machinery of the RNA interference pathway. FEBS Lett 2005; 579(26):5822-5829.
17. Ahlquist P. RNA-dependent RNA polymerases, viruses and RNA silencing. Science 2002; 296(5571):1270-1273.
18. Ruiz-Ferrer V, Voinnet O. Roles of plant small RNAs in biotic stress responses. Annu Rev Plant Biol 2009; 60:485-510.
19. Tomari Y, Zamore PD. Perspective: machines for RNAi. Genes Dev 2005; 19(5):517-529.
20. Hamilton A, Voinnet O, Chappell L et al. Two classes of short interfering RNA in RNA silencing. EMBO J 2002; 21(17):4671-4679.
21. Henderson IR, Zhang X, Lu C et al. Dissecting Arabidopsis thaliana DICER function in small RNA processing, gene silencing and DNA methylation patterning. Nat Genet 2006; 38(6):721-725.
22. Bouche N, Lauressergues D, Gasciolli V et al. An antagonistic function for Arabidopsis DCL2 in development and a new function for DCL4 in generating viral siRNAs. EMBO J 2006; 25(14):3347-3356.
23. Deleris A, Gallego-Bartolome J, Bao J et al. Hierarchical action and inhibition of plant Dicer-like proteins in antiviral defense. Science 2006; 313(5783):68-71.
24. Fusaro AF, Matthew L, Smith NA et al. RNA interference-inducing hairpin RNAs in plants act through the viral defence pathway. EMBO Rep 2006; 7(11):1168-1175.
25. Diaz-Pendon JA, Li F, Li WX et al. Suppression of antiviral silencing by cucumber mosaic virus 2b protein in Arabidopsis is associated with drastically reduced accumulation of three classes of viral small interfering RNAs. Plant Cell 2007; 19(6):2053-2063.
26. Du QS, Duan CG, Zhang ZH et al. DCL4 targets Cucumber mosaic virus satellite RNA at novel secondary structures. J Virol 2007; 81(17):9142-9151.
27. Simon AE, Roossinck MJ, Havelda Z. Plant virus satellite and defective interfering RNAs: new paradigms for a new century. Annu Rev Phytopathol 2004; 42:415-437.
28. Xie Z, Johansen LK, Gustafson AM et al. Genetic and functional diversification of small RNA pathways in plants. PLoS Biol 2004; 2(5):E104.
29. Molnar A, Csorba T, Lakatos L et al. Plant virus-derived small interfering RNAs originate predominantly from highly structured single-stranded viral RNAs. J Virol 2005; 79(12):7812-7818.
30. Szittya G, Moxon S, Pantaleo V et al. Structural and functional analysis of viral siRNAs. PLOS Pathogens 2010, 6(4):e1000838.

31. Ho T, Pallett D, Rusholme R et al. A simplified method for cloning of short interfering RNAs from Brassica juncea infected with Turnip mosaic potyvirus and Turnip crinkle carmovirus. J Virol Methods 2006; 136(1-2):217-223.
32. Donaire L, Barajas D, Martinez-Garcia B et al. Structural and genetic requirements for the biogenesis of tobacco rattle virus-derived small interfering RNAs. J Virol 2008; 82(11):5167-5177.
33. Qi X, Bao FS, Xie Z. Small RNA deep sequencing reveals role for Arabidopsis thaliana RNA-dependent RNA polymerases in viral siRNA biogenesis. PLoS One 2009; 4(3):e4971.
34. Akbergenov R, Si-Ammour A, Blevins T et al. Molecular characterization of geminivirus-derived small RNAs in different plant species. Nucleic Acids Res 2006; 34(2):462-471.
35. Blevins T, Rajeswaran R, Shivaprasad PV et al. Four plant Dicers mediate viral small RNA biogenesis and DNA virus induced silencing. Nucleic Acids Res 2006; 34(21):6233-6246.
36. Moissiard G, Voinnet O. RNA silencing of host transcripts by cauliflower mosaic virus requires coordinated action of the four Arabidopsis Dicer-like proteins. Proc Natl Acad Sci USA 2006; 103(51):19593-19598.
37. Rodriguez-Negrete EA, Carrillo-Tripp J, Rivera-Bustamante RF. RNA silencing against geminivirus: complementary action of post-transcriptional gene silencing and transcriptional gene silencing in host recovery. J Virol 2009; 83(3):1332-1340.
38. Aliyari R, Ding SW. RNA-based viral immunity initiated by the Dicer family of host immune receptors. Immunol Rev 2009; 227(1):176-188.
39. Nakazawa Y, Hiraguri A, Moriyama H et al. The dsRNA-binding protein DRB4 interacts with the Dicer-like protein DCL4 in vivo and functions in the trans-acting siRNA pathway. Plant Mol Biol 2007; 63(6):777-785.
40. Kobayashi K, Tsuge S, Nakayashiki H et al. Requirement of cauliflower mosaic virus open reading frame VI product for viral gene expression and multiplication in turnip protoplasts. Microbiol Immunol 1998; 42(5):377-386.
41. Haas M, Geldreich A, Bureau M et al. The open reading frame VI product of Cauliflower mosaic virus is a nucleocytoplasmic protein: its N terminus mediates its nuclear export and formation of electron-dense viroplasms. Plant Cell 2005; 17(3):927-943.
42. Love AJ, Laird J, Holt J et al. Cauliflower mosaic virus protein P6 is a suppressor of RNA silencing. J Gen Virol 2007; 88(Pt 12):3439-3444.
43. Haas G, Azevedo J, Moissiard G et al. Nuclear import of CaMV P6 is required for infection and suppression of the RNA silencing factor DRB4. EMBO J 2008; 27(15):2102-2112.
44. Qu F, Ye X, Morris TJ. Arabidopsis DRB4, AGO1, AGO7 and RDR6 participate in a DCL4-initiated antiviral RNA silencing pathway negatively regulated by DCL1. Proc Natl Acad Sci USA 2008; 105(38):14732-14737.
45. Papp I, Mette MF, Aufsatz W et al. Evidence for nuclear processing of plant micro RNA and short interfering RNA precursors. Plant Physiol 2003; 132(3):1382-1390.
46. Hiraguri A, Itoh R, Kondo N et al. Specific interactions between Dicer-like proteins and HYL1/DRB-family dsRNA-binding proteins in Arabidopsis thaliana. Plant Mol Biol 2005; 57(2):173-188.
47. Mackenzie J. Wrapping things up about virus RNA replication. Traffic 2005; 6(11):967-977.
48. Sanfacon H, Zhang G. Analysis of interactions between viral replicase proteins and plant intracellular membranes. Methods Mol Biol 2008; 451:361-375.
49. Salonen A, Ahola T, Kaariainen L. Viral RNA replication in association with cellular membranes. Curr Top Microbiol Immunol 2005; 285:139-173.
50. Netherton C, Moffat K, Brooks E et al. A guide to viral inclusions, membrane rearrangements, factories and viroplasm produced during virus replication. Adv Virus Res 2007; 70:101-182.
51. Kopek BG, Perkins G, Miller DJ et al. Three-dimensional analysis of a viral RNA replication complex reveals a virus-induced mini-organelle. PLoS Biol 2007; 5(9):e220.
52. Ahlquist P. Parallels among positive-strand RNA viruses, reverse-transcribing viruses and double-stranded RNA viruses. Nat Rev Microbiol 2006; 4(5):371-382.
53. Cheung AY, de Vries SC. Membrane trafficking: intracellular highways and country roads. Plant Physiol 2008; 147(4):1451-1453.
54. Yu B, Yang Z, Li J et al. Methylation as a crucial step in plant microRNA biogenesis. Science 2005; 307(5711):932-935.
55. Yang L, Liu Z, Lu F et al. SERRATE is a novel nuclear regulator in primary microRNA processing in Arabidopsis. Plant J 2006; 47(6):841-850.
56. Huang Y, Ji L, Huang Q et al. Structural insights into mechanisms of the small RNA methyltransferase HEN1. Nature 2009; 461(7265):823-827.
57. Li J, Yang Z, Yu B et al. Methylation protects miRNAs and siRNAs from a 3′-end uridylation activity in Arabidopsis. Curr Biol 2005; 15(16):1501-1507.
58. Ramachandran V, Chen X. Degradation of microRNAs by a family of exoribonucleases in Arabidopsis. Science 2008; 321(5895):1490-1492.

59. Alefelder S, Patel BK, Eckstein F. Incorporation of terminal phosphorothioates into oligonucleotides. Nucleic Acids Res 1998; 26(21):4983-4988.
60. Lozsa R, Csorba T, Lakatos L et al. Inhibition of 3′ modification of small RNAs in virus-infected plants require spatial and temporal co-expression of small RNAs and viral silencing-suppressor proteins. Nucleic Acids Res 2008; 36(12):4099-4107.
61. Kubota K, Tsuda S, Tamai A et al. Tomato mosaic virus replication protein suppresses virus-targeted post-transcriptional gene silencing. J Virol 2003; 77(20):11016-11026.
62. Csorba T, Bovi A, Dalmay T et al. The p122 subunit of Tobacco mosaic virus replicase is a potent silencing suppressor and compromises both siRNA and miRNA mediated pathways. J Virol 2007; 81(21):11768-80.
63. Vogler H, Akbergenov R, Shivaprasad PV et al. sRNA modification associated with suppression of RNA silencing by tobamovirus replicase protein. J Virol 2007; 81(19):10379-88..
64. Vaucheret H. Plant ARGONAUTES. Trends Plant Sci 2008; 13(7):350-358.
65. Wang Y, Juranek S, Li H et al. Nucleation, propagation and cleavage of target RNAs in Ago silencing complexes. Nature 2009; 461(7265):754-761.
66. Fagard M, Boutet S, Morel JB et al. AGO1, QDE-2 and RDE-1 are related proteins required for post-transcriptional gene silencing in plants, quelling in fungi and RNA interference in animals. Proc Natl Acad Sci USA 2000; 97(21):11650-11654.
67. Morel JB, Godon C, Mourrain P et al. Fertile hypomorphic ARGONAUTE (ago1) mutants impaired in post-transcriptional gene silencing and virus resistance. Plant Cell 2002; 14(3):629-639.
68. Vaucheret H, Vazquez F, Crete P et al. The action of ARGONAUTE1 in the miRNA pathway and its regulation by the miRNA pathway are crucial for plant development. Genes Dev 2004; 18(10):1187-1197.
69. Baumberger N, Baulcombe DC. Arabidopsis ARGONAUTE1 is an RNA Slicer that selectively recruits microRNAs and short interfering RNAs. Proc Natl Acad Sci USA 2005; 102(33):11928-11933.
70. Qi Y, Denli AM, Hannon GJ. Biochemical specialization within Arabidopsis RNA silencing pathways. Mol Cell 2005; 19(3):421-428.
71. Zhang X, Yuan YR, Pei Y et al. Cucumber mosaic virus-encoded 2b suppressor inhibits Arabidopsis Argonaute1 cleavage activity to counter plant defense. Genes Dev 2006; 20(23):3255-3268.
72. Pazhouhandeh M, Dieterle M, Marrocco K et al. F-box-like domain in the polerovirus protein P0 is required for silencing suppressor function. Proc Natl Acad Sci USA 2006; 103(6):1994-1999.
73. Baumberger N, Tsai CH, Lie M et al. The polerovirus silencing suppressor P0 targets ARGONAUTE proteins for degradation. Curr Biol 2007; 17(18):1609-14.
74. Pantaleo V, Szittya G, Burgyan J. Molecular bases of viral RNA targeting by viral small interfering RNA-Programmed RISC. J Virol 2007; 81(8):3797-3806.
75. Shen B, Goodman HM. Uridine addition after microRNA-directed cleavage. Science 2004; 306(5698):997.
76. Okamura K, Ishizuka A, Siomi H et al. Distinct roles for Argonaute proteins in small RNA-directed RNA cleavage pathways. Genes Dev 2004; 18(14):1655-1666.
77. Tomari Y, Du T, Haley B et al. RISC assembly defects in the Drosophila RNAi mutant armitage. Cell 2004; 116(6):831-841.
78. Jones-Rhoades MW, Bartel DP, Bartel B. MicroRNAS and their regulatory roles in plants. Annu Rev Plant Biol 2006; 57:19-53.
79. Mi S, Cai T, Hu Y et al. Sorting of small RNAs into Arabidopsis argonaute complexes is directed by the 5′ terminal nucleotide. Cell 2008; 133(1):116-127.
80. Takeda A, Iwasaki S, Watanabe T et al. The mechanism selecting the guide strand from small RNA duplexes is different among argonaute proteins. Plant Cell Physiol 2008; 49(4):493-500.
81. Doi N, Zenno S, Ueda R et al. Short-interfering-RNA-mediated gene silencing in mammalian cells requires Dicer and eIF2C translation initiation factors. Curr Biol 2003; 13(1):41-46.
82. Siolas D, Lerner C, Burchard J et al. Synthetic shRNAs as potent RNAi triggers. Nat Biotechnol 2005; 23(2):227-231.
83. Kim VN. MicroRNA biogenesis: coordinated cropping and dicing. Nat Rev Mol Cell Biol 2005; 6(5):376-385.
84. Bhattacharjee S, Zamora A, Azhar MT et al. Virus resistance induced by NB-LRR proteins involves Argonaute4-dependent translational control. Plant J 2009; 58(6):940-951.
85. Brodersen P, Sakvarelidze-Achard L, Bruun-Rasmussen M et al. Widespread translational inhibition by plant miRNAs and siRNAs. Science 2008; 320(5880):1185-1190.
86. Lanet E, Delannoy E, Sormani R et al. Biochemical evidence for translational repression by arabidopsis MicroRNAs. Plant Cell 2009; 21(6):1762-1768.
87. Wassenegger M, Krczal G. Nomenclature and functions of RNA-directed RNA polymerases. Trends Plant Sci 2006; 11(3):142-151.
88. Voinnet O. Use, tolerance and avoidance of amplified RNA silencing by plants. Trends Plant Sci 2008; 13(7):317-328.
89. Curaba J, Chen X. Biochemical activities of Arabidopsis RNA-dependent RNA polymerase 6. J Biol Chem 2008; 283(6):3059-3066.

90. Makeyev EV, Bamford DH. Cellular RNA-dependent RNA polymerase involved in post-transcriptional gene silencing has two distinct activity modes. Mol Cell 2002; 10(6):1417-1427.
91. Li CF, Henderson IR, Song L et al. Dynamic regulation of ARGONAUTE4 within multiple nuclear bodies in Arabidopsis thaliana. PLoS Genet 2008; 4(2):e27.
92. Peragine A, Yoshikawa M, Wu G et al. SGS3 and SGS2/SDE1/RDR6 are required for juvenile development and the production of trans-acting siRNAs in Arabidopsis. Genes Dev 2004; 18(19):2368-2379.
93. Allen E, Xie Z, Gustafson AM et al. microRNA-directed phasing during trans-acting siRNA biogenesis in plants. Cell 2005; 121(2):207-221.
94. Xie Z, Allen E, Wilken A et al. DICER-LIKE 4 functions in trans-acting small interfering RNA biogenesis and vegetative phase change in Arabidopsis thaliana. Proc Natl Acad Sci USA 2005; 102(36):12984-12989.
95. Borsani O, Zhu J, Verslues PE et al. Endogenous siRNAs derived from a pair of natural cis-antisense transcripts regulate salt tolerance in Arabidopsis. Cell 2005; 123(7):1279-1291.
96. Elmayan T, Balzergue S, Beon F et al. Arabidopsis mutants impaired in cosuppression. Plant Cell 1998; 10(10):1747-1758.
97. Mourrain P, Beclin C, Elmayan T et al. Arabidopsis SGS2 and SGS3 genes are required for post-transcriptional gene silencing and natural virus resistance. Cell 2000; 101(5):533-542.
98. Kumakura N, Takeda A, Fujioka Y et al. SGS3 and RDR6 interact and colocalize in cytoplasmic SGS3/RDR6-bodies. FEBS Lett 2009; 583(8):1261-1266.
99. Schwach F, Vaistij FE, Jones L et al. An RNA-dependent RNA polymerase prevents meristem invasion by potato virus X and is required for the activity but not the production of a systemic silencing signal. Plant Physiol 2005; 138(4):1842-1852.
100. Vaistij FE, Jones L. Compromised virus-induced gene silencing in RDR6-deficient plants. Plant Physiol 2009; 149(3):1399-1407.
101. Yu D, Fan B, MacFarlane SA et al. Analysis of the involvement of an inducible Arabidopsis RNA-dependent RNA polymerase in antiviral defense. Mol Plant Microbe Interact 2003; 16(3):206-216.
102. Yang SJ, Carter SA, Cole AB et al. A natural variant of a host RNA-dependent RNA polymerase is associated with increased susceptibility to viruses by Nicotiana benthamiana. Proc Natl Acad Sci USA 2004; 101(16):6297-6302.
103. Garcia-Ruiz H, Takeda A, Chapman EJ et al. Arabidopsis RNA-dependent RNA polymerases and dicer-like proteins in antiviral defense and small interfering RNA biogenesis during turnip mosaic virus infection. Plant Cell 2010;(22)2:482-496.
104. Lucas WJ. Plant viral movement proteins: agents for cell-to-cell trafficking of viral genomes. Virology 2006; 344(1):169-184.
105. Havelda Z, Hornyik C, Crescenzi A et al. In situ characterization of cymbidium ringspot tombusvirus infection-induced post-transcriptional gene silencing in Nicotiana benthamiana. J Virol 2003; 77(10):6082-6086.
106. Himber C, Dunoyer P, Moissiard G et al. Transitivity-dependent and -independent cell-to-cell movement of RNA silencing. EMBO J 2003; 22(17):4523-4533.
107. Voinnet O. Noncell autonomous RNA silencing. FEBS Lett 2005; 579(26):5858-5871.
108. Ding SW, Voinnet O. Antiviral Immunity Directed by Small RNAs. Cell 2007; 130(3):413-426.
109. Dunoyer P, Voinnet O. Movement of RNA silencing between plant cells: is the question now behind us? Trends Plant Sci 2009; 14(12):643-644.
110. Mlotshwa S, Voinnet O, Mette MF et al. RNA silencing and the mobile silencing signal. Plant Cell 2002; 14 Suppl:S289-301.
111. Kobayashi K, Zambryski P. RNA silencing and its cell-to-cell spread during Arabidopsis embryogenesis. Plant J 2007; 50(4):597-604.
112. Waigmann E, Lucas WJ, Citovsky V et al. Direct functional assay for tobacco mosaic virus cell-to-cell movement protein and identification of a domain involved in increasing plasmodesmal permeability. Proc Natl Acad Sci USA 1994; 91(4):1433-1437.
113. Vogler H, Kwon MO, Dang V et al. Tobacco mosaic virus movement protein enhances the spread of RNA silencing. PLoS Pathog 2008; 4(4):e1000038.
114. Yoo BC, Kragler F, Varkonyi-Gasic E et al. A systemic small RNA signaling system in plants. Plant Cell 2004; 16(8):1979-2000.
115. Palauqui JC, Elmayan T, Pollien JM et al. Systemic acquired silencing: transgene-specific post-transcriptional silencing is transmitted by grafting from silenced stocks to nonsilenced scions. EMBO J 1997; 16(15):4738-4745.
116. Voinnet O, Vain P, Angell S et al. Systemic spread of sequence-specific transgene RNA degradation in plants is initiated by localized introduction of ectopic promoterless DNA. Cell 1998; 95(2):177-187.
117. Tournier B, Tabler M, Kalantidis K. Phloem flow strongly influences the systemic spread of silencing in GFP Nicotiana benthamiana plants. Plant J 2006; 47(3):383-394.

118. Ratcliff F, Harrison BD, Baulcombe DC. A similarity between viral defense and gene silencing in plants. Science 1997; 276(5318):1558-1560.
119. Ratcliff FG, MacFarlane SA, Baulcombe DC. Gene silencing without DNA. rna-mediated cross-protection between viruses. Plant Cell 1999; 11(7):1207-1216.
120. Bennett CW. Recovery of plants from dodder latent mosaic. Phytopathology 1949; 39(II):637-646.
121. Bennett CW. Recovery of water pimpernel from curly top and the reaction of recovered plants to reinoculation with different virus strains. Phytopathology 1955; 45(II):531-536.
122. Voinnet O, Lederer C, Baulcombe DC. A viral movement protein prevents spread of the gene silencing signal in Nicotiana benthamiana. Cell 2000; 103(1):157-167.
123. Brosnan CA, Mitter N, Christie M et al. Nuclear gene silencing directs reception of long-distance mRNA silencing in Arabidopsis. Proc Natl Acad Sci USA 2007; 104(37):14741-14746.
124. Csorba T, Pantaleo V, Burgyan J. RNA silencing: an antiviral mechanism. Adv Virus Res 2009; 75:35-71.
125. Chapman EJ, Prokhnevsky AI, Gopinath K et al. Viral RNA silencing suppressors inhibit the microRNA pathway at an intermediate step. Genes Dev 2004; 18(10):1179-1186.
126. Lakatos L, Csorba T, Pantaleo V et al. Small RNA binding is a common strategy to suppress RNA silencing by several viral suppressors. EMBO J 2006; 25(12):2768-2780.
127. Chen HY, Yang J, Lin C et al. Structural basis for RNA-silencing suppression by Tomato aspermy virus protein 2b. EMBO Rep 2008; 9(8):754-760.
128. Vargason J, Szittya G, Burgyan J et al. Size selective recognition of siRNA by an RNA silencing suppressor. Cell 2003; 115(7):799-811.
129. Bortolamiol D, Pazhouhandeh M, Marrocco K et al. The polerovirus F box protein P0 targets ARGONAUTE1 to suppress RNA Silencing. Curr Biol 2007; 17(18):1615-21.
130. Glick E, Zrachya A, Levy Y et al. Interaction with host SGS3 is required for suppression of RNA silencing by tomato yellow leaf curl virus V2 protein. Proc Natl Acad Sci USA 2008; 105(1):157-161.
131. Lu R, Folimonov A, Shintaku M et al. Three distinct suppressors of RNA silencing encoded by a 20-kb viral RNA genome. Proc Natl Acad Sci USA 2004; 101(44):15742-15747.
132. Vanitharani R, Chellappan P, Pita JS et al. Differential roles of AC2 and AC4 of cassava geminiviruses in mediating synergism and suppression of post-transcriptional gene silencing. J Virol 2004; 78(17):9487-9498.
133. Yamada K, Lim J, Dale JM et al. Empirical analysis of transcriptional activity in the Arabidopsis genome. Science 2003; 302(5646):842-846.
134. Margis R, Fusaro AF, Smith NA et al. The evolution and diversification of Dicers in plants. FEBS Lett 2006; 580(10):2442-2450.
135. Merai Z, Kerenyi Z, Kertesz S et al. Double-stranded RNA binding may be a general plant RNA viral strategy to suppress RNA silencing. J Virol 2006; 80(12):5747-5756.
136. St-Pierre P, Hassen IF, Thompson D et al. Characterization of the siRNAs associated with peach latent mosaic viroid infection. Virology 2009; 383(2):178-182.
137. Di Serio F, Gisel A, Navarro B et al. Deep sequencing of the small RNAs derived from two symptomatic variants of a chloroplastic viroid: implications for their genesis and for pathogenesis. PLoS One 2009; 4(10):e7539.
138. Navarro B, Pantaleo V, Gisel A et al. Deep sequencing of viroid-derived small RNAs from grapevine provides new insights on the role of RNA silencing in plant-viroid interaction. PLoS One 2009; 4(11):e7686.

CHAPTER 4

MicroRNAs AS POST-TRANSCRIPTIONAL MACHINES AND THEIR INTERPLAY WITH CELLULAR NETWORKS

Sarath Chandra Janga* and Swathi Vallabhaneni

Institute for Genomic Biology, University of Illinois at Urbana-Champaign, Urbana, Illinois, USA
**Corresponding Author: Sarath Chandra Janga—Email: sarath@igb.uiuc.edu*

Abstract: Gene expression is a highly controlled process which is known to occur at several levels in eukaryotic organisms. Although RNAs have been traditionally viewed as passive molecules in the pathway from transcription to translation, there is increasing evidence that their metabolism is controlled by a class of small noncoding RNAs called MicroRNAs (miRNAs). MicroRNAs (miRNAs) control essential gene regulatory pathways in both plants and animals however our understanding about their function, evolution and interplay with other cellular components is only beginning to be elucidated. In this chapter, we provide an overview of the recent developments in our understanding of this class of RNAs from diverse perspectives including biogenesis, mechanism of their function, evolution of their clusters, and discuss the approaches currently available for the construction of post-transcriptional networks governed by them. We also present our current understanding on these post-transcriptional networks in the context other cellular networks. We finally argue that such developments would not only allow us to gain a deeper understanding of regulation at a level that has been under-appreciated over the past decades, but would also allow us to use the newly developed high-throughput approaches to interrogate the prevalence of these phenomena in different states, and thereby exploit the functions of these RNA molecules for therapeutic advantage in higher eukaryotes.

RNA Infrastructure and Networks, edited by Lesley J. Collins.

INTRODUCTION

For a long time, RNA was thought to have a supplementary role in genome function—an intermediate in the flow of biological information from gene to protein. In the past decade, however, the centrality of RNA as a key player in gene expression has become clear; it is far from being merely an intermediate between gene and protein. Instead, new insights into RNA structure and function have revolutionized our view of molecular biology, and the role of RNA in the evolution of life itself. According to the central dogma of molecular biology, RNAs are passive messengers and only take charge of transferring genetic information or carrying out DNA instructions, or code, for protein production in cells. However, this central dogma is getting challenged by the findings that tiny fragments of noncoding RNA typically ~22 nucleotides in length, namely microRNA (miRNA), are able to negatively regulate protein-coding genes by interfering with mRNA's original instructions. Recent studies indicate that miRNAs have emerged as central post-transcriptional repressors of gene expression. MicroRNAs suppress gene expression via imperfect base pairing to the 3′ untranslated region (3′ UTR) of target mRNAs leading to repression of protein production or mRNA degradation.[1-4] These noncoding regulatory RNA molecules have been found in diverse plants, animals, some viruses and even algae, and it now seems likely that all multicellular eukaryotes and perhaps some unicellular eukaryotes utilize them to regulate gene expression. It is currently estimated that miRNAs account for about ~1% of the predicted genes in higher eukaryotes and up to 10-30% of their genes might be regulated by miRNAs.[1]

MicroRNAs regulate many key biological processes, including cell growth, development, proliferation, differentiation, apoptosis and metabolism by controlling the expression of the genes they bind to at the post-transcriptional level. Indeed, animals lacking some mature miRNAs are unable to survive or reproduce.[5-7] For instance, a single malfunctioning microRNA can be sufficient to cause cancer in mice.[8] These discoveries in the recent past have offered new insights into another layer of gene regulation and at the same time underscore the important role noncoding mRNAs play in the cells. MicroRNAs have biological functions in many aspects, with targets ranging from signaling proteins, metabolic enzymes to transcription factors. This diversity and abundance in their target repertoire offers an enormous level of combinatorial possibilities, and suggest that miRNAs and their targets can be densely interconnected with other cellular networks. Thus, it is reasonable to think that miRNAs exert their function by regulating and integrating cellular networks at different levels in the cell.

In this chapter, we provide a comprehensive overview of this quickly developing area of post-transcriptional regulatory networks formed by miRNAs. We organize it into the following major sections: (1) biogenesis, genomic organization and evolution of miRNAs; (2) methods used for the identification of miRNAs and their targets; (3) interplay of miRNAs with other cellular networks and finally; (4) expression dynamics of miRNAs in the context of other regulatory factors. With increasing amount of evidence in support of the roles of miRNAs in diverse cellular processes in a variety of model systems, it is possible to envisage that miRNAs can form promising therapeutic targets in years to come.

BIOGENESIS, GENOMIC ORGANIZATION AND EVOLUTION OF miRNAs

RNA molecules have been thought to be important players in regulating gene expression in metazoans since a vast class of endogenously expressed small RNAs, termed microRNAs, were identified. The first miRNAs were discovered via genetic screening of the nematode *Caenorhabditis elegans*.[9,10] Genes encoding miRNAs contain many similarities and distinctions compared to protein-coding genes. Typically, one miRNA gene can encode only one particular mRNA; however, there are frequent exceptions where the coding sequences of some groups of miRNAs are clustered and expressed as a single polycistronic transcript.[11,12] MicroRNAs in the same cluster may not necessarily show sequence similarities or regulate the identical target genes, but they are often co-expressed.[13,14] Due to their co-expression, they function collaboratively to orchestrate the appropriate dosage of multiple gene products. In addition, if a new miRNA sequence should appear within an existing transcription unit, it immediately expresses its new product without invention or duplication of enhancers and promoters. This enables new miRNA genes to more easily appear without complete gene/transcript duplication and may account for the abundance of miRNA gene clusters. Indeed, many new animal miRNAs are thought to arise from accumulation of nucleotide sequence changes and not from gene duplication supporting this notion.[15] Most miRNAs are located in either the intergenic regions or the intron regions of genes encoding proteins or other noncoding RNAs in the antisense orientation, and expressed as independent transcription units.[11,12] Some intronic miRNA genes, however, are arranged in the sense orientation with respect to the protein or RNA-coding genes, and are cotranscribed with their precursor mRNAs.[14] This type of transcript is organized such that the miRNA sequence is located within an intron.

MicroRNAs are transcribed by RNA polymerase II as long precursor transcripts, which are called primary miRNAs (pri-miRNAs). The pri-miRNAs are capped and polyadenylated, and can reach several kilobases in length.[16] A single pri-miRNA might contain one, or up to several, miRNAs. Several sequential steps of transcript processing are required to produce mature miRNAs from pri-miRNAs. In the nucleus, there is a microprocessor complex in which the major components are the RNase-III enzyme Drosha and its partner DGCR8/Pasha,[17,18] which initially recognize pri-miRNAs and then excise the precursor miRNA (pre-miRNA), a 60-100 nucleotide intermediate that makes imperfect stem-loop hairpin structure containing the mature miRNA. Exportin-5, a nuclear export factor, recognizes and transports the pre-miRNAs to cytoplasm.[19,20] In the cytoplasm, Dicer, a second RNase-III enzyme, cleaves the pre-miRNAs to generate double-stranded 18-24 nucleotide long RNA molecules—mature miRNAs.[16,21,22] RNA-induced silencing complex (RISC), the core component of which is the argonaute protein,[16] incorporates one of these two strands—the guide strand of miRNAs. Finally, the miRNA guides the RISC complex to the target mRNA to suppress gene expression via imperfect base pairing to the 3′ UTR of target mRNAs, leading to repression of protein production or mRNA degradation.[1,4] Targeting by miRNAs is known to affect an mRNA's activity via different mechanisms. The earliest model claimed that an RNAi-like degradation mechanism occurs when the complementarity within miRNA-target duplex is nearly perfect, whereas imperfect pairing results in the suppression of translation possibly via the disruption of elongation[23] or 5′ capping of the mRNA.[24,25] This dichotomy has been severely challenged by miRNA-induced destabilization of mRNA half-life where the miRNA-mRNA target interaction is imperfect. These imperfect miRNA-target effects have indeed been documented in both humans and *C. elegans*.[26-29] However, a recent study showed that changes in mRNA levels closely reflect the impact of miRNAs on gene expression and indicate that destabilization of target mRNAs is the predominant reason for reduced protein output in humans, with 84% of the decreased protein production accounted by the lowered mRNA levels.[30]

Evaluating conservation among species is a robust approach to find critical features within biological sequences and to understand the contribution they make to the existence of an organism. In this context, sequence homology can be found with very high frequency in most miRNAs across closely related species, and some miRNAs are ubiquitous in all the metazoans.[31] In addition, most miRNAs share almost identical biosynthetic pathways, precursor structure and effector mechanism. It is due to this high degree of similarity from biosynthesis to mechanism of action, conservation of miRNAs and their gene clusters have been studied from diverse perspectives across organisms. Members of a given miRNA family can be recovered from genome sequences due to the high sequence conservation of the mature miRNA and the characteristic stable hairpin structure of the precursor. Such a systematic analysis can be used for not only identifying miRNA homologs but to construct phylogenies of the family to understand the likely time of evolutionary origin. Several studies have used the continuously increasing repertoire of miRNAs in eukaryotes, documented in databases,[32] to show that miRNA evolution correlates with major body plan innovations in animal phylogeny.[33-35] These studies also propose that the diversity of the microRNA repertoire, complexity of their expression patterns and the diversity of the miRNA targets are correlated with an animal's morphological complexity. It is possible to imagine that miRNA's can act as important players in controlling large fractions of gene networks in a co-ordinated manner as they are known to influence almost all cellular processes. It is also increasingly becoming clear that lineage-specific microRNAs may account for phenotypic variation in closely related species.[33,36,37] For instance, in a large scale study comparing the miRNA repertoires of human and chimpanzee brains, Plasterk and colleagues[37] have shown that many of the then identified novel miRNAs are not conserved beyond primates, indicating their recent origin. They also found a significant fraction to be species-specific, whereas others are expanded in one of the species through duplication events. These studies support the notion that evolution of miRNAs is an ongoing process and that along with a small fraction of ancient, highly conserved miRNAs, there are a number of emerging miRNAs in the current genome sequences.[15]

The evolution of microRNAs is characterized not only by the continuing innovation of novel families but also by the diversification of established families spawning additional paralogous family members. Animal miRNAs are often organized in genomic clusters, usually indicating a single polycistronic primary precursor transcript, which may carry members of several distinct miRNA families. A second type of cluster organization comprising of many miRNAs which are transcribed independently or possibly in small groups has also been found. An example of this type is the C19MC cluster,[38] whose members are independently transcribed by pol-III. It is now becoming clear that miRNA families evolve through gene duplication and gene loss, among other mechanisms such as accumulation of mutations in the neighborhood of established miRNA gene clusters (discussed above).[15,39-41] Two distinct modes of duplication events have been proposed: (1) local duplications leading to additional copies on the same primary transcript, and (2) nonlocal duplications which eventually place the paralogs under different transcriptional control. Evidently, the cause for nonlocal duplications is mostly the whole-genome duplication events in early vertebrate evolution.[41] Mature miRNA paralogs usually acquire minimal substitutions, suggesting that the differences in the functions among members of a miRNA family might be associated with the differences in their expression and/or downstream processing in order to affect their targets. In contrast to the small-sized but abundant number of miRNA families in the animal genomes, plants have been found to have fewer but larger miRNA families.[42] In addition, plants have also been found to harbor a large number of species-specific miRNAs, which often outnumber the conserved fraction.[41,43-45] Many of these species-specific miRNAs are single copy genes and show significant sequence similarity with their putative targets supporting the view that these miRNAs are indeed evolutionarily recent.

METHODS TO IDENTIFY miRNAs AND THEIR TARGETS

Computational methods have played an important role in the prediction of miRNAs from the very beginning.[11] Traditionally, some major features such as the hairpin-shaped stem loop structure, minimal folding free-energy and high evolutionary conservation have been used in the computational identification of miRNAs. In general, approaches for the identification of miRNAs can be broadly classified into three major classes: Approaches based on the features in the sequence and their conservation, approaches based on thermodynamic stability of the miRNAs and experimental data-driven approaches.

Most simple strategies employed for the identification of miRNAs fall into the first category and use sequence homology to experimentally known miRNAs as well as the characteristic hairpin structure of the pre-miRNA.[11] However, conservation is high at both sides of the stem region and is decreasing towards the unpaired region of the apical loop. If only one mature miRNA is produced from the precursor, the region encoding the mature sequences is best conserved. In some cases, both sides of the hairpin produce mature sequences in which case both are conserved equally well. Several software tools have been designed to take into account these features of miRNA sequences and their conservation (see Table 1). These approaches are typically used for the identification of miRNAs in a newly sequenced genome. For closely related species, phylogenetic shadowing has also been proposed to be a powerful approach to identify regions that are under stabilizing selection and hence exhibit the characteristic variations in sequence conservation between stems, loop and mature miRNA (see Chapter 2, Figure 3, page 24 for an example of an RNA comparative alignment).[46] Some approaches such as Mirscan-II have also exploited the genomic context of a miRNA under the notion that miRNAs usually appear in genomic clusters.[47,48] More complicated approaches in this category employ a number of different features contained in bona fide miRNAs, in order to predict novel ones.[11] MicroRNA detection without the aid of comparative sequence analysis is a very hard task but unavoidable when species-specific miRNAs are of prime interest. This is because a typical idea of the above approaches using comparative genomics is to filter out hairpins that are not evolutionary conserved in related species. Such a filtering step makes the methods unable to identify novel miRNAs when there is no known close homology either because of the limitation of the current data or due to the rapid evolution of miRNAs. The mir-abela approach, first searches for hairpins that are robust against changes in the folding windows and are also thermodynamically stable, and then uses a support vector machine to identify miRNAs among these filtered candidates.[49] A similar approach has been proposed by Xue et al.[50] These approaches fall into the second group of methods for predicting miRNAs. Some approaches use both

Table 1. Different resources and methods available for finding miRNA genes

Algorithm	Description of Features	Data Sets	Download/Web Server	References
Mir-abela	A support vector machine approach which computes 16 different statistics from the entire hair pin structure	Human	http:www.mirz.unibas.ch/cgi/pred_ miRNA_genes.cgi	49
miPred	This approach uses 32 global and intrinsic hairpin folding attributes based on sequence, structural and folding thermodynamics	Human	http://bioinfo.au.tsinghua.edu. cn/mirnasvm/	93
RNAmicro	12 features based on structure, sequence, composition, thermodynamic stability and structural conservation are included in this approach	Animal	http://www.bioinf.uni-leipzig.de/ ~jana/index.php/jana-hertel-software/65-ja-na-hertel-rnamicro	94
Micro-ProcessorSVM	This approach includes 686 features from structure and sequence; 7 additional Drosha processing site features are also included	Human	http://demo1.interogan.com/miRNA/	95
BayesmiRNAfind	Secondary structure and sequence features are included in this approach	Mouse	https://bioinfo.wistar.upenn.edu/ miRNA/miRNA/login.php	96
miPred	This approach uses structure—sequence composition and free energy of the secondary structure	Human	http://www.bioinfo.seu.edu.cn/miRNA	51
MiRFinder	This approach uses 18 parameters including secondary structure, differences of the stem region of miRNA and secondary structures of pre-miRNAs	Human	http://www.bioinfomatics.org/mirfinder/	97

sequence and thermodynamic aspects of miRNA to build integrated models to predict miRNAs.[51,52] Finally, experimental driven-approaches can be considered to be the result of the recent explosion in the deep-sequencing techniques, which allow the unambiguous identification of miRNAs in different populations. Since these technologies result in the production of large amount of sequence data, bioinformatics tools are essential to analyze and understand the resulting sequences. For instance, Mirdeep uses a probabilistic model of miRNA biogenesis to score the compatibility of the position and frequency of sequenced RNA with the secondary structure of the miRNA precursor. The main idea being to detect miRNAs by analyzing how sequenced RNAs are compatible with miRNA precursors processed in the cell, by comparing with published datasets.[53] Other approaches such as Miranalyzer fall into this category[54] (Table 2). More recent studies also report the accumulation of the processed next generation sequencing datasets into unified databases for efficient downstream analysis of post-transcriptional regulation in model organisms.[55,56]

MicroRNAs act as guide molecules that program the RISC complex to recognize a target miRNA, so it is essential to understand the mechanism by which miRNAs recognize their targets in order to predict target mRNAs for a given miRNA sequence. Currently, the number of miRNA-mRNA interactions is still small which makes it difficult for computational approaches to accurately predict the targets. For instance, Tarbase,[57] a database dedicated to the collection of experimentally identified miRNA-mRNA interactions currently reports 1333 mRNAs targeted by ~150 animal miRNAs. These known interactions have been used by a number of groups to develop rules for the prediction of miRNA targets in the form of a variety of computational methods (see Table 2). In plants, many targets can be predicted with confidence simply by searching for the messages with extensive complementarity to the miRNAs.[58] However, in animals, extensive complementarity is not a common phenomenon.[59]

Current prediction methods are diverse in both the adopted approach and performance (see Table 2). However, some major guiding principles have emerged over the years. These include (1) absence of a perfect complementarity between miRNA and mRNA; (2) miRNA:mRNA duplexes are asymmetric with the 5′ end of the miRNA binding more strongly than the 3′ end of the miRNA; (3) the region comprising positions 2-7 on the miRNA often exhibit perfect complementarity and is therefore commonly referred to as the seed region of miRNA; (4) target sites with evolutionary conserved seed regions tend to be more likely true sites and show stronger regulatory impact than nonconserved ones, and (5) highly conserved miRNAs have many conserved targets (see Fig. 1).[60-66]

MicroRNA INTERPLAY WITH OTHER CELLULAR NETWORKS

MicroRNAs have been shown to have biological functions in many aspects. Their targets range from signaling proteins, enzymes to transcription factors and RNA binding proteins. The diversity and abundance of miRNA targets offer an enormous level of combinatorial possibilities and suggest that miRNAs with their targets might form a rather complex regulatory network intertwined with other cellular networks such as signal transduction, metabolic, gene regulatory and protein interaction networks. Therefore, in order to understand the global principles of miRNA regulation, it is imperative to understand how they take part in different cellular processes at a systems level. Four

Table 2. Different methods and software implementations available for predicting miRNA targets from genomic sequences

Tool	Supported Organisms	Website	References
TargetScan, TargetScanS	Vertebrates, flies and worms	http://hollywood.mit.edu/burgelab/software	65
miRanda	Flies, Vertebrates	http://www.microrna.org/	98
DIANA-microT	Vertebrates	http://diana.cslab.ece.ntua.gr	99
RNA hybrid	Flies	http://bibiserv.techfak.uni-bielefeld.de/rnahybrid/	100
PicTar	Nematodes, flies and vertebrates	http://pictar.mdc-berlin.de/	101
Rna22	Nematodes, flies and vertebrates	http://cbcsrv.watson.ibm.com/rna22.html	102
EIMMo	Vertebrates, flies and worms	http://www.mirz.unibas.ch./EIMMo2	103
PITA Top	Vertebrates, flies and worms	http://genie.weizmannac.il/pubs/mir07/mir07_data.html	104
TargetBoost	Worm and fruit fly	http://demo1.interagon.com/demo	105
mirWIP	*C. elegans*	http://www.psi.toronto.edu/genmir	106
miRDeep	Vertebrates, flies and worms	http://www.mdc-berlin.de/en/research/research_teams/systems_biology_of_gene_regulatory_elements/projects/miRDeep/index.html	53
miRanalyzer	Vertebrates, flies and worms	http://web.bioinformatics.cicbiogune.es/microRNA/	54

different kinds of networks are commonly studied within the context of a cell, namely gene regulatory, metabolic, signaling and protein interaction networks. In what follows, we will summarize our current understanding of the influence of miRNAs on these cellular networks.

Signaling network of a cell can be considered as a complex system responding to stimuli, signals and messages from other cells and environment. Once a cell receives signals, it processes the information and the signals reach transcription factors, thereby triggering responses of gene regulatory networks. In other words, the signaling network integrates the extra- and intra-cellular signals for the proper functioning of a cell. Signaling networks are typically represented as graphs containing both directed and undirected edges, with the nodes representing proteins. In such a representation directed links represent activation or inactivation relationships between proteins, while the undirected links represent physical interactions between proteins. In a large-scale analyses performed by Cui et al,[67] the authors collected publicly available dataset on signaling interactions in human[68] and extended it by manually curating additional signaling interactions to construct the largest known signaling interactome. By employing already available miRNA target predictions in human, Cui and coworkers

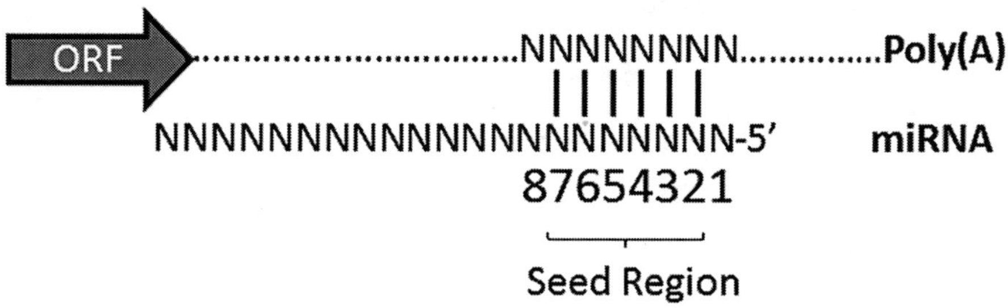

Figure 1. Representation of a miRNA-mRNA duplex with the seed region labeled. Nucleotide stretch numbered from 2 to 7 on the 5′ end of the miRNA is considered as its seed region and is often found to be completely complementary to the 3′ untranslated region (UTR) on the target transcript, where the miRNA binds and represses its activity by translational inhibition or by destabilization. Base complementarity is also found between position 1 and 8 on the miRNA and the target transcript in addition to the match on the seed region. The vertical dashes represent the match (complementarity) between the seed region and mRNA transcript using Watson-Crick base pairing and is commonly referred to as the canonical site. Other supplementary sites (in conjunction with the match on the seed region) on the 3′ region of the miRNA have also been found where there is complementarity between the transcripts however these are far less common than the canonical sites alone.[60] Multiple miRNAs can bind to the same UTR of the mRNA with members of the same miRNA family binding to identical sites, to work in a combinatorial fashion.

showed that the fraction of miRNA targets increases with the signal information flow from the upstream to the downstream in this network e.g., from ligands, cell surface receptors, intracellular signaling proteins to nuclear proteins. In other words, only 9.1% of the ligands were miRNA targets while half of the nuclear proteins, most of which were transcription factors were found to be miRNA targets. In addition, they noted that highly linked scaffold proteins such as CRK (sarcoma virus CT10 oncogene homolog) and SNAP25 (synaptosome-associated protein) have higher probability to be targeted by miRNAs. More generally, this study[67] showed that miRNAs avoid targeting common components of cellular pathways, possibly to avoid disturbing basic cellular processes because these common proteins are highly shared by basic cellular machines and should be frequently used in various cellular conditions. A network analysis of the signaling interactions allowed them to infer that miRNAs frequently target positively linked network motifs compared to negatively linked ones.[67] By selectively regulating positive regulatory motifs, scaffolding proteins and networks' downstream components, miRNAs may provide a mechanism to terminate the pre-existing messages and facilitate quick and robust transitions for responses to new signals.

Gene regulatory networks comprise of the regulatory interconnections between transcription factors (TFs), RNA binding proteins (RBPs), noncoding RNAs and their target genes or transcripts. In a simpler version of these networks only one type of regulatory molecule (either TF or RBP or RNA) is considered to target genes or transcripts to construct transcriptional or post-transcriptional networks, while more complex representations involve the integration of multiple layers of regulation. In one of the preliminary studies on the impact of miRNAs on the transcriptional regulatory network of a eukaryote, the authors accumulated publicly available TF binding data for three embryonic stem cell specific TFs—NANOG, SOX2 and OCT4,[69] to show that miRNA targets are significantly enriched in the genes that are regulated by many transcription factors. They extended these observations to the complete genome by using computationally predicted binding sites in human to show that a gene that is regulated by a large number of TFs is also more likely to be regulated by many miRNAs.[70] This observation suggests that genes which are more complexly regulated at transcriptional level are required to be turned on more frequently and hence are more likely to be controlled under different conditions. Therefore, they are also more likely to be regulated by many miRNAs. Another area of intense research is to dissect the integrated network of transcriptional and post-transcriptional regulatory interactions, to unravel patterns of interconnections which appear more often than expected by chance—often referred to network motifs analogous to sequence motifs in genomes.[71,72] For instance, a commonly occurring network motif which was identified in a number of biological networks including transcriptional regulatory networks, where in TFs form the regulatory nodes and the genes transcriptionally controlled by them form the target nodes, is the Feed Forward Loop (FFL).[72,73] In this motif, a TF, X, regulates another TF, Y and a target gene, Z while the second TF, Y also regulates the activity of Z forming a loop. Initial efforts to study miRNA network motifs have focused on analyzing the co-operation between TFs and miRNAs by using computational approaches.[74] These computational techniques involved the identification of pairs of TF-miRNAs, which show high co-occurrence of their sequence motifs in a number of target genes and resulted in our understanding that miRNAs and TFs work together to form network motifs.[74-76] These studies have shown that two kinds of FFLs are prevalent in the integrated miRNA-TF regulatory networks: (1) Type I comprising of a TF activating/repressing both the miRNA and the target gene, while the miRNA represses the target transcript; and (2) Type II comprising of a TF activating the miRNA but repressing the expression of the target gene (or the other way around) with the miRNA also repressing the expression of the target. One example has been characterized experimentally for type I FFL—MicroRNA, miR-17-5p represses the target gene, E2F1 and both the miRNA and E2F1 are transcriptionally activated by c-Myc in human cells.[77] This type of motif has been proposed to prevent proliferation of noise-driven transitions.[77] In addition to these motif types, superposed or composite FFL motifs, where in the interacting miRNA-TF pairs have a bi-directional relationship between them and indirect feed-forward motifs have also been observed (Fig. 2). In the former, miRNA represses a TF and a target gene while the TF activates both the miRNA and the target. The later motif constitutes a TF activating another TF and a target gene; with the second TF activating a miRNA, which itself represses the target gene. However, the functions of these motifs have not been studied experimentally.

Protein interaction networks provide a snapshot of the proteome's functional organization by elucidating the links between pairs of proteins in a cell. Large-scale determination of these maps during the last decade in various model organisms has enabled us to exploit them to understand the cell's functional architecture and their interplay with the rest of the cellular constituents. In the context of miRNAs, one study showed that analogous to the transcriptional networks discussed above, proteins with more number of interacting partners are preferentially targeted by miRNAs.[78] In other words, protein connectivity in the human protein interaction network was shown to be positively correlated with the number of miRNA target-site types. This observation indicates that when a protein has more number of interacting partners, it will be regulated by many regulatory molecules. The study also showed that inter-modular hubs i.e., proteins which

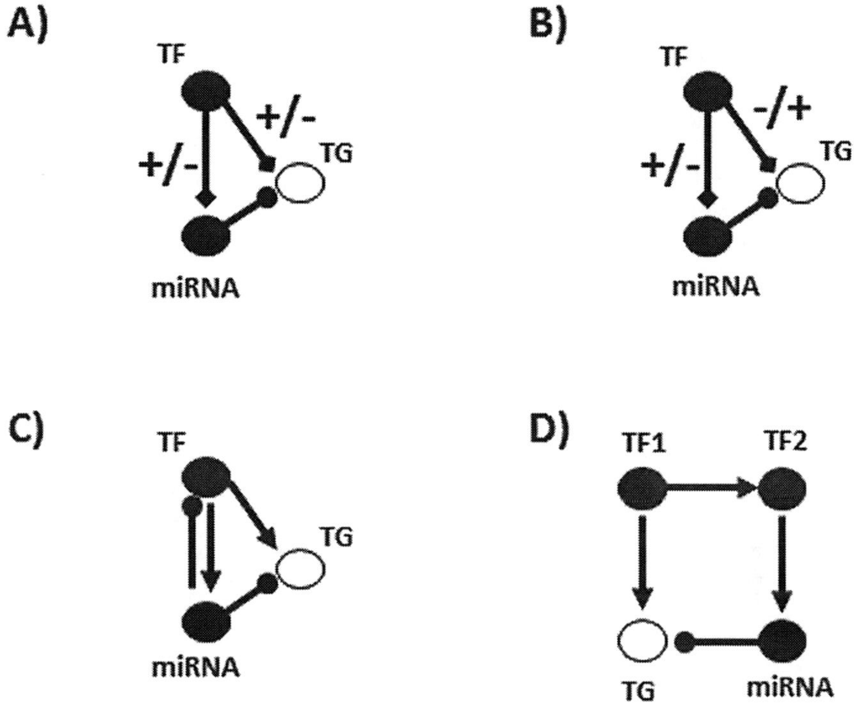

Figure 2. Different types of network motifs identified in the integrated miRNA-TF regulatory networks. Darker nodes correspond to regulators, either Transcription Factors (TFs) or miRNAs, while light colored nodes correspond to the target genes controlled by them. Oval arrows always correspond to the repression mediated by miRNAs while diamond arrows shown for TFs correspond to both activation and repression. When only an arrow is shown for a TF, it corresponds to activation mediated by the TF. A) In a type 1 Feed Forward Loop (FFL), TF sends either an activation or inhibition signal, at the same time to both miRNA and the Target Gene (TG), thereby causing an incoherent effect on target genes' expression. B) In type 2 FFL, likewise there are two possibilities, one where TF activates miRNA and inhibits TG at the same time and the other where TF inhibits miRNA and activates TG. This results in a net coherent effect on the TG. C) In superposed/composite FFL, TF activates miRNA and TG while the miRNA represses both of them. D) In an indirect feed forward loop, feedforward effect is achieved when TF1 activates a TG and TF2, which in succession activates miRNA. MicroRNA then represses TG to bring in a cascading effect.

link different modules/complexes in the protein interaction network, are more likely to be regulated by miRNAs.

EXPRESSION DYNAMICS AND CROSSTALK OF miRNAs WITH OTHER REGULATORY FACTORS

Analogous to transcriptional regulatory networks, wherein TFs regulate their target genes in a combinatorial manner,[79] most miRNAs fine tune the expression of hundreds of genes in a combinatorial manner at the post-transcriptional level.[1] This combinatorial regulation among miRNAs is believed to result in the following outcomes on their targets:

(1) several miRNAs have been found to regulate a single mRNA target by binding its transcript preferentially in the 3′ UTR region although regulation by multiple miRNAs in coding sequences has also been reported;[80,81] (2) MicroRNAs which are clustered and are often co-expressed could regulate functionally related proteins resulting in the targeting of same or functionally related pathways.[82-85] These computational and experimental studies suggest that co-ordinate regulation by miRNAs is a flexible and efficient strategy to regulate cellular processes in a condition or tissue-specific manner. However, whether the combinatorial interplay is conserved across species or there would have been a conserved set of coregulating pairs which maintain this phenomenon in specific tissues or conditions in related organisms is unclear. MicroRNAs, in addition to their interplay with other miRNAs, are also known to combinatorially control genes with the help of TFs (discussed above), RNA-binding proteins and chromatin associated factors to control important functions during development and differentiation.[86-90] Precise roles and mechanisms of most of which remain unclear to date.

In addition to the key roles played by miRNAs in the context of other regulatory proteins in the cell, they are also known to work as regulators for buffering the protein expression noise. Since miRNAs can tune a protein's expression level more rapidly by targeting the encoded transcript at the post-transcriptional level compared to transcription factors at the transcriptional level, they have been proposed to significantly shorten the response delay and in turn provide more effective noise buffering. For example, in flies miR-9a is suggested to set up a threshold for signals in a positive feedback loop so that it can filter out noise.[91] During fly sensory organ development, a fly gene, *senseless's* expression is activated by proneuronal proteins and feeds back positively to reinforce proneural gene expression. If *senseless,* the target of miR-9a is highly expressed, flies produce extra sense organs. MicroRNA miR-9a has been suggested to set a threshold that *senseless* expression has to overcome to induce the normal developmental program. These observations are strongly supported by the fact that most of the developmental genes are highly conserved targets of miRNAs, indicating the importance of preserving their regulatory circuitry across organisms with similar developmental programs. It is also evident from genome-wide studies that miRNAs might act as genetic buffers to constrain gene expression divergence between organisms, as was shown in a study comparing the expression divergences of miRNA targets across diverse organisms.[92] It is possible that the abundance of miRNAs in higher eukaryotes minimizes the expression divergence in a population by decreasing the detrimental effects of errors in gene regulation. Such a buffering phenomenon might also provide the cell with a way to silence-accumulating mutations without being subjected to extreme selective forces and hence contribute to the evolvability of an organism.

CONCLUSION

The emerging picture of miRNA regulation is getting far richer and increasingly complex than the simple linear pathways that one would have imagined these micromanagers to be controlling a decade ago. It is now becoming clear that miRNAs not only control a significant fraction of the protein coding genes in higher eukaryotes, but also function as key regulators in development, proliferation, differentiation, apoptosis, signalling and metabolism. In addition to their core roles in controlling cellular mRNA pool by managing the protein output, miRNAs have also been shown to work in mediating cross-talk between different cellular networks by integrating various signals. Given the enormous ability these tiny RNAs have in managing the interactome of higher eukaryotes, the stage is set to manipulate their precise roles in regulating RNAs for therapeutic benefit in years to come.

ACKNOWLEDGEMENTS

SCJ acknowledges financial support from Institute for Genomic Biology, University of Illinois at Urbana-Champaign. We would also like to thank Magdy Alabady and Nitish Mittal for providing helpful comments on a previous version of this manuscript.

REFERENCES

1. Bartel DP. MicroRNAs: genomics, biogenesis, mechanism, and function. Cell 2004; 116(2):281-297.
2. Carthew RW. Gene regulation by microRNAs. Curr Opin Genet Dev 2006; 16(2):203-208.
3. Carthew RW, Sontheimer EJ. Origins and Mechanisms of miRNAs and siRNAs. Cell 2009; 136(4):642-655.
4. Valencia-Sanchez MA, Liu J, Hannon GJ et al. Control of translation and mRNA degradation by miRNAs and siRNAs. Genes Dev 2006; 20(5):515-524.
5. Cao X, Yeo G, Muotri AR et al. Noncoding RNAs in the mammalian central nervous system. Annu Rev Neurosci 2006; 29:77-103.
6. Plasterk RH. Micro RNAs in animal development. Cell 2006; 124(5):877-881.
7. Wienholds E, Koudijs MJ, van Eeden FJ et al. The microRNA-producing enzyme Dicer1 is essential for zebrafish development. Nat Genet 2003; 35(3):217-218.
8. Costinean S, Zanesi N, Pekarsky Y et al. Pre-B cell proliferation and lymphoblastic leukemia/high-grade lymphoma in E(mu)-miR155 transgenic mice. Proc Natl Acad Sci U S A 2006; 103(18):7024-7029.
9. Lee RC, Feinbaum RL, Ambros V. The C. elegans heterochronic gene lin-4 encodes small RNAs with antisense complementarity to lin-14. Cell 1993; 75(5):843-854.
10. Wightman B, Ha I, Ruvkun G. Posttranscriptional regulation of the heterochronic gene lin-14 by lin-4 mediates temporal pattern formation in C. elegans. Cell 1993; 75(5):855-862.
11. Lee RC, Ambros V. An extensive class of small RNAs in Caenorhabditis elegans. Science 2001; 294(5543):862-864.
12. Lagos-Quintana M, Rauhut R, Lendeckel W et al. Identification of novel genes coding for small expressed RNAs. Science 2001; 294(5543):853-858.
13. Baskerville S, Bartel DP. Microarray profiling of microRNAs reveals frequent coexpression with neighboring miRNAs and host genes. RNA 2005; 11(3):241-247.
14. Rodriguez A, Griffiths-Jones S, Ashurst JL et al. Identification of mammalian microRNA host genes and transcription units. Genome Res 2004; 14(10A):1902-1910.
15. Lu J, Shen Y, Wu Q et al. The birth and death of microRNA genes in Drosophila. Nat Genet 2008; 40(3): 351-355.
16. Kim VN. MicroRNA biogenesis: coordinated cropping and dicing. Nat Rev Mol Cell Biol 2005; 6(5):376-385.
17. Denli AM, Tops BB, Plasterk RH et al. Processing of primary microRNAs by the Microprocessor complex. Nature 2004; 432(7014):231-235.
18. Gregory RI, Yan KP, Amuthan G et al. The Microprocessor complex mediates the genesis of microRNAs. Nature 2004; 432(7014):235-240.
19. Bohnsack MT, Czaplinski K, Gorlich D. Exportin 5 is a RanGTP-dependent dsRNA-binding protein that mediates nuclear export of pre-miRNAs. RNA 2004; 10(2):185-191.
20. Lund E, Guttinger S, Calado A et al. Nuclear export of microRNA precursors. Science 2004; 303(5654):95-98.
21. Bernstein E, Kim SY, Carmell MA et al. Dicer is essential for mouse development. Nat Genet 2003; 35(3):215-217.
22. Hutvagner G, McLachlan J, Pasquinelli AE et al. A cellular function for the RNA-interference enzyme Dicer in the maturation of the let-7 small temporal RNA. Science 2001; 293(5531):834-838.
23. Olsen PH, Ambros V. The lin-4 regulatory RNA controls developmental timing in Caenorhabditis elegans by blocking LIN-14 protein synthesis after the initiation of translation. Dev Biol 1999; 216(2):671-680.
24. Humphreys DT, Westman BJ, Martin DI et al. MicroRNAs control translation initiation by inhibiting eukaryotic initiation factor 4E/cap and poly(A) tail function. Proc Natl Acad Sci U S A 2005; 102(47):16961-16966.

25. Pillai RS, Bhattacharyya SN, Artus CG et al. Inhibition of translational initiation by Let-7 MicroRNA in human cells. Science 2005; 309(5740):1573-1576.
26. Jing Q, Huang S, Guth S et al. Involvement of microRNA in AU-rich element-mediated mRNA instability. Cell 2005; 120(5):623-634.
27. Lim LP, Lau NC, Garrett-Engele P et al. Microarray analysis shows that some microRNAs downregulate large numbers of target mRNAs. Nature 2005; 433(7027):769-773.
28. Sood P, Krek A, Zavolan M et al. Cell-type-specific signatures of microRNAs on target mRNA expression. Proc Natl Acad Sci U S A 2006; 103(8):2746-2751.
29. Bagga S, Bracht J, Hunter S et al. Regulation by let-7 and lin-4 miRNAs results in target mRNA degradation. Cell 2005; 122(4):553-563.
30. Guo H, Ingolia NT, Weissman JS et al. Mammalian microRNAs predominantly act to decrease target mRNA levels. Nature 2010; 466(7308):835-840.
31. Lim LP, Lau NC, Weinstein EG et al. The microRNAs of Caenorhabditis elegans. Genes Dev 2003; 17(8): 991-1008.
32. Kozomara A, Griffiths-Jones S. miRBase: integrating microRNA annotation and deep-sequencing data. Nucleic Acids Res 2011; 39(Database issue):D152-D157.
33. Lee CT, Risom T, Strauss WM. Evolutionary conservation of microRNA regulatory circuits: an examination of microRNA gene complexity and conserved microRNA-target interactions through metazoan phylogeny. DNA Cell Biol 2007; 26(4):209-218.
34. Heimberg AM, Sempere LF, Moy VN et al. MicroRNAs and the advent of vertebrate morphological complexity. Proc Natl Acad Sci U S A 2008; 105(8):2946-2950.
35. Prochnik SE, Rokhsar DS, Aboobaker AA. Evidence for a microRNA expansion in the bilaterian ancestor. Dev Genes Evol 2007; 217(1):73-77.
36. Bentwich I, Avniel A, Karov Y et al. Identification of hundreds of conserved and nonconserved human microRNAs. Nat Genet 2005; 37(7):766-770.
37. Berezikov E, Thuemmler F, van Laake LW et al. Diversity of microRNAs in human and chimpanzee brain. Nat Genet 2006; 38(12):1375-1377.
38. Borchert GM, Lanier W, Davidson BL. RNA polymerase III transcribes human microRNAs. Nat Struct Mol Biol 2006; 13(12):1097-1101.
39. Tanzer A, Amemiya CT, Kim CB et al. Evolution of microRNAs located within Hox gene clusters. J Exp Zool B Mol Dev Evol 2005; 304(1):75-85.
40. Tanzer A, Stadler PF. Evolution of microRNAs. Methods Mol Biol 2006; 342:335-350.
41. Fahlgren N, Howell MD, Kasschau KD et al. High-throughput sequencing of Arabidopsis microRNAs: evidence for frequent birth and death of MIRNA genes. PLoS One 2007; 2(2):e219.
42. Li A, Mao L. Evolution of plant microRNA gene families. Cell Res 2007; 17(3):212-218.
43. Lindow M, Krogh A. Computational evidence for hundreds of nonconserved plant microRNAs. BMC Genomics 2005; 6:119.
44. Barakat A, Wall PK, Diloreto S et al. Conservation and divergence of microRNAs in Populus. BMC Genomics 2007; 8:481.
45. Fattash I, Voss B, Reski R et al. Evidence for the rapid expansion of microRNA-mediated regulation in early land plant evolution. BMC Plant Biol 2007; 7:13.
46. Berezikov E, Guryev V, van de Belt J et al. Phylogenetic shadowing and computational identification of human microRNA genes. Cell 2005; 120(1):21-24.
47. Ohler U, Yekta S, Lim LP et al. Patterns of flanking sequence conservation and a characteristic upstream motif for microRNA gene identification. RNA 2004; 10(9):1309-1322.
48. Altuvia Y, Landgraf P, Lithwick G et al. Clustering and conservation patterns of human microRNAs. Nucleic Acids Res 2005; 33(8):2697-2706.
49. Sewer A, Paul N, Landgraf P et al. Identification of clustered microRNAs using an ab initio prediction method. BMC Bioinformatics 2005; 6:267.
50. Xue C, Li F, He T et al. Classification of real and pseudo microRNA precursors using local structure-sequence features and support vector machine. BMC Bioinformatics 2005; 6:310.
51. Jiang P, Wu H, Wang W et al. MiPred: classification of real and pseudo microRNA precursors using random forest prediction model with combined features. Nucleic Acids Res 2007; 35(Web Server issue):W339-W344.
52. Ding J, Zhou S, Guan J. MiRenSVM: towards better prediction of microRNA precursors using an ensemble SVM classifier with multi-loop features. BMC Bioinformatics 2010; 11 Suppl 11:S11.
53. Friedlander MR, Chen W, Adamidi C et al. Discovering microRNAs from deep sequencing data using miRDeep. Nat Biotechnol 2008; 26(4):407-415.
54. Hackenberg M, Sturm M, Langenberger D et al. miRanalyzer: a microRNA detection and analysis tool for next-generation sequencing experiments. Nucleic Acids Res 2009; 37(Web Server issue):W68-W76.
55. Yang JH, Li JH, Shao P et al. starBase: a database for exploring microRNA-mRNA interaction maps from Argonaute CLIP-Seq and Degradome-Seq data. Nucleic Acids Res 2011; 39(Database issue):D202-D209.

56. Yang JH, Shao P, Zhou H et al. deepBase: a database for deeply annotating and mining deep sequencing data. Nucleic Acids Res 2010; 38(Database issue):D123-130.
57. Papadopoulos GL, Reczko M, Simossis VA et al. The database of experimentally supported targets: a functional update of TarBase. Nucleic Acids Res 2009; 37(Database issue):D155-158.
58. Rhoades MW, Reinhart BJ, Lim LP et al. Prediction of plant microRNA targets. Cell 2002; 110(4):513-520.
59. Davis E, Caiment F, Tordoir X et al. RNAi-mediated allelic trans-interaction at the imprinted Rtl1/Peg11 locus. Curr Biol 2005; 15(8):743-749.
60. Bartel DP. MicroRNAs: target recognition and regulatory functions. Cell 2009; 136(2):215-233.
61. Baek D, Villen J, Shin C et al. The impact of microRNAs on protein output. Nature 2008; 455(7209):64-71.
62. Selbach M, Schwanhausser B, Thierfelder N et al. Widespread changes in protein synthesis induced by microRNAs. Nature 2008; 455(7209):58-63.
63. Brennecke J, Stark A, Russell RB et al. Principles of microRNA-target recognition. PLoS Biol 2005; 3(3):e85.
64. Krek A, Grun D, Poy MN et al. Combinatorial microRNA target predictions. Nat Genet 2005; 37(5):495-500.
65. Lewis BP, Burge CB, Bartel DP. Conserved seed pairing, often flanked by adenosines, indicates that thousands of human genes are microRNA targets. Cell 2005; 120(1):15-20.
66. Xie X, Lu J, Kulbokas EJ et al. Systematic discovery of regulatory motifs in human promoters and 3′ UTRs by comparison of several mammals. Nature 2005; 434(7031):338-345.
67. Cui Q, Yu Z, Purisima EO et al. Principles of microRNA regulation of a human cellular signaling network. Mol Syst Biol 2006; 2:46.
68. Ma'ayan A, Jenkins SL, Neves S et al. Formation of regulatory patterns during signal propagation in a Mammalian cellular network. Science 2005; 309(5737):1078-1083.
69. Lee TI, Jenner RG, Boyer LA et al. Control of developmental regulators by Polycomb in human embryonic stem cells. Cell 2006; 125(2):301-313.
70. Cui Q, Yu Z, Pan Y et al. MicroRNAs preferentially target the genes with high transcriptional regulation complexity. Biochem Biophys Res Commun 2007; 352(3):733-738.
71. Shen-Orr SS, Milo R, Mangan S et al. Network motifs in the transcriptional regulation network of Escherichia coli. Nat Genet 2002; 31(1):64-68.
72. Alon U. Network motifs: theory and experimental approaches. Nat Rev Genet 2007; 8(6):450-461.
73. Janga SC, Collado-Vides J. Structure and evolution of gene regulatory networks in microbial genomes. Res Microbiol 2007; 158(10):787-794.
74. Shalgi R, Lieber D, Oren M et al. Global and local architecture of the mammalian microRNA-transcription factor regulatory network. PLoS Comput Biol 2007; 3(7):e131.
75. Zhou Y, Ferguson J, Chang JT et al. Inter- and intra-combinatorial regulation by transcription factors and microRNAs. BMC Genomics 2007; 8:396.
76. Wang J, Lu M, Qiu C et al. TransmiR: a transcription factor-microRNA regulation database. Nucleic Acids Res 2010; 38(Database issue):D119-D122.
77. O'Donnell KA, Wentzel EA, Zeller KI et al. c-Myc-regulated microRNAs modulate E2F1 expression. Nature 2005; 435(7043):839-843.
78. Liang H, Li WH. MicroRNA regulation of human protein protein interaction network. RNA 2007; 13(9):1402-1408.
79. Balaji S, Babu MM, Iyer LM et al. Comprehensive analysis of combinatorial regulation using the transcriptional regulatory network of yeast. J Mol Biol 2006; 360(1):213-227.
80. Fontana L, Pelosi E, Greco P et al. MicroRNAs 17-5p-20a-106a control monocytopoiesis through AML1 targeting and M-CSF receptor upregulation. Nat Cell Biol 2007; 9(7):775-787.
81. Tay Y, Zhang J, Thomson AM et al. MicroRNAs to Nanog, Oct4 and Sox2 coding regions modulate embryonic stem cell differentiation. Nature 2008; 455(7216):1124-1128.
82. Dews M, Homayouni A, Yu D et al. Augmentation of tumor angiogenesis by a Myc-activated microRNA cluster. Nat Genet 2006; 38(9):1060-1065.
83. Ivanovska I, Cleary MA. Combinatorial microRNAs: working together to make a difference. Cell Cycle 2008; 7(20):3137-3142.
84. Kim YK, Yu J, Han TS et al. Functional links between clustered microRNAs: suppression of cell-cycle inhibitors by microRNA clusters in gastric cancer. Nucleic Acids Res 2009; 37(5):1672-1681.
85. Yuan X, Liu C, Yang P et al. Clustered microRNAs' coordination in regulating protein-protein interaction network. BMC Syst Biol 2009; 3:65.
86. Hafner M, Landthaler M, Burger L et al. PAR-CliP—a method to identify transcriptome-wide the binding sites of RNA binding proteins. J Vis Exp 2010(41).
87. Khalil AM, Guttman M, Huarte M et al. Many human large intergenic noncoding RNAs associate with chromatin-modifying complexes and affect gene expression. Proc Natl Acad Sci U S A 2009; 106(28):11667-11672.
88. Alvarez-Saavedra M, Antoun G, Yanagiya A et al. miRNA-132 orchestrates chromatin remodeling and translational control of the circadian clock. Hum Mol Genet 2010.

89. van Wolfswinkel JC, Ketting RF. The role of small non-coding RNAs in genome stability and chromatin organization. J Cell Sci 2010; 123(Pt 11):1825-1839.
90. Laporte P, Merchan F, Amor BB et al. Riboregulators in plant development. Biochem Soc Trans 2007; 35(Pt 6):1638-1642.
91. Li Y, Wang F, Lee JA et al. MicroRNA-9a ensures the precise specification of sensory organ precursors in Drosophila. Genes Dev 2006; 20(20):2793-2805.
92. Cui Q, Yu Z, Purisima EO et al. MicroRNA regulation and interspecific variation of gene expression. Trends Genet 2007; 23(8):372-375.
93. Ng KL, Mishra SK. De novo SVM classification of precursor microRNAs from genomic pseudo hairpins using global and intrinsic folding measures. Bioinformatics 2007; 23(11):1321-1330.
94. Hertel J, Stadler PF. Hairpins in a Haystack: recognizing microRNA precursors in comparative genomics data. Bioinformatics 2006; 22(14):e197-e202.
95. Helvik SA, Snove O Jr., Saetrom P. Reliable prediction of Drosha processing sites improves microRNA gene prediction. Bioinformatics 2007; 23(2):142-149.
96. Yousef M, Jung S, Kossenkov AV et al. Naive Bayes for microRNA target predictions—machine learning for microRNA targets. Bioinformatics 2007; 23(22):2987-2992.
97. Huang JC, Babak T, Corson TW et al. Using expression profiling data to identify human microRNA targets. Nat Methods 2007; 4(12):1045-1049.
98. Enright AJ, John B, Gaul U et al. MicroRNA targets in Drosophila. Genome Biol 2003; 5(1):R1.
99. Kiriakidou M, Nelson PT, Kouranov A et al. A combined computational-experimental approach predicts human microRNA targets. Genes Dev 2004; 18(10):1165-1178.
100. Rehmsmeier M, Steffen P, Hochsmann M et al. Fast and effective prediction of microRNA/target duplexes. RNA 2004; 10(10):1507-1517.
101. Grun D, Wang YL, Langenberger D et al. microRNA target predictions across seven Drosophila species and comparison to mammalian targets. PLoS Comput Biol 2005; 1(1):e13.
102. Miranda KC, Huynh T, Tay Y et al. A pattern-based method for the identification of MicroRNA binding sites and their corresponding heteroduplexes. Cell 2006; 126(6):1203-1217.
103. Gaidatzis D, van Nimwegen E, Hausser J et al. Inference of miRNA targets using evolutionary conservation and pathway analysis. BMC Bioinformatics 2007; 8:69.
104. Kertesz M, Iovino N, Unnerstall U et al. The role of site accessibility in microRNA target recognition. Nat Genet 2007; 39(10):1278-1284.
105. Saetrom O, Snove O Jr., Saetrom P. Weighted sequence motifs as an improved seeding step in microRNA target prediction algorithms. RNA 2005; 11(7):995-1003.
106. Hammell M, Long D, Zhang L et al. mirWIP: microRNA target prediction based on microRNA-containing ribonucleoprotein-enriched transcripts. Nat Methods 2008; 5(9):813-819.

CHAPTER 5

SPATIOTEMPORAL ASPECTS OF MicroRNA-MEDIATED GENE REGULATION

Joris Pothof and Dik C. van Gent

Department of Genetics, Erasmus MC, University Medical Center, Netherlands
Emails: j.pothof@erasmusmc.nl, d.vangent@erasmusmc.nl

Abstract: MicroRNA-mediated modulation of translation has been recently discovered as a new dimension in gene expression regulation. In this chapter we review how this regulation operates in time between the fast protein modification and degradation steps on the one hand and the slow transcriptional reprogramming associated with more stable changes in gene expression patterns on the other hand. We also discuss the additional layer of complexity associated with spatial redistribution of the RNA silencing machinery in subcellular structures. Various stress conditions induce both a transient change in microRNA expression within the first few hours after exposure and a redistribution of the RNA silencing machinery from P-bodies to Stress Granules, which differ in their function and protein content. Insight into the spatiotemporal aspects of the microRNA response will be indispensable for a full understanding of this level of gene regulation.

INTRODUCTION

Regulation of gene expression has been studied for many decades. The need for such regulation was clear as soon as the gene concept had been formulated. Cells have to be able to respond to changes in the environment (such as lack of nutrients or intracellular stress) or to stimuli that are part of a developmental program. Various parameters determine the robustness of the response: a stable change (e.g., differentiation of a stem cell due to an extracellular signal) requires stable changes in the cell's molecular make-up, but transient changes in the environment (e.g., exposure to toxins) should cause a transient response, which reverts to its original situation when the environmental stress disappears. All these situations trigger various regulation mechanisms in the cell.

RNA Infrastructure and Networks, edited by Lesley J. Collins.

Stable changes in gene expression generally require transcriptional reprogramming, while protein modifications that alter activity or stability affect gene regulation in a more transient way.

More recently, an additional layer of gene regulation has been discovered in the emerging world of noncoding RNAs. This RNA infrastructure consists of a variety of noncoding RNAs.[1] Among all these small noncoding RNA classes, microRNAs have been found to be particularly important for regulation of gene expression. MicroRNAs are endogenous, small, noncoding RNAs that are encoded in the genome. They are either regulated by their own promoters or encoded within introns of protein-coding genes. They are transcribed by RNA polymerase II as primary microRNAs,[2] which are subsequently processed into precursor microRNAs by a complex containing the Drosha and DGCR8 proteins[3] (Fig. 1). Precursor microRNAs are then exported from the nucleus to the

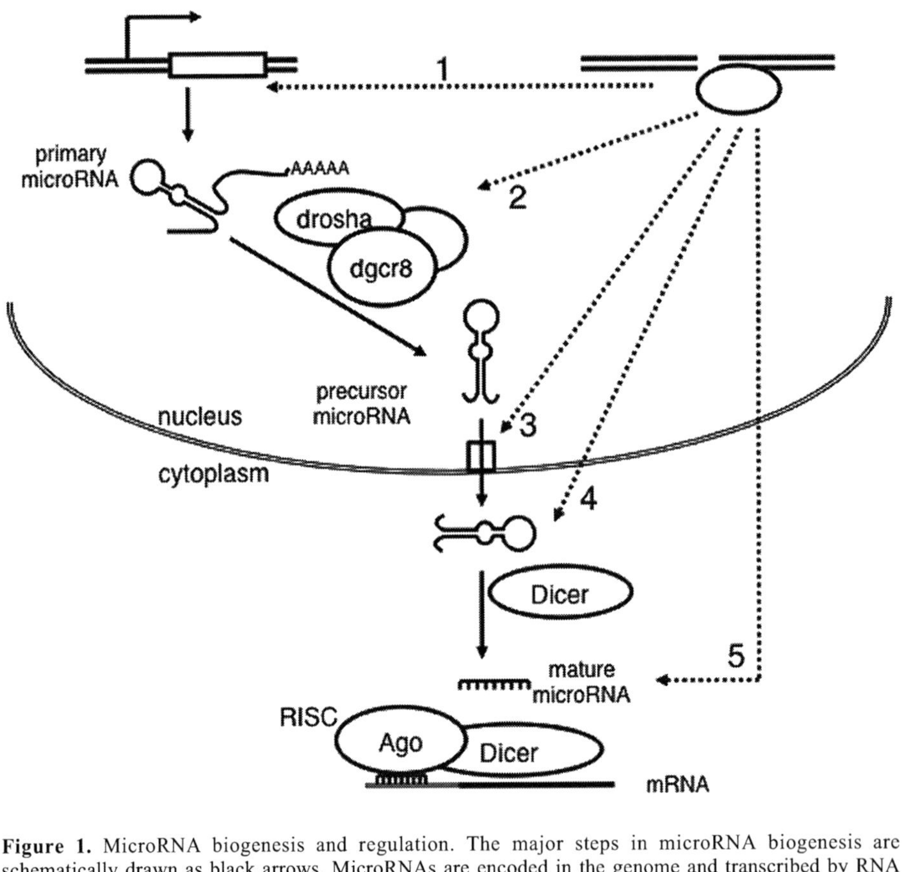

Figure 1. MicroRNA biogenesis and regulation. The major steps in microRNA biogenesis are schematically drawn as black arrows. MicroRNAs are encoded in the genome and transcribed by RNA polymerase II. These primary microRNAs are further processed by the Drosha-DGCR8 complex into precursor microRNAs, which are then exported to the cytoplasm. Subsequently DICER processes these precursor microRNAs into mature microRNAs that are then incorporated in RISC. Dashed arrows with numbers indicate possible microRNA regulation mechanisms after DNA damage: 1) Transcriptional regulation, 2) Regulation of primary microRNA processing by modulating the Drosha-DGCR8 complex, 3) Regulation of nuclear export, 4) Regulation of precursor microRNA processing and RISC activity and 5) microRNA degradation.

cytoplasm and processed by Dicer into mature ~22 nucleotide long microRNAs.[4] Mature microRNAs are then incorporated in the RNA Induced Silencing Complex (RISC), which contains one of the Argonaute (AGO) proteins that bind the microRNA and are essential for mRNA silencing or degradation.[5] Of the four AGO proteins in mammals, only AGO2 contains an intact nuclease site and is therefore the only AGO protein capable of mRNA degradation.[6,7] It is possible that the other AGO proteins facilitate mRNA silencing without RNA degradation. MicroRNA loaded RISC complexes bind complementary sites in mRNAs, primarily located in 3′ untranslated regions (3′UTR), thereby inducing mRNA degradation or translation inhibition. Currently, more than 1000 mammalian microRNAs have been identified for which it is predicted that they regulate ~60% of all genes in the genome.[8] These numbers suggest that most, if not all cellular processes will be under microRNA control. In this chapter, we will highlight the spatiotemporal aspects of microRNA regulation. We use the DNA damage response as a convenient example to introduce several aspects of this regulation, as the timing of DNA damage induction is easy to control in cells. However, it is to be expected that similar mechanisms also apply to other areas of gene regulation.

DNA DAMAGE RESPONSE AS AN EXAMPLE OF MULTILAYERED REGULATION

The integrity of our DNA is continuously threatened by a variety of endogenous factors, such as reactive oxygen species and reactive metabolites and exogenous factors such as ultraviolet (UV) radiation in sunlight and genotoxic chemicals. Failure to repair DNA damage correctly results in mutations and gross chromosomal rearrangements, which can eventually give rise to tumor formation.[9,10] On the other hand, persistent unrepaired DNA damage is a driving force of aging.[11,12] It is thus important for a cell to deal with DNA damage in a coordinated set of reactions, involving DNA repair, cell cycle control and regulation of apoptosis or senescence (see Fig. 2 for schematic overview). Collectively, these processes are referred to as the DNA damage response (DDR). Table 1 summarizes and clarifies terms, genes and processes involved in the DDR.

Cells have various DNA repair mechanisms (each repairing a specific set of lesions), which operate directly to repair the DNA damage.[9] On top of this damage reversal reaction, an elaborate system of cell cycle checkpoints is activated to arrest cell cycle progression to allow time for repair of the damage.[10] Moreover, inhibiting cell proliferation prevents the formation of mutations during DNA synthesis or the mis-segregation of broken chromosomes during mitosis. To establish a cell cycle block, the checkpoint kinases ATM and/or ATR phosphorylate the CDC25a protein, which is then rapidly degraded, thereby preventing specific CDK activation and cell cycle progression.[13] In another branch of the DDR, the TP53 protein is stabilized, which activates its transcriptional program.[14,15] Initially, TP53 induces cell cycle control genes, such as p21, that maintain the cell cycle block. At later stages, various critical pro-apoptotic genes are induced, which in the end induces cell death when damage is beyond repair. Alternatively, some cell types do not go into apoptosis, but may suffer premature senescence in the case of excessive DNA damage.[16] However, if DNA damage is repaired correctly, cells continue the normal cell cycle.

It is critical for a cell to tightly regulate its DDR: specific defects in one of the processes described above results in impaired repair fidelity, which increases the frequency of

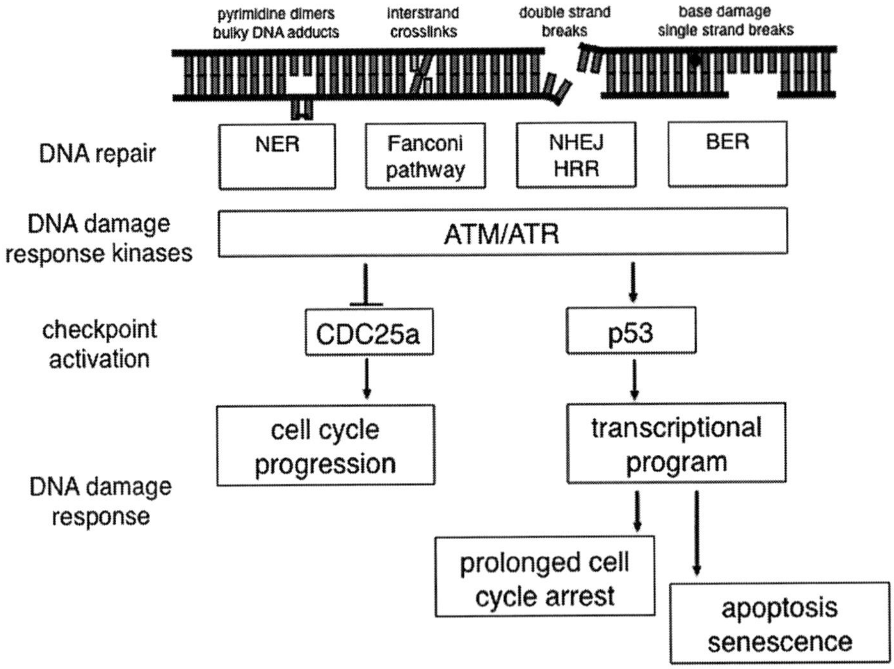

Figure 2. Schematic overview of the DNA damage response. After DNA damage, DNA repair mechanisms sense and repair the damage. Each type of DNA damage has its own repair mechanism. In addition, DNA damage checkpoint proteins (ATM/ATR) bind (in close proximity of) the DNA lesion and contribute to its repair, but also activate signal transduction pathways that halt cell proliferation or, in case of damage beyond repair, cell death or senescence pathways are activated.

mutations and chromosomal aberrations, which may eventually lead to tumor formation. A prime example of such a DDR defect in humans is Xeroderma Pigmentosum, in which a defect in Nucleotide Excision Repair causes highly increased mutation frequencies and cancer in UV-exposed areas of the skin. Hyperactivity of the DDR, on the other hand, may lead to increased cell death or senescence, which contributes to the aging process. The DDR is regulated at multiple levels. First, a fast response involves posttranslational modifications of various proteins,[17] thereby altering their activity or expression levels. At later time points, regulation of gene expression is taken over by transcriptional regulation.[10,18] However, an additional layer of gene regulation, involving microRNAs, was recently discovered to operate mainly at intermediate time points.

MicroRNA-MEDIATED GENE SILENCING IN THE DDR

One of the first clues for involvement of microRNA-mediated gene silencing in the DDR was the observation that knockdown of Dicer or AGO2, which interferes with microRNA biogenesis and mRNA silencing, results in sensitivity to UV light and altered cell cycle progression after damage induction.[19] These results indicate that microRNAs are important for proper DDR execution. Indeed, various microRNAs are regulated after

Table 1. Description of genes, processes and terms commonly used in the DNA damage response

Gene/Process	Description
DDR	DNA damage response; the combination of processes induced by DNA damage (DNA repair, cell cycle checkpoints and apoptosis or senescence).
NER	Nucleotide Excision Repair. Repairs UV-induced 6-4 photoproducts and cyclobutane pyrimidine dimers, but also other DNA helix distorting bulky lesions. NER defects can cause Xeroderma Pigmentosum, Cockayne's Syndrome or Trichothiodystrophy.
HRR	Homologous Recombination Repair. Repairs double stranded DNA breaks primarily in the G2 phase of the cell cycle. HRR defects can cause hereditary breast cancer.
NHEJ	Nonhomologous End Joining. Repairs double stranded DNA breaks in all phases of the cell cycle. NHEJ defects can cause Severe Combined Immuno Deficiency (because of defective immunoglobulin and T-cell receptor gene recombination).
Fanconi pathway	Pathway involved in repairing DNA interstrand crosslinks. Defects in this pathway cause Fanconi Anemia.
BER	Base Excision Repair. Repairs damaged bases.
ATM	Ataxia Telangiectasia Mutated. Initial DNA damage signal transducing protein.
ATR	Ataxia Telangiectasia-mutated and Rad3-related. Initial DNA damage signal transducing protein. ATR defects can cause Seckle Syndrome.
CDC25a	Cell Division Cycle 25a. Cell cycle control protein.
CDK	Cyclin Dependent Kinase. Cell cycle protein.
TP53	P53. Central checkpoint gene that regulates prolonged cell cycle arrest, programmed cell death and senescence.
P21	Key p53 target gene involved in cell cycle control.
BRCA1	BReast CAncer 1. Involved in HRR and the DNA damage checkpoint.
DNA damage Checkpoints	Signal transduction pathway that signals to the cell cycle control machinery to halt the cell cycle upon DNA damage.
Cell cycle arrest	Transient halt of the cell cycle to allow cells time to repair the damage.
Apoptosis	Programmed cell death.
Senescence	Irreversible state of cell cycle arrest induced by short telomeres, DNA damage and other stresses.

DNA damage induction by UV, ionizing radiation or topoisomerase inhibitors (such as etoposide).[19,20] These DNA damage regulated microRNAs target several genes that have important functions in the DDR, e.g., in the regulation of the cell cycle, cell death or resistance to DNA damage. For example, UV-inducible microRNAs miR-16, 21, 34a and 221 have been shown to regulate both cell proliferation and cell death.[21]

The majority of specific changes in microRNA expression was differentially regulated early after UV damage in human fibroblasts (4 hours after exposure), whereas the expression levels of many microRNAs were almost back at basal (unirradiated) levels within 24 hours.[19] This early regulation of most microRNAs was faster than the relatively slow transcriptional response following TP53 stabilization, one of the most prominent DNA damage responsive transcription factors. This leads to a model in which

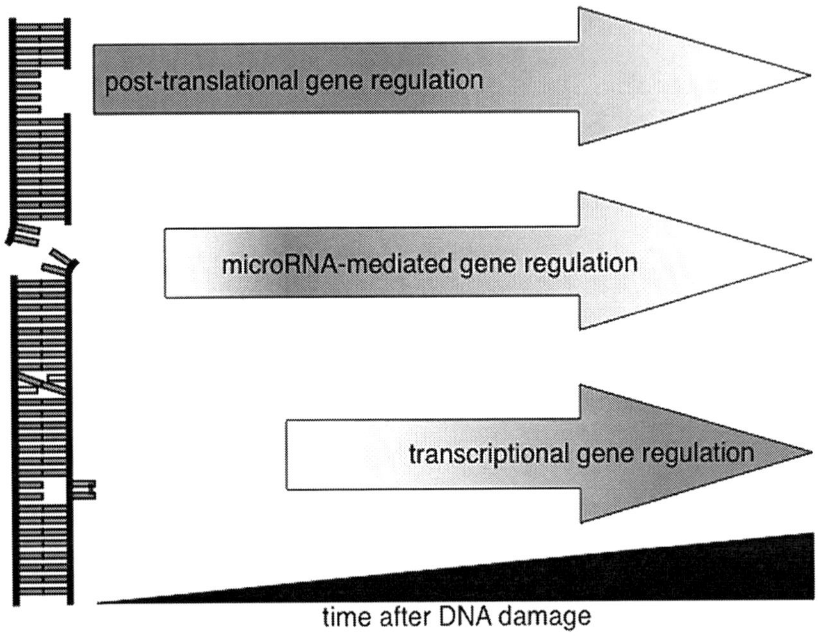

Figure 3. Model for microRNA-mediated gene regulation in the DNA damage response. Post-transcriptional gene silencing by microRNAs appear in time after the fast protein modifications, but earlier than most gene transcriptional responses.

microRNAs generally act in time between the fast posttranslational modifications and the relatively slow transcriptional response (Fig. 3).

Additional evidence for this model was provided by the regulation of the cell cycle checkpoint gene *CDC25a*. Within minutes after DNA damage, the CDC25a protein is phosphorylated by the DNA damage kinases ATM or ATR, leading to its degradation.[13] Several hours later, the CDC25a promoter is silenced by p21, whose expression after DNA damage depends on TP53.[22,23] However, CDC25a mRNA downregulation was already observed in the first hours after DNA damage,[24] a time frame in which p21 is not yet upregulated. In the intermediate time frame, between protein degradation and promoter silencing, microRNA-mediated gene silencing is responsible for CDC25a downregulation. In this case, miR-16 upregulation resulted in degradation of CDC25a mRNA and abrogation of cell cycle progression.

This type of intermediate regulation has several advantages. First, quick regulation of a few microRNAs can influence the expression of hundreds to thousands of genes, which enables a coordinated response in gene expression. Later, transcriptional regulation can take over and establish more permanent changes. Furthermore, quick regulation at the level of mRNA translation prevents futile rounds of protein synthesis and degradation, which would otherwise result from the protein modifications, e.g., phosphorylation and degradation of CDC25a after DNA damage induction. Finally, it may be easier to resume normal gene expression when only microRNA levels have been altered without changes in promoter activity.

MECHANISM OF FAST MicroRNA REGULATION

The observation that the DDR is partly regulated by microRNAs leads to the question how microRNA expression is regulated. Several steps in the microRNA biogenesis could be regulated, but little is known at this moment about this subject. We will quickly mention the various possible regulation mechanisms as summarized in Figure 1 and go over the limited evidence that exists to date.

1. MicroRNA expression might be regulated at the level of transcription. Many primary microRNAs have small transcripts (e.g., the miR-15a—miR-16 cluster contains less than 1000 nucleotides,[25] while the average transcript of a gene is >20 kilobases). This size difference could be responsible for the observed fast responses in microRNA expression levels. However, this will not apply to microRNAs that are located within introns and are thus coregulated (at least transcriptionally) with their "host" gene. Transcriptional regulation of miR-34a, which has a >30 kilobase primary transcript,[26] by TP53 as part of the DDR is a late reaction (not yet visible at four hours after UV irradiation, but clearly observed at 24 hours), suggesting that transcriptional regulation may be primarily important for the late changes in expression of target genes.
2. Processing of primary microRNA transcripts into precursor microRNAs might be regulated. Recently, such post-transcriptional regulation of microRNAs has been demonstrated. This type of microRNA regulation has the distinct advantage that it bypasses the relatively slow process of transcription. For example, within the DDR, TP53 has been shown to regulate microRNAs at the post-transcriptional level by binding to dead-box helicase DDX5.[27] This RNA helicase is present in a protein complex associated with Drosha and DGCR8.[28] The TP53 dependent activation of DDX5 results in the processing of a subset of specific primary microRNA species into precursor microRNAs. This type of regulation bypasses transcription, as the nuclear pool of primary microRNAs is selectively processed. This type of regulation is not unique to the DDR: TGF-β signaling causes the rapid upregulation of miR-21 via SMAD dependent regulation of DDX5 in human vascular smooth muscle cells.[29] In addition, the Drosha-DGCR8 subcomplex with DDX5 contains several other proteins, which may also be involved in similar regulatory functions. For example, KSRP also regulates a subset of specific microRNAs.[30]
3. Export of precursor microRNAs from the nucleus to the cytoplasm might be regulated. These RNA molecules are transported by a specific transport system, which makes use of Exportin-5.[31] Although there is no positive evidence for such a regulation at this moment, modulation of Exportin-5 levels or activity could influence the level of microRNAs in the cytosol, where they have their function.
4. Processing of precursor microRNAs into mature microRNAs by Dicer or RISC activity could also be regulated. An example of such regulation involves the let-7 family of microRNAs by LIN28 and LIN28b proteins, which bind to the hairpin loop of precursor microRNAs thereby blocking further processing by Dicer.[32] Interestingly, the let-7 family of microRNAs is rapidly downregulated after ionizing radiation,[33] while both overexpression of let-7 and knockdown of LIN28 sensitize cells to ionizing radiation.[34] It is tempting to speculate that a LIN28—let-7 regulatory circuit is important to confer radioresistance by

regulating precursor microRNAs. Although there is no evidence that Dicer and/or AGO2 activity are directly regulated upon specific cues in living cells (which could induce specific microRNA expression alterations), it is possible that interacting factors play a regulatory role. For example, RNA Helicase A (RHA) specifically interacts with siRNAs, DICER and AGO2 and functions as a siRNA loading factor in RISC.[35] Since RHA interacts with the DNA damage checkpoint gene BRCA1,[36] it is possible that this step in the microRNA pathway is subject of regulation as well.

5. MicroRNA degradation is also an important factor for the level of mature microRNAs in the cell. Little if anything is currently known about their degradation, but the general regulation of microRNA stability by controlling the level or activity of specific ribonucleases could be important for overall microRNA levels. It is conceivable that not all microRNAs are equally sensitive to degradation, which could be exploited as a mechanism for differential regulation of microRNA classes.

These types of post-transcriptional regulation of microRNAs provide cells with a tool to react rapidly via protein-protein interactions to DNA damage or other cues, because it bypasses the relatively slow transcriptional process. It is also more flexible, since abrogation of the signal that induces the selective post-transcriptional regulation of microRNAs directly influences gene expression levels.

LOCALIZATION OF THE MicroRNA-MEDIATED GENE SILENCING MACHINERY

MicroRNA function is not only regulated in time, but also in space. In order to carry out their function in mRNA silencing or degradation, microRNAs need to bind to one of the AGO proteins in the RNA-induced silencing complex (RISC).[5] As only AGO2 has an intact active site, most experiments on localization of the RISC complex have used this protein as a marker. AGO2-containing RISC exists as a soluble complex throughout the cytosol, but it is enriched in Processing Bodies (PBs; also called GW bodies, because they contain Gly/Trp rich proteins), which are putative sites of mRNA degradation and/or silencing. These PBs are present in all cells, irrespective of microRNA expression or external factors. However, upon certain stress conditions, RISC complexes accumulate in another type of cytosolic bodies, called Stress Granules (SGs).[37] PBs and SGs are found in all eukaryotes from yeast to mammals. Other types of RNA granules have been found in specialized cell types, such as germ cell granules and neuronal granules (see Chapter 8, Table 1, page 124 for definitions of cellular components).[38] These granules also contain elements of the microRNA machinery, suggesting that they also have functions in post-transcriptional gene regulation. SGs have a different protein composition than PBs.[39] It is not yet clear how different localization of mRNA and microRNA complexes exactly influences mRNA silencing and degradation, but it is generally assumed that PBs are more involved in degradation, while SGs may be mainly involved in translational silencing.[40] PBs and SGs have often been found to be very close together and may even touch each other, suggesting that mRNAs might be transferred between both types of granules.[41] Therefore, SGs have been described as site of 'RNA triage'.

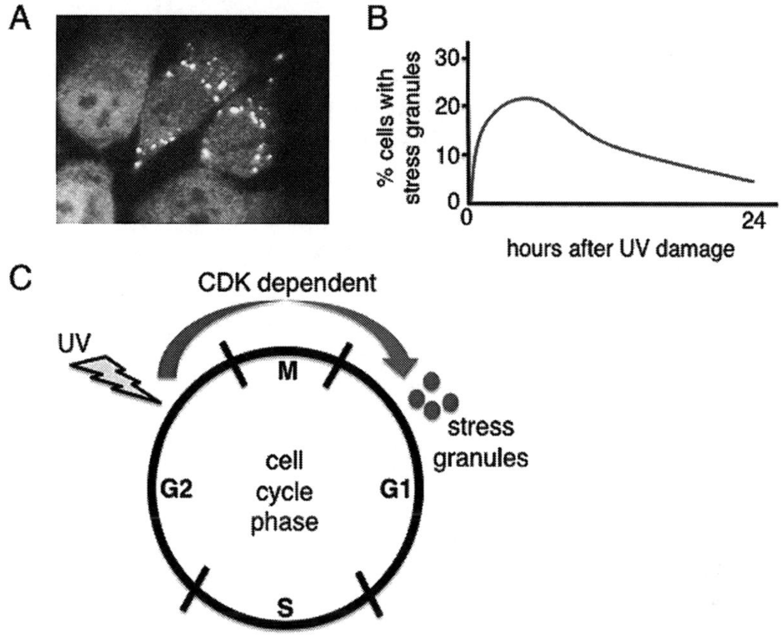

Figure 4. Stress granule formation after UV treatment. A) Immunofluorescent staining for stress granule marker TIA1 shows SG formation after UV treatment. B) Kinetics of SG formation after UV. C) Schematic overview of cell cycle dependent assembly of SGs after UV treatment.

SGs have been most extensively characterized under conditions of translational stress, which can be induced by Arsenite or the translation inhibitor hippuristanol.[37,42] Under these conditions all cells develop SGs with translationally inhibited mRNAs within 30 minutes. However, SGs can also be induced by exposure to DNA damaging agents, such as UV light or H_2O_2[19,21] (Fig. 4A). Interestingly, SG formation kinetics are quite different under these conditions: SGs reach a maximum after several hours and not all cells acquire these granules (~20% in HeLa cells) (Fig. 4B). Unexpectedly, this phenomenon was cell cycle dependent: cells that were irradiated in the G2 phase of the cell cycle developed SGs in the next G1 phase. Although this phenomenon is not yet understood, SG formation after UV irradiation requires Cyclin Dependent Kinase (CDK) activity, suggesting that this process is under direct control of the cell cycle machinery (Fig. 4C). It is therefore tempting to speculate that mRNA silencing and/or degradation depends not only on microRNA expression, but also on the cell cycle phase.

The timing of SG formation after UV irradiation is very similar to the timing of most changes in microRNA expression pattern: maximum induction is within a few hours and most microRNAs are back to pretreatment levels within 24 hours. Therefore, it is expected that accumulation of microRNA-loaded RISC complexes and translationally repressed mRNAs in SGs may have significant impact on gene expression levels. However, more research will be required to address this point experimentally.

CONCLUSION

Currently, most studies concerning microRNAs are aimed at their identification and function in the process of interest. Although more than 1000 mammalian microRNAs have been experimentally verified, the regulation of their expression is known for only a handful of microRNAs. It will be important to get a much more complete insight into the various mechanisms of gene expression regulation in time and space. It will be especially important to take into account the kinetics of microRNA expression regulation, protein modification and transcriptional responses. As all these layers of regulation increase the complexity of the cellular response to specific cues tremendously, a broad time range should be studied for adequate understanding of these regulatory mechanisms.

On top of this temporal aspect of gene expression regulation, the spatial organization of various regulatory mechanisms within a cell may also contribute to the final outcome. It will be important to investigate how incorporation of silencing complexes into PBs or SGs influences regulation of mRNA expression. Although these structures most probably are the sites that attract mRNAs that are targeted for silencing or degradation, it is not yet clear whether this localization is necessary for proper expression regulation. At this moment it is even questionable whether SGs that are formed after different types of stress are functionally similar. Their shape and timing differs between treatments, but their content appears similar upon first inspection. However, more careful analysis is required to fully address this issue.

In conclusion, the impact of microRNA-mediated gene regulation in time and space is just beginning to surface. It will be important to address these aspects in various model systems and experimental set-ups to verify the general nature of the concepts described in this chapter.

REFERENCES

1. Collins LJ, Penny D. The RNA infrastructure: dark matter of the eukaryotic cell? Trends Genet 2009; 25(3):120-128.
2. Lee Y, Kim M, Han J et al. MicroRNA genes are transcribed by RNA polymerase II. EMBO J 2004; 23(20):4051-4060.
3. Lee Y, Ahn C, Han J et al. The nuclear RNase III Drosha initiates microRNA processing. Nature 2003; 425(6956):415-419.
4. Bernstein E, Caudy AA, Hammond SM et al. Role for a bidentate ribonuclease in the initiation step of RNA interference. Nature 2001; 409(6818):363-366.
5. Hutvagner G, Simard MJ. Argonaute proteins: key players in RNA silencing. Nat Rev Mol Cell Biol 2008; 9(1):22-32.
6. Liu J, Carmell MA, Rivas FV et al. Argonaute2 is the catalytic engine of mammalian RNAi. Science 2004; 305(5689):1437-1441.
7. Meister G, Landthaler M, Patkaniowska A et al. Human Argonaute2 mediates RNA cleavage targeted by miRNAs and siRNAs. Mol Cell 2004; 15(2):185-197.
8. Bartel DP. MicroRNAs: target recognition and regulatory functions. Cell 2009; 136(2):215-233.
9. Hoeijmakers JHJ. Genome maintenance mechanisms for preventing cancer. Nature 2001; 411(6835):366-374.
10. Jackson SP, Bartek J. The DNA-damage response in human biology and disease. Nature 2009; 461(7267):1071-1078.
11. de Boer J, Andressoo JO, de Wit J et al. Premature aging in mice deficient in DNA repair and transcription. Science 2002; 296(5571):1276-1279.
12. Niedernhofer LJ, Garinis GA, Raams A et al. A new progeroid syndrome reveals that genotoxic stress suppresses the somatotroph axis. Nature 2006; 444(7122):1038-1043.

13. Mailand N, Falck J, Lukas C et al. Rapid destruction of human Cdc25A in response to DNA damage. Science 2000; 288(5470):1425-1429.
14. Colman MS, Afshari CA, Barrett JC. Regulation of p53 stability and activity in response to genotoxic stress. Mutat Res 2000; 462(2-3):179-188.
15. Kruse J-P, Gu W. Modes of p53 Regulation. Cell 2009; 137(4):609-622.
16. Busuttil RA, Doll M, Campisi J et al. Genomic Instability, Aging and Cellular Senescence. Ann N Y Acad Sci 2004; 1019(Strategies for Engineered Negligible Senescence: Why Genuine Control of Aging May Be Foreseeable):245-255.
17. Lukas J, Bartek J. Cell division: The heart of the cycle. Nature 2004; 432(7017):564-567.
18. Bartek J, Lukas J. Pathways governing G1/S transition and their response to DNA damage. FEBS Letters 2001; 490(3):117-122.
19. Pothof J, Verkaik NS, van Ijcken W et al. MicroRNA-mediated gene silencing modulates the UV-induced DNA-damage response. EMBO J 2009; 28(14):2090-2099.
20. Simone NL, Soule BP, Ly D et al. Ionizing Radiation-Induced Oxidative Stress Alters miRNA Expression. PLoS ONE 2009; 4(7):e6377.
21. Pothof J, Verkaik NS, Hoeijmakers JH et al. MicroRNA responses and stress granule formation modulate the DNA damage response. Cell Cycle 2009; 8(21):3462-3468.
22. Vigneron A, Cherier J, Barré B et al. The cell cycle inhibitor p21waf1 binds to the myc and cdc25A promoters upon DNA damage and induces transcriptional repression. J Biol Chem 2006; 281(46):34742-34750.
23. Rother K, Kirschner R, Sanger K et al. p53 downregulates expression of the G1//S cell cycle phosphatase Cdc25A. Oncogene 2006; 26(13):1949-1953.
24. Garinis GA, Mitchell JR, Moorhouse MJ et al. Transcriptome analysis reveals cyclobutane pyrimidine dimers as a major source of UV-induced DNA breaks. EMBO J 2005; 24(22):3952-3962.
25. Calin GA, Ferracin M, Cimmino A et al. A MicroRNA signature associated with prognosis and progression in chronic lymphocytic leukemia. N Engl J Med 2005; 353(17):1793-1801.
26. Chang T-C, Wentzel EA, Kent OA et al. Transactivation of miR-34a by p53 broadly influences gene expression and promotes apoptosis. Molecular Cell 2007; 26(5):745-752.
27. Suzuki HI, Yamagata K, Sugimoto K et al. Modulation of microRNA processing by p53. Nature 2009; 460(7254):529-533.
28. Gregory RI, Yan K-p, Amuthan G et al. The Microprocessor complex mediates the genesis of microRNAs. Nature 2004; 432(7014):235-240.
29. Davis BN, Hilyard AC, Lagna G et al. SMAD proteins control DROSHA-mediated microRNA maturation. Nature 2008; 454(7200):56-61.
30. Trabucchi M, Briata P, Garcia-Mayoral M et al. The RNA-binding protein KSRP promotes the biogenesis of a subset of microRNAs. Nature 2009; 459(7249):1010-1014.
31. Yi R, Qin Y, Macara IG, Cullen BR. Exportin-5 mediates the nuclear export of premicroRNAs and short hairpin RNAs. Genes Dev 2003; 17(24):3011-3016.
32. Viswanathan SR, Daley GQ, Gregory RI. Selective blockade of microRNA processing by Lin28. Science 2008; 320(5872):97-100.
33. Weidhaas JB, Babar I, Nallur SM et al. MicroRNAs as potential agents to alter resistance to cytotoxic anticancer therapy. Cancer Res 2007; 67(23):11111-11116.
34. Oh J-S, Kim J-J, Byun J-Y et al. Lin28-let7 Lin28-let7 modulates radiosensitivity of human cancer cells with activation of K-Ras. Int J Radiat Oncol Biol Phys 2010; 76(1):5-8.
35. Robb GB, Rana TM. RNA Helicase A Interacts with RISC in Human Cells and Functions in RISC Loading. Mol Cell 2007; 26(4):523-537.
36. Anderson SF, Schlegel BP, Nakajima T et al. BRCA1 protein is linked to the RNA polymerase II holoenzyme complex via RNA helicase A. Nat Genet 1998; 19(3):254-256.
37. Leung AKL, Calabrese JM, Sharp PA. Quantitative analysis of Argonaute protein reveals microRNA-dependent localization to stress granules. Proc Natl Acad Sci U S A 2006; 103(48):18125-18130.
38. Buchan JR, Parker R. Eukaryotic stress granules: the ins and outs of translation. Mol Cell 2009; 36(6):932-941.
39. Anderson P, Kedersha N. RNA granules: post-transcriptional and epigenetic modulators of gene expression. Nat Rev Mol Cell Biol 2009; 10(6):430-436.
40. Anderson P, Kedersha N. Stress granules: the Tao of RNA triage. Trends Biochem Sci 2008; 33(3):141-150.
41. Kedersha N, Anderson P. Stress granules: sites of mRNA triage that regulate mRNA stability and translatability. Biochem Soc Trans 2002; 30(Pt 6):963-969.
42. Kedersha N, Anderson P, Jon L. Mammalian stress granules and processing bodies. Methods Enzymol 2007; 431:61-81.

CHAPTER 6

SPLICEOSOMAL RNA INFRASTRUCTURE:

The Network of Splicing Components and Their Regulation by miRNAs

Lesley J. Collins
Institute of Fundamental Sciences, Massey University, Palmerston North, New Zealand
Email: l.j.collins@massey.ac.nz

Abstract: The RNA infrastructure model highlights the major roles played by RNA–based networks in cellular biology. One of the principle concepts behind the RNA-infrastructure is that proteins shared between RNP machineries network their processes in a temporal (over the cell cycle) and spatial (across the cell, or intercellular) manner. In order to dig deeper into the RNA-infrastructure we need to examine the networking aspects of RNPs in a more detailed manner. The eukaryotic spliceosome is an excellent example of an RNA machine that contains RNA-Protein and RNA-RNA interactions, as well as temporal and spatial networking to other processes. This chapter will examine some different types of spliceosomal networks that involve RNPs and illustrate how current networking tools can be used to dissect the many faces of the RNA-infrastructure.

INTRODUCTION

RNA-based metabolism underpins the regulation and action of proteins in eukaryotic cells.[1-3] Studies continue to reveal enormous genetic complexity through processes such as RNA interference (RNAi), alternative splicing and epigenetics, but we now face a massive challenge in mapping this RNA complexity onto already known molecular biology.[2,4] We have regulatory RNAs (e.g., miRNAs) targeting specific genes, or in some cases thousands of genes; and we have processing RNAs (e.g., snRNAs) which interact with proteins in large complexes (for example snRNAs interact with multiple proteins to form snRNPs which themselves interact to form the spliceosome). Associated with these

RNA Infrastructure and Networks, edited by Lesley J. Collins.

processes are modifying ncRNAs such as snoRNAs. These were once thought to modify only rRNA, snRNA and tRNA, but they are now suspected to be more important in the regulation of many other genes. There are also long ncRNAs (>200 nucleotides) implicated in epigenetic marking (reviewed in refs.4-6). All of these different types of RNAs work with proteins and together they make up a network called the 'RNA-Infrastructure'.[1]

Until recently, investigation of genetic networks was largely restricted to proteins, since protein-protein interactions (PPI) are important indicators of metabolic pathways,[7] with networks often used for drug discovery,[8] plant development[9] and cancer studies.[10,11] However, increases in the use of high-throughput technologies to gather gene expression and regulation information (e.g., see refs. 12, 13) place an increasing importance in connecting RNA information to existing protein information.[14] Instead of just protein-protein interactions, we need to also consider RNA-protein and RNA-RNA interactions in order to use gene networks to make more accurate predictions. Understanding the networking properties of miRNA-target information for example can reveal the complicated facets of RNA-based regulation. For example, the Snu13p spliceosomal protein of the yeast *Saccharomyces cerevisiae* binds to U4 snRNA, but is also involved in the processing of the U3 snoRNA, having different interactions depending on the complex in which it is present.[15]

The spliceosome comprised of general and specific stage proteins (many of which are regulated by miRNAs), medium length snRNAs (some of which are modified by medium length snoRNAs) makes an ideal model for studying RNA-based networks. For model species such as humans and *S. cerevisiae* we can examine networks produced from Protein-Protein interactions (e.g., from BioGrid[16]), RNA-RNA and RNA-protein (miRNA-target and other) information. This chapter will use examples from the literature and our own studies to highlight how network analysis is even more relevant to understanding biological function when RNA-based regulation is added.

THE SPLICEOSOME AS AN EXAMPLE OF THE RNA-INFRASTRUCTURE

The macromolecular spliceosome is responsible for the removal of introns from gene transcripts (pre-mRNA) prior to translation.[17] About 200-300 proteins are part of the spliceosome during the splicing cycle,[17,18] but it is the inclusion of 5 critical small nuclear RNAs (snRNAs) which allow catalysis in the actual splicing reactions.[19,20] These catalytic snRNAs must first bind to specific proteins to form the U-class snRNPs (small nuclear ribonucleoproteins) before the spliceosome assembles. As shown in Figure 1, during splicing, the U1 and U2 snRNPs recognize the intron-exon boundaries and form the prespliceosome (called the A complex). In the next stage the U4/U6.U5 snRNP tri-complex interacts and the B complex is formed. The U6 snRNA then breaks its interactions with the U4 snRNA and instead binds to the U2 snRNA. Both the U1 snRNP and the U4 snRNP leave the spliceosome allowing the formation of the catalytic unit. The mRNA is spliced by the action of the U2/U6 snRNAs then the exons are ligated. At this stage a protein complex called the 'Exon Junction Complex' (EJC) is deposited on the ligated exons which signals that splicing is complete. The spliced mRNA then goes through further checking before it is exported from the nucleus to the cytoplasm where translation takes place. We can readily see different types of RNA interactions taking place within the splicing cycle, some of them between the mRNA and splicing components (e.g., U1, U2, U5 interacting with the mRNA) and some between supporting components (e.g., snRNAs and their proteins).

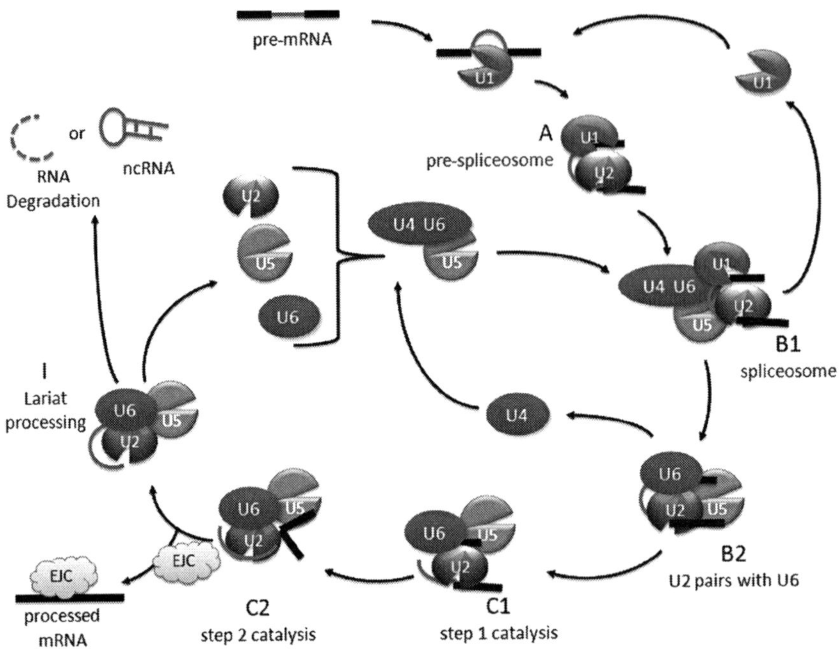

Figure 1. The spliceosome cycle (based on the major spliceosomal cycle in ref. 17) The major cycle is shown here and although the minor cycle is believed to be very similar due to many of the proteins being shared, one important difference in the minor cycle is that the U11 and U12 snRNPs interact before recognising the minor splice sites. A color version of this figure is available at www.landesbioscience.com/curie.

'Major splicing' (or U2-splicing) as described above, is only one form of splicing and over the last decade the importance of so-called 'minor' splicing has become clearer.[21] Although the cycle is similar, other snRNAs (U11, U12, U4atac and U6atac) replace their counterparts (U1, U2, U4, U6 respectively) and recognise a different set of introns. Although the U5 snRNA is a part of both splicing systems it is very likely that there are some different interactions. One example is that the U11 and U12 form a di-snRNP before interacting with the mRNA (in contrast, U1 and U2 snRNPs interact independently during major splicing). Minor splicing first came to light in plants but since then, it has been characterised in many groups of eukaryotes and may play an important role in alternative splicing.

The spliceosome (both major and minor versions) is a dynamic 'organelle' that displays characteristics of the RNA-infrastructure. Interactions between the RNA and protein components have both temporal and spatial aspects, especially in the biogenesis of different sub-complex components. The dynamic network of the spliceosome means that it cannot presently be investigated in its entirety. Instead we have to break it into either temporal (e.g., A, B, C complexes) or spatial (U2 snRNP-based, U4/U6 tri snRNP-based) partitions. In a similar manner we can also break our network analysis into physical interactions (spatial) and regulatory interactions (temporal). Many proteomic studies (e.g., refs. 22-24) have investigated parts of the spliceosome using mass spectrometry and microscopy giving us a more detailed look at how different components interact. However, we are still limited in how we can peek into the working spliceosome as most of these methods are not compatible with intact RNA. The integration of proteomic (i.e., mass spectrometry) and transcriptomics/expression (i.e., from mRNA sequencing) is a powerful combination of two genomic scale technologies is opening up further analysis into macromolecular complexes such as the spliceosome. Unfortunately, many of the RNA interactions are not collated into any database (especially the non-mRNA interactions) so the literature remains the largest source for RNA-RNA and modifying RNA-protein interactions. For this reason many macromolecular studies begin with a few proteins from a sub-complex and work up from there, because as will be seen further in this chapter, RNP networks can get very complicated.

RNA-PROTEIN INTERACTION NETWORKS

Before examining the complex RNA-protein interaction networks within the spliceosome, let us first look at RNA interaction networks in general. Interaction networks imply a physical connection between interacting partners. We can find many such networks in general life (e.g., subway networks, computer networks) and mapping biological networks has become one way in which we can investigate different cellular connections. One of the most common forms is the protein-protein interaction (PPI) network.

Data for PPI networks can be gathered from experimental data by a variety of methods including large scale tandem affinity purification coupled to mass spectrometry (TAP-MS) and yeast two-hybrid methods.[25] Currently the most studied PPI organism is the yeast *S.cerevisiae*,[7] but large datasets are now available for other eukaryotes such as humans, mice, *Drosophila mealanogaster* (fruitfly) and nematode (*Caenorhabditis elegans*),[16] and also from some bacteria such as *Bacillus anthracis*, *Francisella tularensis* and *Yersina pestis*.[26]

A gene interaction network is made up of the following parts (Fig. 2). Each node (a gene or in PPIs specifically a node represents a protein) is connected by edges representing the interaction with other nodes. The connectivity (or degree, k) or a node is the number of connections a node makes with other nodes. Nodes which connect to numerous other nodes (i.e., have a high degree or connectivity e.g., k in Fig. 2A) are often called hubs. Network edges can be weighted in different ways (with numerical or colour differences) representing the experimental procedure used to determine the interaction. This distinction between different experiments is important as interactions may differ due to experimental biases. This can be applied to RNP networks where RNA-protein interactions are included. However, generally in these cases the RNA-associated nodes and edges are treated as 'special' types of protein nodes meaning that RNA-specific features (such as specific target areas) may be lost during network construction. Current PPI networks are also rife with noise in that they contain large numbers of false positives and false negatives.[27]

PPI networks in general can have different topologies (reviewed in ref. 25). Scale-free networks (Fig. 2A) are characterised by a power law degree distribution; that is a few hubs have a very high degree (many interactions) distinct from the majority of hubs that have few interactions.[25,28] Random networks (Fig. 2C) are often constructed for comparison with PPI networks to show significance of hubs. The nodes in these random networks tend to have similar numbers of links, no hubs and a Poisson degree distribution. Hierarchical

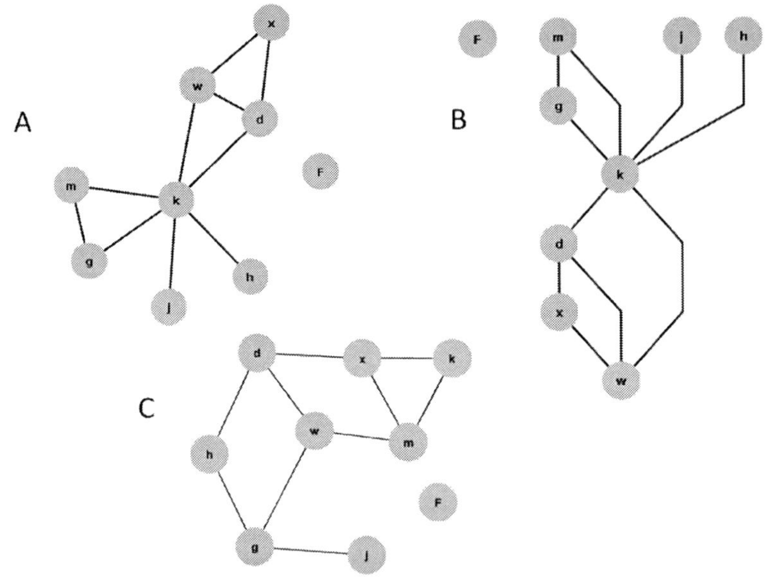

Figure 2. Components of PPI networks. x, w, d, k, j, g, m are nodes representing genes which are connected by edges (solid lines) indicating an interaction between them. F is an unconnected node. In network A, a scale-free network, node k has a high degree of connectivity while nodes j and h have a low degree of connectivity. In this network node k represents a hub. The same network can be redrawn as a hierarchical network (B) indicating again how central node k is to the network. This representation can also aid in finding paths between nodes such as those between nodes d and w. A random network (C) is drawn from the same nodes with different interactions. In this network there are no hubs and the majority of nodes have an average degree of 3. Networks visualised with Cytoscape 2.7.0.[31,32] A color version of this figure is available at www.landesbioscience.com/curie.

networks (Fig. 2B) contain hubs and defined modules and are considered to more accurately reflect biological systems.[25] Hierarchical networks can be considered a different representation of a scale-free network but can be more useful in indicating interaction pathways between different protein complexes. There are also many statistics that can be used to describe PPI networks and they are also applicable to RNP networks. These are described in Table 1.

From some studies (e.g., see ref. 28) RNP networks follow a power law degree distribution and hence fit into the scale-free and hierarchical topologies. Randomly generated RNP networks (summarized in ref. 28) display the expected Poisson degree distribution where most nodes carry an average degree and hence it could be surmised that the graph properties shown by RNP networks offer some selective evolutionary advantage. Another difference is that RNP graphs show a clustering coefficient of 0.05 (largely unclustered, see Table 1 for definition of scoring) whereas 'real world' networks (those with hierarchical and scale-free topologies) generally show higher clustering (~0.3). On our example in Figure 2 the scale-free network (Fig. 2A) has a clustering coefficient of 0.496, whereas the random network (Fig. 2C) has a clustering coefficient of 0.185, although both networks contain the same number of nodes and edges and isolated nodes (statistics calculated with Network Analyzer[29]).

It is interesting in how adding ncRNAs such as regulatory RNAs, change PPI networks. Figure 3 shows a network of some human U2 proteins and their miRNA

Table 1. Network terms and characteristics used in this chapter

Network Term	Definition
Node	Represents a gene (generally protein or RNA)
Edge	Represents the interaction between nodes. Edges are graphically displayed by lines connecting nodes. Different experimental conditions can be displayed either by a number on the edge or by edge colour
Hub	A node that has many connections to other nodes
Undirected	This term can apply to an edge or the entire network. It implies that nodes interact but not regulation or direction
Directed	This term can apply to an edge or the entire network. It implies direction to the connection between nodes (e.g., regulation)
Scale-Free network	Network characterised by a power law degree distribution. I.e. there are a small number of highly connected hubs and the probability of these hubs is statistically more significant than in a random network
Random network	Each pair of nodes is connected with a probability (p). Node degrees follow a Poisson distribution where most nodes have about the same number of links
Hierarchical network	A type of scale free network allowing for hubs but also for modular sub-networks within the entire network
Degree (k)	The number of neighbours of a node. Describes connectiveness
Clustering coefficient (C)	The average ratio of the proportion of edges between connected nodes and the possible number of edges between them (i.e., the number of connections if the graph is fully connected), values between 0 and 1. 1 being fully connected and 0 being fully unconnected.
Assortativity (NC)	The average degree of the nearest neighbours of a node. A negative correlation indicates that hubs tend to interact with nodes with a low degree. A positive correlation suggests a modular mode with hubs linking to other hubs
Shortest path (SP)	The shortest path between each two nodes (i.e., the number of edges connecting two nodes)
Betweenness (B)	The frequency with which a node is located on the shortest path between all other nodes. Nodes with a high betweenness control the flow of information across the network
Proteomic seed	A protein significantly differentially expressed between conditions as identified by proteomic screening
Network crosstalk	The degree of network proximity and connectivity between (groups of) proteins, modelled as the amount of information flow between these proteins n a PPI network
Sub-graph or Graphlets	A small sub-motif of a network occurring more than would be expected by random
RGF Distance	Identifies all subgraphs with 3-5 nodes in two networks and compares the frequency of their appearance
GGDA statistic	Calculates node specific permutation groups (called automorphism orbits) within each of the 29 (2-5 nodes) possible subgraphs of the two networks being compared

Definitions taken from references 11, 28 and 40.

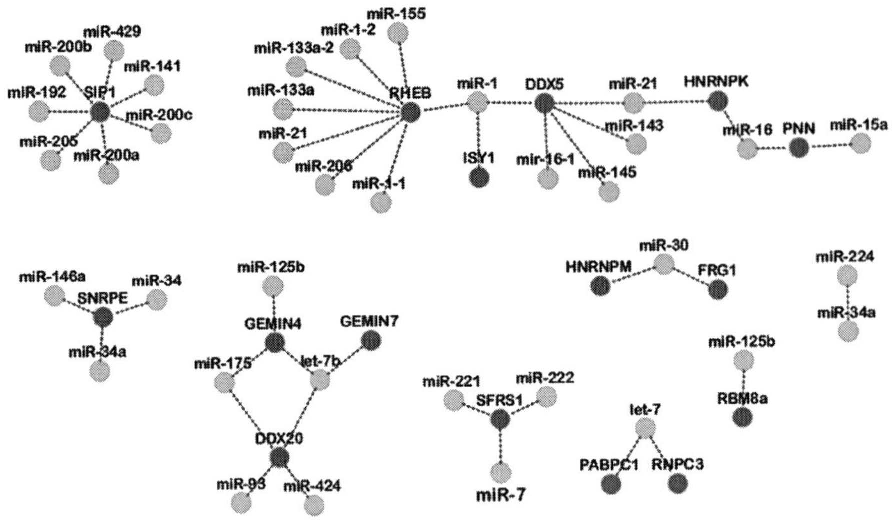

Figure 3. Networks of some splicing proteins connected by miRNAs. Some miRNA-target interactions have a simple one to one relationship whereas other proteins are regulated by many miRNAs and some miRNAs regulate many proteins. miRNA interaction information was gathered from miRsel,[35] miRbase,[36] Rfam[37,38] and miRecords.[39] miRNAs are in light nodes and proteins are in dark nodes. PPI interactions are solid lines and RNA-Protein interactions are dashed lines. Network visualised with Cytoscape 2.7.0.[31,32] A color version of this figure is available at www.landesbioscience.com/curie.

connections visualised with the network software Cytoscape.[30-32] We can see here many patterns common to miRNA-based regulation. Some proteins (e.g., RBM8a) are regulated by a single miRNA whereas others (e.g., SIP1, SNRPE and SFRS1) are regulated by many miRNAs. It should be noted that these networks indicate overall interactions and do not factor in that some miRNAs may regulate their targets only in specific tissues or at specific times (i.e., these networks are not graphed temporally or spatially). We can see, however that some miRNAs can connect proteins in a network (e.g., DDX5, ISY1, RHEB, HNRNPK and PNN) whereas these proteins may remain unconnected nodes in a strictly PPI network (this is not really the case with these proteins as connecting nonsplicing proteins have been removed for this exercise).

One downside with these networks is that computational approaches are presently limited in their ability to resolve the connections in an RNA-centric manner. Typically miRNAs are known to target the 3′ UTR region of a gene, but others may target the 5′ UTR region.[33] Although these are still all miRNA-mRNA interactions, they are quite different in their nature and hence their biological relevance. Our way to resolve this matter presently is to weight the edges for different connection types in a manner similar to determining different experiments in PPI networks.

On a larger scale, regulated protein networks (e.g., RNP networks consisting of proteins and their miRNA regulators, but not other ncRNAs) display characteristics similar to that of PPIs from transcription factor proteins.[34] This study[34] compared data from an early database of RNA-protein interactions from six species (bacteria-*E.coli*, yeast-*S.cerevisae*, nematode-*C.elegans*, Fruitfly-*D.melanogaster*, mouse and human) to transcription factor interactions. It showed that these networks showed a power-law behaviour (a few miRNAs regulate many proteins) in a scale-free fashion, but that there appeared to be a maximal degree meaning that very highly connected nodes were not present (for human there was a maximum of 10). Current miRNA informational databases (e.g., miRsel,[35] miRbase,[36] Rfam,[37,38] miRecords[39]) have since swelled especially with newly reported interactions from high-throughput sequencing reactions. It will be interesting to see if the degree restriction holds on analysis from these much larger datasets.

Network comparisons can often be difficult because of the network size,[27] because the size includes both the number of nodes but also the (multiple) interactions between them. Because cellular biology is hierarchical we can examine a network by splitting it into modules and analysing for network motifs. One example of a sub-graph motif is where two transcription factors regulate two target genes in parallel possibly due to gene duplication (or perhaps genome duplication). For RNP networks, a duplicated protein may in fact be regulated by the original miRNA or a second miRNA may evolve. With genome duplication a more complicated situation may occur where miRNAs from both copies may cross-regulate until such time as the individual miRNA and their targets co-evolve away from each other. The identification of other sub-graph motifs[40] that represent different modes of RNA-RNA and RNA-protein interactions is an important area for future research. With sub-graph motifs we can begin to model the occurrence and distribution of the different interactions for different organisms and perhaps gain insight into how such networks represent the way that each organism has evolved.

We can take the network motif concept further with the extraction of small sub-graphs or motifs from the larger network. Sub-graphs (or graphlets) are defined as sections of the network with a defined topology that are present in a higher abundance than expected from a random network with the same degree distribution.[34,40] Counting small connected subgraphs especially in large PPI networks can be computationally demanding as the number of possible subgraphs of n-nodes increases exponentially with n. For this reason, calculations tend to use subgraphs with five to seven nodes.[27] Additional statistics such as the Relative Graphlet Frequency (RGF) and the Graphlet Degree Distribution Agreement (GDDA) score (described in Table 1) can be used to count the occurrence of these subgraphs which in turn can be used to compare networks.[27,41] These local approaches to network analysis and comparison are most often more successful than complete network analyses because of the incomplete and noisy nature of biological networks.[40,41]

REGULATORY (EXPRESSION MODULATED) NETWORKS

Biological networks can also represent a pathway of events be used in a 'directed' fashion to indicate. In RNP metabolism in particular we can use directed networks to indicate the immediate and downstream effects of RNA regulation. This means that the interaction between the RNA and the target goes one way (e.g., the miRNA affects the regulation of the target protein).

We know that a gene may be regulated by RNA, but it is only after we connect that protein to its interacting partners (both protein and RNA) can we see possible large scale and even phenotypic effects. Regulation is not only by RNA interference (i.e., miRNA or siRNAs) but can also include alternative splicing, nucleotide modification (e.g., methylation, pseudouridylation), alternative transcription initiation and termination.[2] We concentrate here on miRNA-based regulation, but similar RNA-target and network issues apply for these other types of RNA regulation. Most methods of discovering miRNA-mRNA interactions (experimental and computational) focus on down-regulatory interactions where

RNA silences the gene of interest. However, there is also RNA-mediated up-regulation and mix-regulation where a gene is up-regulated in one instance but down-regulated in another.[42] An example of this is let7 and the synthetic miRcxcr4 which up-regulate target mRNAs upon cell cycle arrest, but down-regulate in proliferating cells.[42,43]

In the past, regulatory RNA-target interactions were typically discovered firstly at a single gene level by painstaking (or sometimes accidental) molecular biology experiments. As with all ncRNAs classes, once a type of miRNA or siRNA was characterised then further members of that group could then be discovered computationally, quickly building up a more genomic (i.e., multi-gene) view of the regulation network for that particular regulatory RNA. However, computational approaches are limited in their resolution of the true connections between the RNA and the target in a sensitive or specific way. This may be because computational algorithms for RNA-target connections are either written for specific species or specific RNAs and thus cannot perform perfectly in a general situation. Nowadays we have on the experimental side, high-throughput sequencing which can quickly gather genome-wide small RNA information [(i.e., from 21-25 nucleotide long RNA sequences including miRNAs, siRNAs, piRNAs and other short classes)(refer to Chapter 16, page 239 for more detail on computational and experimental discovery of small RNAs)]. So long as we have an accurate genome to map this data against then establishing more accurate connection edges becomes a little more accurate. In one example MacLean et al 2010[28] generated networks from data collected from high-throughput sequencing of *Arabidopsis* short RNAs. This directed graph of 39,994 short RNA nodes, 18,054 long RNA nodes (primary transcripts) was connected by 38,149 source edges (a match to the positive strand of the genome) and 140,035 target edges (a match to the complementary strand of the genome). As we would expect the network showed a power-law (i.e., scale free) property and a high clustering coefficient (0.32 compared with random network expectations of 0.05). One other interesting property was that of disassortativity[44] where high degree nodes connect preferentially with low degree nodes (as opposed to assorted graphs were high degree nodes are connected to other high degree nodes, as seen in internet social networks.[44]

In plants, double stranded RNA can initiate a range of sequence-specific gene silencing pathways.[14] For example, short small interfering RNAs (siRNAs) are 21 nucleotides (nt) long while the ‘long’ siRNAs are 24 nt long with each being produced from different pathways.[14] High-throughput sequencing does not distinguish between the pathways used to generate the sequences, it just produces the sequence. When these RNAs are connected to protein targets, length, target specificity and location are often key factors in estimating which process was used to produce each sequence.

However, when we take some smaller networks of splicing proteins we can see a network that displays more assortive behaviour. The SF3B complex is a multi-protein assembly that is an integral part of the U2 snRNP and also the U11/U12 snRNP in minor splicing. It consists of seven proteins (SF3B1, SF3B2, SF3B3, SF3B4, SF3B5, PHF5A and SF3B14, also known as SF3b155, SF3b145, SF3b130, SF3b49, SF3b10, SF3b14b and P14 respectively). When a network of these proteins is drawn including known first neighbour interacting proteins and miRNAs (Fig. 4) we can see that all but one protein has a host of miRNAs, most of which associate with only one of the SF3B proteins. There are some miRNAs and a few proteins (SF3A2, TCERG1, CDC5L, TCA1, SMNDC1 and RNPS1) which interact with more than one SF3B protein, whereas the SF3B proteins themselves are highly interacting. This assorted behaviour is perhaps indicative of a tightly bound multi-protein complex as opposed to a general protein interaction network.

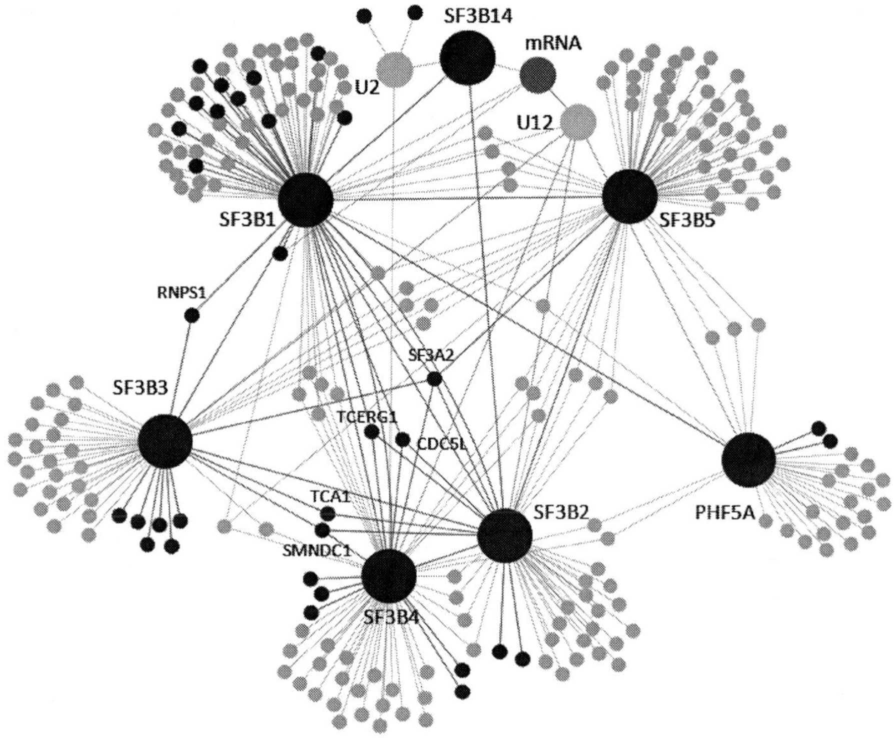

Figure 4. The seven proteins of the SF3B complex (large circles) and their regulatory miRNAs. Six of the proteins SF3B1, SF3B2, SF3B3, SF3B4, SF3B5 and PHF5A have interactions with many miRNAs and some other proteins. However, SF3B14 has interactions only with proteins SF3B1 and SF3B2 and no miRNAs have yet been found to regulate it. SF3B14 plays an important part in the complex by interacting both with the U2 snRNA and the mRNA (any interaction with the U12 snRNA is not known). Other proteins that have interactions with more than one SF3B protein are also indicated. miRNAs are in light nodes and proteins are in dark nodes. PPI interactions are dark lines and RNA-Protein interactions are light lines. In miRNA interaction information was gathered from miRsel,[35] miRbase,[36] Rfam[37,38] and miRecords.[39] Network visualised with Cytoscape 2.7.0.[31,32] A color version of this figure is available at www.landesbioscience.com/curie.

It is interesting that SF3B14 is the only one of the seven SF3B proteins not to have any known regulating miRNAs (as yet). This protein is thought to be positioned within the inner cage of the SF3B complex structure which has very few openings.[45] It is thus likely that there is a conformational change that allows the SF3B14 protein to interact with the U2 (or U12) snRNA and the mRNA (Table 2), two extremely critical functions. It is as yet unknown as to what this means in terms of the regulation of SF3B14 and the SF3B complex. It is suggested that SF3B plays a critical role near or at the spliceosome catalytic core,[46] since SF3a and SF3b reside at around the intron branchpoint are present prior to the first step and completely absent prior to the second step of splicing arrestment.[47] One belief is that SF3 functions to restrain the branchpoint hydroxyl from the cell's chemistry until the catalytic core of the spliceosome is properly assembled.

With a more comprehensive picture of the regulatory aspects behind our RNP interaction networks becoming available, we are discovering that there is a high level

Table 2. Protein and RNA interactions within the U2-SF3B complex. pp: number of protein-protein interactions, pr: number of protein RNA interactions. U2 interactions are taken from 60; U12 interactions from BioGrid16 (version 3.0.68)

Protein	Alias	pp	pr miRNA	U2 snRNA	mRNA
SF3B1	SF3b155	23	53	-	+
SF3B2	SF3b145	10	26	-	-
SF3B3	SF3b130	10	30	-	-
SF3B4	SF3b49	11	41	+	-
SF3B5	SF3b10	2	70	-	-
SF3B14	P14	2	0	+	+
PHF5A	SF3b14b	2	28	-	-

of regulatory feedback through multiple mechanisms in many species. We can also use directed networks to illustrate feedback loops between miRNAs and the targets they regulate. Take for example the splicing factor SF2/ASF (the SFRS1 gene) and one of its 40 miRNAs, the up-regulated hsa-miR-7 (miR-7).[48] Figure 5A is a simple representation of how SFRS1 promotes the Drosha cleavage step of miR-7 maturation and the mature miR-7 represses SFRS1 by binding to the 3′UTR of its transcript. Another gene EGFR (epidermal growth factor receptor), is also repressed by miR-7 leading to the downstream repression of the AKT pathway.[49] Both activation and repression of the genes are indicated graphically and it is not hard to understand the pathways involved. However, the real world is very different. If we add in just a couple of other genes that are known to be repressed by miR-7 including BL2 and MAPK3 (both very prominent in cancer studies) (Fig. 5B), we can see that many other miRNAs are involved and that some of these miRNAs regulate more than one of these genes. In the case of SFRS1 it is known that three other miRNAs (miR-29b, miR-221 and miR-222) are up-regulated by SFRS1 and not all at the Drosha-cleavage step.[48] Increased levels of mature miR-29 but not pri-miR-29 were observed upon SFRS1 induction indicating that this regulation occurs at the postDrosha stage. SFRS1 is also known to be feedback regulated through other means including unproductive alternative splicing and inhibition of translation initiation (summarized in ref. 48). We can also see two upstream regulators of the Akt pathway (IRS1 and IRS2)[49] are also regulated by miR-7 and have connections to the BCL2 and AKT1 proteins. Here in this network we can find connections to SFRS1 and the other splicing proteins to which it connects, as well as key components of the AKT pathway. With all these mechanisms to take into account our network graphs could get even more complex if we added in 'everything'. However, as we learn more about how each type of RNA regulation affects others we can trim down our representations to give us a less detailed, but clearer picture.

RNA interaction networks have been studied using Bayesian analysis, integrating miRNA targeting information with expression profiles.[42] This is because predictions based on sequence may not be sufficient to determine the complex interactions of miRNA-mRNA pairs. Adding in information about the expression state (i.e., the conditions in which the RNA sample was taken) and grouping different expression profiles based on these states aids in network accuracy. In Bayesian network analysis,[42] the interactions between miRNAs and mRNAs are defined as dependencies of their states encoded in a graphical representation with miRNA and mRNA nodes connected by directed edges. The

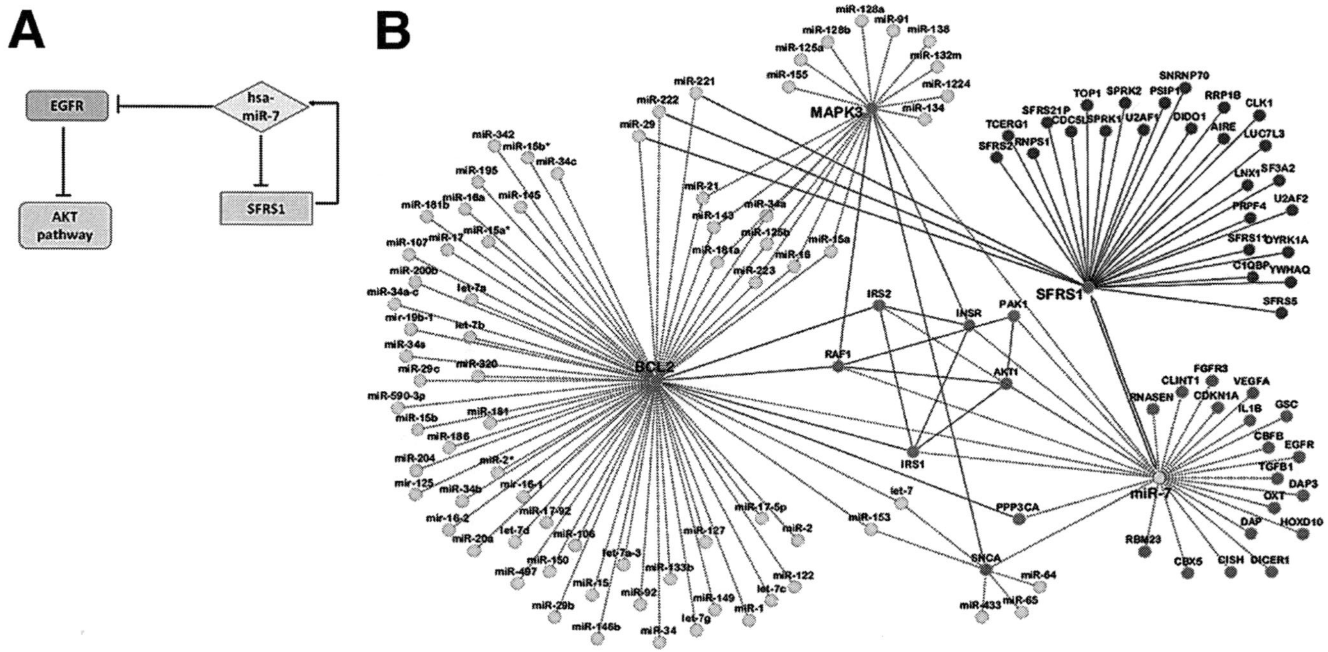

Figure 5. The relationship between hsa-miR-7 (miR-7) miRNA and two of its targets can be displayed in a simple manner as in (A), showing how miR-7 regulates the SFRS1 protein but in return the SFRS1 protein activates production of miR-7. Another protein EGFR is repressed by miR-7 which in return represses the AKT pathway showing a downstream effect of regulation. Network B shows how very quickly networks can get very complicated when other miR-7 regulated proteins are added into the network. The proteins connected to SFRS1 do not have their regulatory miRNAs attached but three other proteins BCL2, MAPK3 and SNCA have their regulatory miRNAs mapped showing how other miRNAs may also regulate some of the same proteins. There are many shared interactions between miRNAs that regulate EFFR and BCL2 which are not shown on this network. miRNAs are in light nodes and proteins are in dark nodes(spliceosomal proteins are darkest). PPI interactions are solid lines and RNA-Protein interactions are dashed lines. Interactions where SFRS1 enhances miRNA expression are herringbone lines. The feedback interaction between SFRS1 and miR-7 has a double line. miRNA interaction information was gathered from miRsel,[35] miRbase,[36] Rfam[37,38] and miRecords.[39] Network B visualised with Cytoscape 2.7.0.[31,32] miRNA names have been shortened by removing the ‘ha- in many cases for visualisation. A color version of this figure is available at www.landesbioscience.com/curie.

presence or absence of a directed edge from a miRNA to an mRNA indicates the state of the mRNA as dependent or independent on that of the miRNA implying a regulatory relationship. Observations of different relationship states (either A and B are independent or A regulates B) are taken from the expression data and the one that receives the highest score is used to represent the relationship. This method uses expression values between conditions, so we can determine if A is up, down, or mixed between states (up in one and down in another). This work was done with microarray data but can also be used for mRNA sequencing data, giving an example where incorporating other biological knowledge into protein-RNA networks can aid in the discovery of both strong and subtle interactions.[42]

SPATIOTEMPORAL DYNAMIC RNA NETWORKS

In order to investigate how RNAs really interact in cellular mechanisms, networks need to be analysed in a spatiotemporal manner (i.e., they must take in both the cellular location of the interaction and the timing/ordering of the interactions). This is especially so when studying multicellular organisms, where there are many signals passed between the cells. This extracellular signalling not only permits the cells to be organised, it also permits signals from the environment to be processed and acted upon. However, spatiotemporal organisation is not limited to multicellular situations. The eukaryotic cell is organised into numerous sub-compartments, some of them membrane bound (e.g., a mitochondria and nucleus) and others considered more dynamic (e.g., nucleolus and spliceosome). Prokaryotes also have sub-cellular localisation of RNP complexes [e.g., Ribosome and SRP (Signal recognition particle) which targets signal peptides and conducts them to the protein-conducting channel on the plasma membrane (or endoplasmic reticulum membrane in eukaryotes)].[50] Dynamic compartments in particular offer interesting insights into biological networks since they must form and disassociate at a particular points in the cell cycle or under set cellular conditions (e.g., the formation of a nucleolus around a nucleolar organiser region).

The spliceosome is also a dynamic compartment in which initial components recognise the exon-intron boundaries then other splicing components are recruited to complete the splicing catalysis steps to free the ligated exon. If we analyse interactions from large databases, many of them will have gene ontology, a library of terms describing function and cell localisation. However, these definitions are 'cell-wide' in which they tell us that a protein is involved in the spliceosome, but not which interactions are only present during the different splicing stages (Fig. 1). Using data from mass spectrometry experiments (Complex A,[22] Complex B,[23] Complex C[24]), PPI information downloaded from Biogrid[16] and RNA-Protein interactions from splicing-related publications,[17,51] we can examine the dynamic nature of RNP networks (Fig. 6). The initial proteins for the network in Figure 6 are U2-snRNP associated, since U2 snRNP remains within the spliceosome for the majority of the splicing cycle. In the first stage of splicing the exon-intron boundaries are recognised by the U1 and U2 snRNPs (Fig. 6A). The U5 and U6 snRNPs then join the spliceosome in Complex B (Fig. 6B). All but one of the proteins is present in the B complex when the actual working spliceosome is formed. The U4 and U1 snRNPs leave the spliceosome in complex C (Fig. 6C) and with them many of their associated proteins. This can leave a network with supposedly unconnected components. However, in order to keep these networks focused, our network in Figure 6 does not show all the connections that can occur with other splicing and nonsplicing proteins. It is also likely that other connections between splicing proteins and their RNAs have not yet been found

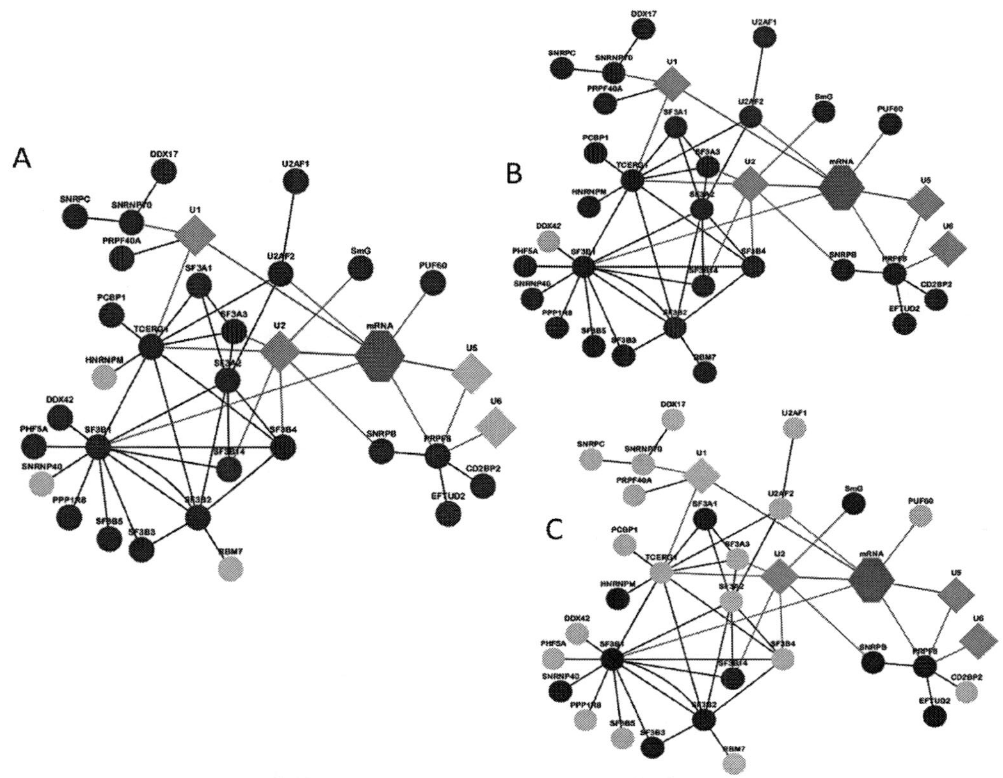

Figure 6. Protein-ncRNA and protein-protein interactions change during the splicing cycle. Inactive nodes are shown in light grey. Complex A (A) is where the splice sites of the mRNA are recognized. The U5 and U6 snRNPs then join the spliceosome in Complex B (B). The U4 and U1 snRNPs leave the spliceosome in complex C (C) and with them many of their interacting proteins. Mass spectrometry has determined which proteins are within the spliceosome at these stages (A,[22,61] B,[23] C[24]). PPI information was downloaded from Biogrid.[16] RNA-Protein interactions were gathered from splicing-based publications.[17,51,60] Networks visualised with Cytoscape 2.7.0.[31,32] A color version of this figure is available at www.landesbioscience.com/curie.

due to the limitations of mass spectrometry with RNA macromolecular complexes. The spliceosome is a massive complex where in humans over 200 proteins can be involved. Many of these proteins link to other transcriptionally related functions such as 5′ capping and 3′ tailing as well as RNA export from the nucleus. Detailed analysis of small sections of the spliceosome such as SF3 can greatly aid in eventually forming an overall picture of the larger macromolecular spliceosome.

CONCLUSION

The spliceosome is of great biological importance as there are direct links between splicing proteins and medical conditions such as Alzheimer's disease,[52] retinal disorders,[53] spinal muscular atrophy[54] and especially cancer.[55] Indeed the spliceosome is already under investigation as a target for anti-cancer treatment[56]). It is also important as a model for studying how RNA interactions influence and enhance our network analysis.

There are two main issues surrounding RNP networks; visualisation and interpretation. Common graphical visualization tools such as Cytoscape and BioLayout have evolved greatly over the years to allow us to connect Gene Ontology (functional definitions) and metabolic pathway (e.g., KEGG) information. It is clearly important to connect as much biological inference to the interactions as possible in order for network comparisons to make sense.[11,40] However when we come to interpretation, in general the larger the network, the harder it is to visualise then make sense of it. For accurate comparisons and predictions, we require accurate and biologically relevant networks that are large enough to describe the cellular situation, but not too large as to obscure the meaningful connections with noise. For the moment we can concentrate on sub-networks such as those with the spliceosome to gather the accuracy we require before expanding to larger networks.

Within the field of systems biology, there is a greater move to improve the accuracy of proteomic and transcriptomics linkage.[57] Issues surrounding transcriptomic analysis[58] and RNA granule sequestering of transcripts,[59] means that the proteins discovered in a sample by mass spectrometry may not match the transcripts deemed to be present by transcriptomic sequencing. One approach has been to break down the networks by designating 'proteomic seeds'[11] (a protein that is differentially expressed between two conditions) and using Synergistic dysregulation[11] (coordinate mRNA-level differential expression of a group of genes in the phenotype). Although this sounds impressive what it deals with is the output from two high-throughput technologies that can be used to tackle biological networks at a truly genomic scale. If for example, we also apply epigenetic and metabolomic[12] information to our networks this can result in layers upon layers of network complexity. Our challenge is not to let the enormity or complexity of these networks overwhelm us, but instead concentrate on future developments for genetic information integration, network construction and graph visualisation.

ACKNOWLEDGEMENTS

This work could not have been done without the support of Prof. David Penny and the support of the Institute of Fundamental Sciences at Massey University. This work was partly funded through the New Zealand Health Research Council.

REFERENCES

1. Collins LJ, Penny D. RNA-infrastructure: Dark Matter of the Eukaryotic Cell? Trends in Genetics 2009; 25(3):120-128.
2. Licatalosi DD, Darnell RB. RNA processing and its regulation: global insights into biological networks. Nat Rev Genet 2010; 11(1):75-87.
3. Amaral PP, Dinger ME, Mercer TR et al. The eukaryotic genome as an RNA machine. Science 2008; 319(5871):1787-1789.
4. Collins LJ, Chen XS, Schonfeld B. The Epigenetics of Non-coding RNA. In: Tollefsbol T, ed. Handbook of Epigenetics. Oxford: Academic Press; 2010:49-61.
5. Collins LJ, Chen XS. Ancestral RNA: The RNA biology of the eukaryotic ancestor. RNA Biology 2009; 6(5):1-8.
6. Costa FF. Non-coding RNAs: Lost in translation? Gene 2007; 386(1-2):1-10.
7. Costanzo M, Baryshnikova A, Bellay J et al. The genetic landscape of a cell. Science 2010; 327(5964):425-431.
8. Janga SC, Tzakos A. Structure and organization of drug-target networks: insights from genomic approaches for drug discovery. Mol Biosyst 2009.
9. MacLean D, Elina N, Havecker ER et al. Evidence for large complex networks of plant short silencing RNAs. PLoS One 5(3):e9901.
10. Li L, Zhang K, Lee J et al. Discovering cancer genes by integrating network and functional properties. BMC Med Genomics 2009; 2:61.
11. Nibbe RK, Koyuturk M, Chance MR. An integrative -omics approach to identify functional sub-networks in human colorectal cancer. PLoS Comput Biol 2010; 6(1):e1000639.
12. Yizhak K, Benyamini T, Liebermeister W et al. Integrating quantitative proteomics and metabolomics with a genome-scale metabolic network model. Bioinformatics 2010; 26(12):i255-260.
13. Tsai ZY, Singh S, Yu SL et al. Identification of microRNAs regulated by activin A in human embryonic stem cells. J Cell Biochem 2010; 109(1):93-102.
14. Jamalkandi SA, Masoudi-Nejad A. Reconstruction of Arabidopsis thaliana fully integrated small RNA pathway. Funct Integr Genomics 2009; 9(4):419-432.
15. Dobbyn HC, McEwan PA, Krause A et al. Analysis of pre-mRNA and prerRNA processing factor Snu13p structure and mutants. Biochem Biophys Res Commun 2007; 360(4):857-862.
16. Breitkreutz BJ, Stark C, Reguly T et al. The BioGRID Interaction Database: 2008 update. Nucleic Acids Res 2008; 36(Database issue):D637-D640.
17. Collins L, Penny D. Complex spliceosomal organization ancestral to extant eukaryotes. Mol Biol Evol 2005; 22(4):1053-1066.
18. Jurica MS, Moore MJ. Pre-mRNA splicing: awash in a sea of proteins. Mol Cell 2003; 12(1):5-14.
19. Valadkhan S. The spliceosome: a ribozyme at heart? Biol Chem 2007; 388(7):693-697.
20. Valadkhan S. Role of the snRNAs in spliceosomal active site. RNA Biol 2010;7(3):345-53.
21. Lorkovic ZJ, Lehner R, Forstner C et al. Evolutionary conservation of minor U12-type spliceosome between plants and humans. RNA 2005; 11(7):1095-1107.
22. Behzadnia N, Golas MM, Hartmuth K et al. Composition and three-dimensional EM structure of double affinity-purified, human prespliceosomal A complexes. EMBO J 2007; 26(6):1737-1748.
23. Deckert J, Hartmuth K, Boehringer D et al. Protein composition and electron microscopy structure of affinity-purified human spliceosomal B complexes isolated under physiological conditions. Mol Cell Biol 2006; 26(14):5528-5543.
24. Ilagan J, Yuh P, Chalkley RJ et al. The role of exon sequences in C complex spliceosome structure. J Mol Biol 2009; 394(2):363-375.
25. Yamada T, Bork P. Evolution of biomolecular networks: lessons from metabolic and protein interactions. Nat Rev Mol Cell Biol 2009; 10(11):791-803.
26. Dyer MD, Neff C, Dufford M et al. The human-bacterial pathogen protein interaction networks of Bacillus anthracis, Francisella tularensis and Yersinia pestis. PLoS One 5(8):e12089.
27. Rito T, Wang Z, Deane CM et al. How threshold behaviour affects the use of subgraphs for network comparison. Bioinformatics 26(18):i611-i617.
28. MacLean D, Elina N, Havecker ER et al. Evidence for large complex networks of plant short silencing RNAs. PLoS One 2010; 5(3):e9901.
29. Assenov Y, Ramirez F, Schelhorn SE et al. Computing topological parameters of biological networks. Bioinformatics 2008; 24(2):282-284.
30. Killcoyne S, Carter GW, Smith J et al. Cytoscape: a community-based framework for network modeling. Methods Mol Biol 2009; 563:219-239.
31. Shannon P, Markiel A, Ozier O et al. Cytoscape: a software environment for integrated models of biomolecular interaction networks. Genome Res 2003; 13(11):2498-2504.

32. Cline MS, Smoot M, Cerami E et al. Integration of biological networks and gene expression data using Cytoscape. Nat Protoc 2007; 2(10):2366-2382.
33. Moretti F, Thermann R, Hentze MW. Mechanism of translational regulation by miR-2 from sites in the 5′ untranslated region or the open reading frame. RNA 2010; 16(12):2493-502.
34. Nacher JC, Araki N. Structural characterization and modeling of ncRNA-protein interactions. Biosystems 2010; 101(1):10-9.
35. Naeem H, Kuffner R, Csaba G et al. miRSel: automated extraction of associations between microRNAs and genes from the biomedical literature. BMC Bioinformatics 2010; 11:135.
36. Griffiths-Jones S. miRBase: microRNA sequences and annotation. Curr Protoc Bioinformatics 2010; Chapter 12:Unit 12 19 11-10.
37. Gardner PP, Daub J, Tate JG et al. Rfam: updates to the RNA families database. Nucleic Acids Res 2009; 37(Database issue):D136-D140.
38. Daub J, Gardner PP, Tate J et al. The RNA WikiProject: community annotation of RNA families. RNA 2008; 14(12):2462-2464.
39. Xiao F, Zuo Z, Cai G et al. miRecords: an integrated resource for microRNA-target interactions. Nucleic Acids Res 2009; 37(Database issue):D105-D110.
40. Rito T, Wang Z, Deane CM et al. How threshold behaviour affects the use of subgraphs for network comparison. Bioinformatics 2010; 26(18):i611-i617.
41. Przulj N, Corneil DG, Jurisica I. Efficient estimation of graphlet frequency distributions in protein-protein interaction networks. Bioinformatics 2006; 22(8):974-980.
42. Liu B, Li J, Tsykin A et al. Exploring complex miRNA-mRNA interactions with Bayesian networks by splitting-averaging strategy. BMC Bioinformatics 2009; 10:408.
43. Vasudevan S, Tong Y, Steitz JA. Switching from repression to activation: microRNAs can up-regulate translation. Science 2007; 318(5858):1931-1934.
44. Redner S. Networks: teasing out the missing links. Nature 2008; 453(7191):47-48.
45. Golas MM, Sander B, Will CL et al. Luhrmann R, Stark H. Molecular architecture of the multiprotein splicing factor SF3b. Science 2003; 300(5621):980-984.
46. Das BK, Xia L, Palandjian L et al. Characterization of a protein complex containing spliceosomal proteins SAPs 49, 130, 145 and 155. Mol Cell Biol 1999; 19(10):6796-6802.
47. Lardelli RM, Thompson JX, Yates JR 3rd et al. Release of SF3 from the intron branchpoint activates the first step of pre-mRNA splicing. RNA 16(3):516-528.
48. Wu H, Sun S, Tu K et al. A splicing-independent function of SF2/ASF in microRNA processing. Mol Cell 2010; 38(1):67-77.
49. Kefas B, Godlewski J, Comeau L et al. microRNA-7 inhibits the epidermal growth factor receptor and the Akt pathway and is down-regulated in glioblastoma. Cancer Res 2008; 68(10):3566-3572.
50. Janda CY, Li J, Oubridge C et al. Recognition of a signal peptide by the signal recognition particle. Nature 465(7297):507-510.
51. Collins L, Penny D. Investigating the intron recognition mechanism in eukaryotes. Mol Biol Evol 2006; 23(5):901-910.
52. Ohe K, Mayeda A. HMGA1a trapping of U1 snRNP at an authentic 5′ splice site induces aberrant exon skipping in sporadic Alzheimer's disease. Mol Cell Biol 30(9):2220-2228.
53. Sun N, Zhao H. Reconstructing transcriptional regulatory networks through genomics data. Stat Methods Med Res 2009; 18(6):595-617.
54. Pedrotti S, Bielli P, Paronetto MP et al. The splicing regulator Sam68 binds to a novel exonic splicing silencer and functions in SMN2 alternative splicing in spinal muscular atrophy. EMBO J 2010; 29(7):1235-1247.
55. Lee JH, Horak CE, Khanna C et al. Alterations in Gemin5 expression contribute to alternative mRNA splicing patterns and tumor cell motility. Cancer Res 2008; 68(3):639-644.
56. van Alphen RJ, Wiemer EA, Burger H et al. The spliceosome as target for anticancer treatment. Br J Cancer 2009; 100(2):228-232.
57. Zdobnov EM, von Mering C, Letunic I et al. Comparative genome and proteome analysis of Anopheles gambiae and Drosophila melanogaster. Science 2002; 298(5591):149-159.
58. Bullard JH, Purdom E, Hansen KD et al. Evaluation of statistical methods for normalization and differential expression in mRNA-Seq experiments. BMC Bioinformatics 2010; 11:94.
59. Anderson P, Kedersha N. RNA granules. J Cell Biol 2006; 172(6):803-808.
60. Dybkov O, Will CL, Deckert J et al. U2 snRNA-protein contacts in purified human 17S U2 snRNPs and in spliceosomal A and B complexes. Mol Cell Biol 2006; 26(7):2803-2816.
61. Behzadnia N, Hartmuth K, Will CL et al. Functional spliceosomal A complexes can be assembled in vitro in the absence of a penta-snRNP. RNA 2006; 12(9):1738-1746.

CHAPTER 7

CONSTRUCTION, STRUCTURE AND DYNAMICS OF POST-TRANSCRIPTIONAL REGULATORY NETWORK DIRECTED BY RNA-BINDING PROTEINS

Sarath Chandra Janga*[,1,3] and Nitish Mittal[2]

[1]MRC Laboratory of Molecular Biology, Cambridge, UK; [2]Department of Biotechnology, National Institute of Pharmaceutical Education and Research, SAS Nagar, Punjab, India; [3]Institute of Genomic Biology, University of Illiniois at Urbana Champaign, Urbana, IL, USA

**Corresponding Author: Sarath Chandra Janga—Email: sarath@mrc-lmb.cam.ac.uk*

Abstract: Gene expression is a highly controlled process which is known to occur at several levels in eukaryotic organisms. Although messenger RNAs have been traditionally viewed as passive molecules in the pathway from transcription to translation, there is increasing evidence that their metabolism is controlled by a class of proteins called RNA-binding proteins (RBPs). In this chapter, we provide an overview of the recent developments in our understanding of the repertoire of RBPs across diverse model systems and discuss the approaches currently available for the construction of post-transcriptional networks governed by them. We also present the first analysis of the network properties of a post-transcriptional system in a model eukaryote using currently available data and discuss the implications of understanding the dynamic properties of this important class of regulatory molecules as more data detailing their dynamic, spatial and tissue-specific maps across diverse model systems accumulates. We argue that such developments would not only allow us to gain a deeper understanding of regulation at a level that has been under-appreciated over the past decades, but would also allow us to use the newly developed high-throughput approaches to interrogate the prevalence of these phenomena in different states and thereby study their relevance to physiology and disease across organisms.

RNA Infrastructure and Networks, edited by Lesley J. Collins.

INTRODUCTION

Gene expression is a highly regulated process controlled at several levels. In all organisms from the prokaryote *Escherichia coli* to the multi-cellular eukaryote human, gene expression is first regulated at the transcriptional level where transcription factors facilitate the RNA synthesis in response to internal or external stimuli.[1-4] On the other hand, at the protein level, several post-translational modifications, such as phosphorylation by kinases, sumoylation by ubiquitin ligases and acetylation by acetyl-transferases, are known to spatially and temporally control the availability of functional protein products within the cell. However, a much less understood level of regulation of gene expression, which occurs between these two layers, is the post-transcriptional control of RNAs. In contrast to prokaryotes where transcription and translation are coupled, in eukaryotes transcription usually takes place in the nucleus and translation in the cytoplasm. This uncoupling of transcription and translation provides an opportunity for an additional layer of gene expression control at post-transcriptional level in eukaryotes. The presence of this post-transcriptional control is also evidenced by a number of studies which showed that in general, there is a poor correlation between the mRNA and protein pools in eukaryotic cells.[5-7] It is now increasingly accepted that this level is controlled by a complex interplay of numerous RNA associated factors with the major players being the RNA-binding proteins (RBPs).[8-11] In this chapter, we provide a comprehensive overview of this quickly developing area of post-transcriptional regulatory networks formed by RBPs. We organize it into three major sections namely, (a) sequence attributes and functional processes associated with RBPs, (b) methods used for the construction of the networks formed by them and, finally, (c) discuss the structure and dynamics of these post-transcriptional networks based on recent publicly available data.

RNA BINDING PROTEINS AND POST-TRANSCRIPTIONAL REGULATION

RNA binding proteins (RBPs) are key regulators of different steps in the metabolism of RNA in eukaryotes. As shown in Figure 1, they participate in the processing of pre-mRNA which includes splicing, poly-adenylation and capping to produce mature mRNA. Following which, they are responsible for mediating the transport of mRNA from nucleus to cytoplasm. RBPs are also found to facilitate and control the localization, translation, stability and degradation of mRNA (see Fig. 1). To regulate the different steps of RNA metabolism, RBPs bind to RNA and form ribonucleoprotein complexes (RNP). Depending upon whether RBPs are bound to pre-mRNA or mRNA, RNPs are classified as Heteronuclear RNPs (hnRNPs) or messenger RNPs (mRNPs) respectively. RNPs are inherently highly dynamic complexes due to their ability to associate and dissociate with various RBPs to mediate different steps of RNA metabolism. Some RBPs associated with RNP complexes are known to remain bound to their target RNA during all the steps of the RNA processing, from splicing to translation. For instance, SF2/ASF, a member of the SR class of RBPs in mammals, is found to facilitate splicing, export and translation initiation of its target RNA.[12,13] Similarly Npl3, a yeast SR protein, has also been shown to interact with pre-mRNA and regulate the events from splicing to translational elongation.[14] Yet another example is that of the neuronal ELAV protein which regulates the fate of its target RNA by mediating the events from poly-adenylation to translation.[15] On the other hand, several RBPs are also responsible for participating

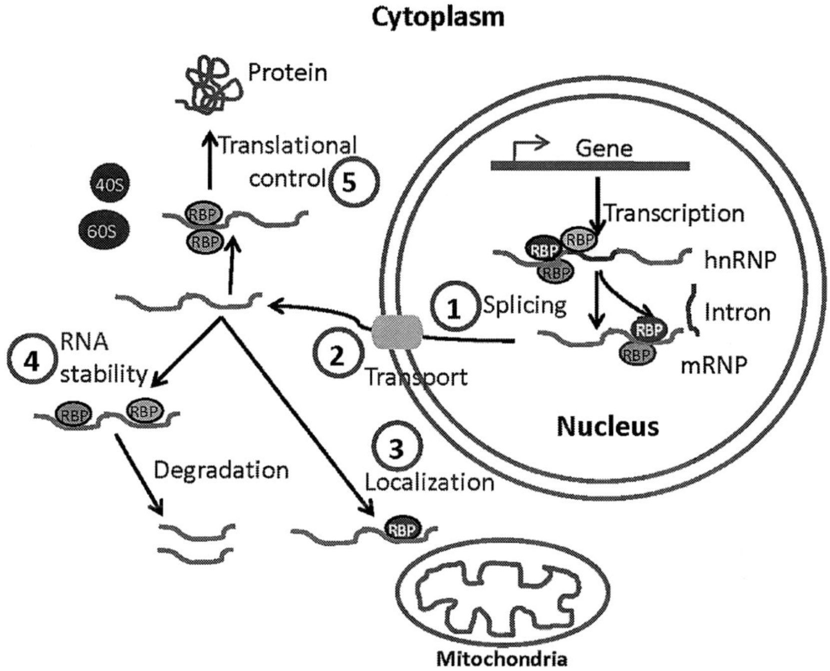

Figure 1. Schematic diagram showing the extensive role of RBPs in various post-transcriptional processes at different locations in eukaryotic cells. The circled number indicates the process in which RBPs are involved. RBPs are major players in splicing pre-mRNAs into mature mRNAs in the nucleus which are then exported into the cytoplasm by various other RBPs. Depending upon whether RBPs are bound to pre-mRNA or mRNA to form a ribonuclear protein (RNP) complex, RNPs are classified as Heteronuclear RNPs (hnRNPs) or messenger RNPs (mRNPs) respectively. In addition, RBPs are responsible for the localization of mRNAs to distinct sub-cellular compartments such as the mitochondria. In the cytoplasm, RBPs are also involved in governing the stability of transcripts by binding the substrate RNAs and in controlling the translation of mRNAs into corresponding protein products. Often multiple RBPs can bind to a single RNA at one or more locations giving rise to a plethora of combinatorial possibilities at every step of post-transcriptional control. For this reason, RBPs have been found to be playing a major role in the cause of several disorders due to changes in regulation they bring about at post-transcriptional level.

in specific steps of RNA metabolism such as the Nova protein, which is associated with splicing in neuronal cells.[16,17] Tap protein, like its yeast homolog Mex67, was reported to be a *bona fide* mRNA nuclear export factor.[18] All these examples highlight (1) the role of RBPs in regulating the expression of genes in multiple steps at post-transcriptional level and (2) the complex combinatorial interplay of different RBPs to integrate various post-transcriptional events to fine tune the availability of transcripts both spatially and temporally.

RBPs bind to their RNA targets with the help of several domains having different specificity and affinity. Some of the most common domains are RRM (RNA recognition motif), KH (K homology domain), SR (serine arginine domain), Zn-finger, Pumilio/FBF (PUF domain) and Sm.[8] Table 1 shows the most frequently occurring RNA binding domains in the yeast, *Saccharomyces cerevisiae*, along with the commonly appearing

Table 1. Common RNA binding domains in the putative list of RBPs from yeast, *S. cerevisiae*[19] along with their frequency of occurrence. Also shown are the domains most often associated with these RNA binding domains according to the Pfam[62] domain database

Domain	Pfam Accession	Description	Protein Frequency	Frequently Occurring Partner Domains
RRM_1	PF00076	RNA recognition motif (RRM). Many eukaryotic proteins containing one or more copies of this putative RNA-binding domain of about 90 amino acids. They are known to bind single-stranded RNAs.	0.105	RRM_1, Lsm_ interact
DEAD	PF00270	DEAD/DEAH box helicase. Members of this family include the DEAD and DEAH box helicases.	0.042	Helicase C,
KH_1	PF00013	K homology (KH) is a domain of 70 amino acids and is present in diverse RBPs.	0.015	KH_1
PUF	PF00806	Pumilio-family RNA binding repeat. Puf domain usually occurs as a tandem repeat of eight domains.	0.013	PUF, RRM_1
WD40	PF00400	WD-40 repeats (also known as WD or beta-transducin repeats) are short ~40 amino acid motifs, often terminating in a Trp-Asp (W-D) dipeptide.	0.013	WD40

partner domains in the conventional list of 560 RBPs reported recently by Hogan and coworkers.[19] A large number of proteins have been predicted to be RBPs in several model organisms including humans on the basis of the presence of these commonly occurring domains. A list of approximate number of RBPs identified in different model organisms is shown in Table 2 along with a reference to the study reporting it. For instance, in *Caenorhabditis elegans* approximately 500 proteins are annotated as RBPs on the basis of the presence of one or more RNA binding domains. In the yeast, *S. cerevisiae* about 560 proteins have been reported as putative RBPs till date (Table 2). In human, more than 1000 proteins are considered to be RBPs, with 497 containing at least one RRM domain.[20] Other than these putative RBPs (on the basis of previously known RNA binding domains), several metabolic enzymes have also been shown to bind to RNA molecules.[21] For example Aco1, a TCA cycle enzyme, in yeast *S. cerevisiae* binds to several RNAs encoded by the mitochondrial genome.[19] Likewise, recent studies have also shown the ability of RBPs to bind to DNA, suggesting that some of the known RBPs might act as unconventional DNA-binding proteins.[22] These examples indicate the potential for the existence of novel classes of RBPs in eukaryotes with yet to be discovered functional roles.

Table 2. Putative number of RBPs reported in different organisms

Organism	Putative RBPs	Approximate Number of Genes	Reference
S. cerevisiae (Yeast)	561	7000	19
C. elegans (Nematode)	500	20000	63
D. melanogaster (Fruitfly)	300	13290	64
Mus musculus (Mouse)	380	28287	65
Homo sapiens (Human)	800	30000	66

METHODS TO IDENTIFY RBPs AND THEIR TARGETS

Although, several RBPs have been identified on the basis of conservation of domains in different organisms, targets of these RBPs are poorly understood. Therefore, several methods have been employed to identify the targets of RBPs, both in vitro and in vivo. A list of some commonly used methods for identification of RBP targets have been described in Table 3. Traditionally, RNA targets for known RBPs have been identified in vitro by using cross-linking immunoprecipitation followed by electromobility shift assays.[23,24] More recently, one hybrid[25] and three hybrid assays[26] have been used to identify the interaction of an RBP and RNA molecule in vivo. But these traditional methods have limitations in their ability to identify new targets on a genomic scale. Therefore, other in vivo assays have been developed to identify the novel targets of a RBP such as ultraviolet (UV) cross-linking and immunoprecipitation (CLIP) and RNP immunoprecipitation-microarray (RIP-ChIP). These high-throughput immunoprecipitation assays usually work on a similar concept where in (i) the RBP complex together with its target RNAs is first extracted and (ii) the target RNA identified. However, they differ in the procedure used for extracting RBP-RNA complexes and identification of target RNAs. For instance, in ultraviolet (UV) cross-linking and immunoprecipitation (CLIP) method, cells are exposed to ultraviolet light to crosslink RBP-RNA molecules inside the cells. Then cells are lysed and cross-linked RBP-RNA complexes are immunoprecipitated using an antibody against the RBP of interest. Further, RNA is isolated from the complexes and identified by RT-PCR. For instance, in a study to discover the targets of the splicing factor Nova, thirty four transcripts were identified by using the CLIP method.[16] In contrast, in the RNP immunoprecipitation-microarray (RIP-ChIP) method, instead of treating cells with UV light to crosslink RBP-RNA complexes, cells are lysed directly and native RBP-RNA complexes for the RBP of interest are purified from the cell lysate using immunoprecipitation. Following this, RNA is isolated from the complexes and identified by using high-density oligonucleotide microarrays. The targets of the Puf family of RBPs and other RBPs in *S. cerevisiae* have been identified by using modified RIP-Chip method, where tandem affinity tagged (TAP) RBPs were used to facilitate the immunoprecipitation step.[19,27] These studies showed that the RNA targets vary from 1-1300 approximately for the studied RBPs in yeast. For instance Nop13, responsible for pre18s rRNA processing, has 2 RNA targets whereas Npl3 and Mex67, both involved in mRNA export, have 1266 and 1150 RNA targets respectively.[19,28]

Another fundamental area of exploration in elucidating post-transcriptional networks is the identification of the repertoire of RBPs across organisms and several approaches both computational and experimental have been developed in recent years. Computational approaches involve the identification of the set of protein-coding genes containing the

Table 3. Different methods to identify novel RBPs, their targets or RBP-RNA interactions

Method	Description	Reference
Three hybrid	In vivo yeast genetic method to detect and analyze the RNA-RBP interaction of known RNA and RBPs. This method is based on the binding of bifunctional RNA to both of the two hybrid proteins which activates the expression of reporter gene.	26
RNA compete	In vitro identification of RNA binding specificity of a RBP. High concentration of RNA is used and incubated with tagged RBP. This high concentration provides competition for binding and hence this technique gets its name. RBP-RNA complexes are purified and a microarray is used to identify the specific binding sites of RBP.	67
RIP-ChIP	In vivo identification of RNA targets of a RBP of interest. Cells are lysed and RBP-RNA complexes are immunoprecipitated in native state. Target RNA is extracted from the RBP-RNA complexes and identified by the microarray method where total RNA of the cell is used as a control.	68
CLIP	In vivo identification of RNA targets for RBP of interest. Cells are treated with ultraviolet light to covalently crosslink RBP-RNA complexes. Cells are then lysed and RBP-RNA complexes are immunoprecipitated and the RNA identified by RT PCR.	16
PAIR	In vivo identification of novel RBPs. A mRNA binding PNA probe is delivered into the cells. Cells are then exposed to ultraviolet light that enables PNA to bind with RBP. Cells are lysed and PNA-RNA-RBP complexes are immunoprecipitated and RBPs are identified by mass spectrometry.	34
SERF	In vitro selection of RNA fragments that bind to RBP. A random pool of fragmented RNA is generated and incubated with RBP in a test tube. The RBP-RNA complex is extracted by filtration on nitrocellulose membrane. Selection cycle is repeated several times and selected RNA fragments are cloned and identified by the consensus sequences binding to RBP.	69
TRAP	In vivo system for identification of RNA-RBP interactions in yeast. This involves the transformation of reporter mRNA encoding GFP protein and expression of RBP of interest. Fluorescence intensity of the GFP is measured to know the binding of the RBP of interest. High affinity interactions lead to low expression and low fluorescence intensity.	70
SNAAP	In vitro method used to identify mRNAs bound to specific RBPs. Purified tagged RBP is treated with cell lysate. This is followed by immunoprecipitation of the mRNP using antibody against tag. Target mRNA is identified by the differential display method	71
Quantitative proteomics	In vitro method to identify RBPs bound to specific RNA sequences. An RNA aptamer is tagged with an RNA sequence then incubated with cell lysate. The RNA aptamer-RNA-RBPs complex is purified and RBPs are identified by using mass spectrometry.	72

bonafide RNA-binding domains. This step is usually followed by manual curation of the collected set to identify a high confidence set of RBPs.[19,29] Experimental techniques comprise of employing the protein chip of an organism of interest to probe for the potential binding of the cellular RNA molecules and are analogous to the attempts at characterizing the repertoire of DNA-binding proteins.[22,30-32] Another strategy developed to identify the RBPs attached to known RNA molecules is the Peptide nucleic acid (PNA)-Assisted Identification of RBPs (PAIR).[33] This assay utilizes a specific mRNA binding probe (PNA) that has the ability to cross the cell membrane and bind to the RNA of interest. This probe also contains the photoactivable amino acid adduct *p*-benzophenylalaline (Bpa) which can covalently crosslink with the RBP, associated with the RNA, on photoactivation. After delivery of PNA, cells are exposed to ultra violet light for crosslinking of PNA to RBPs associated with RNA of interest. Cells are then lysed, treated with RNase and PNA-RBP adducts are isolated by using sense oligo (bound to PNA)-coupled magnetic beads. Following which RBPs are identified by mass spectrometry. This method has been used to identify the RBPs associated with ankylosis (ank) RNA, a panneuronal dendritically localized RNA.[34] The ank RNA encodes for an inorganic pyrophosphate transporter and its mutation is known to cause a generalized, progressive form of arthritis accompanied by mineral deposition, formation of bony outgrowths and joint destruction.[35] The dendritic localization of RNAs is a rare event occurring for only ≈5% of the cellular RNAs.[36] Therefore, given the importance of the ank protein in the periphery, as well as its distinct neuronal subcellular localization, identification of RBPs that complex with the ank mRNA forms an important step in elucidating the post-transcriptional events controlled by them.

RBPs AND POST-TRANSCRIPTIONAL OPERONS

In prokaryotes, it has been long known that the genes involved in similar processes tend to cluster on chromosomes and are transcribed together using the same promoter thus forming operons such as the well studied, Gal, Lac operons. On the other hand, in eukaryotes, chromosomal operons are rare. However, recently the concept of post-transcriptional operons has been proposed in eukaryotes,[37] which has become possible due to the availability of the wealth of information on RBP-RNA interactions. According to this concept, diverse RNAs related to a common biological process are regulated by similar RBPs. For instance, in yeast, a study of the RBP-RNA interactions by modified RIP-Chip method has revealed that each member of Puf family RBPs bind with functionally and cytotopically related RNAs.[27] Puf1 and Puf2 have been shown to bind to mRNAs of membrane associated proteins. Similarly, Puf3 binds to cytoplasmic mRNAs of mitochondrial proteins. Likewise, the Nova protein was found to regulate splicing of pre-mRNA encoding components of inhibitory synapses and a stem loop binding protein (SLBP) was involved solely in splicing and translation of replication dependent histone RNAs.[38] Further examples in support of post-transcriptional operons have been reviewed extensively elsewhere.[10,39] These examples demonstrate the role of RBPs in view of post-transcriptional operons for coordinating the expression of functionally related genes in eukaryotes. Given the advantages involved in spatially and temporally controlling the post-transcriptional events by selectively binding to compartment or location or process specific transcripts, it is possible to speculate that in eukaryotes RBPs act as mediators in facilitating operon-like organization, which by the virtue of polycistronic transcripts and coupled transcription/translation is inherent in prokaryotes.

POST-TRANSCRIPTIONAL NETWORK FORMED BY RBPs

The development of several high throughput approaches such as CLIP and RIP-ChIP has increased the amount of data for targets of RBPs in diverse organisms. This data of RBPs and their targets can be utilized to construct RBP-RNA interaction network which is also typically referred to as post-transcriptional regulatory network. A post-transcriptional network is represented in the form of a directional network with each edge corresponding to a regulatory link between the nodes as shown in Figure 2A. In this directed network, one set of nodes are RBPs forming the regulatory proteins, while the other set of nodes are RNAs encoded by either protein-coding or nonprotein coding genes referred to as the target nodes. These two nodes (regulator node and target node) are joined by an arrow starting at the regulator node and directing towards the target node. The target RNA may belong to diverse functional proteins including other RBPs. This network can also contain loops as a link starting from RBP and targeting itself, typically referred to as autoregulation of an RBP (Fig. 2B). This loop structure suggests that RBP can bind to its own RNA and control its metabolism at transcript level. There are several examples suggesting the auto-regulation of RBPs at post-transcriptional level. For instance, in humans, RBPs such as AUF1, HuR, KSRP, NF90, TIA-1 and TIAR were reported to associate with their own mRNA and other RBPs.[40]

The availability of the network of post-transcriptional interactions for a considerable fraction of RBPs in model systems such as *S. cerevisiae*,[19] makes it possible to address several questions concerning the structure and organization of post-transcriptional networks directed by RBPs. Table 4 summarizes some of the properties which govern the structure of this network. It is evident from this table that the majority of the mRNA transcriptome encoded by about ~70% of the genes, has significant associations with at least one of the RBPs screened for RNA interactions. In fact, on average, each distinct yeast mRNA was found to interact with three of the RBPs, suggesting the potential for

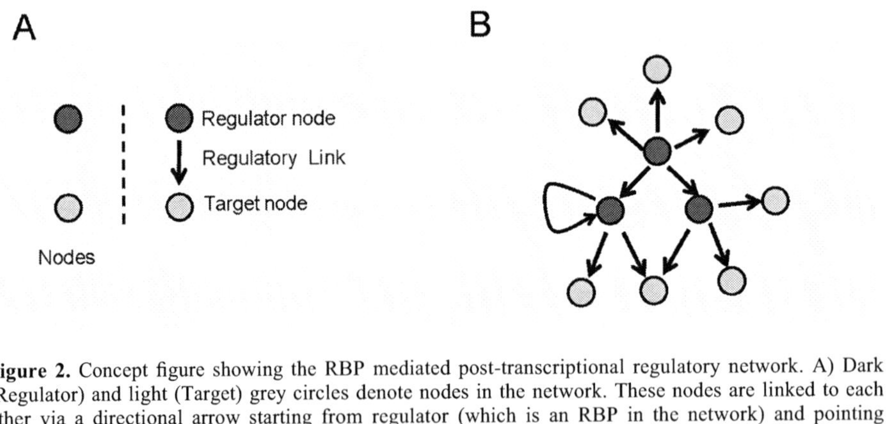

Figure 2. Concept figure showing the RBP mediated post-transcriptional regulatory network. A) Dark (Regulator) and light (Target) grey circles denote nodes in the network. These nodes are linked to each other via a directional arrow starting from regulator (which is an RBP in the network) and pointing towards target (which may be an RNA or miRNA) in the directional network. These linked nodes simply indicate that RBP (Dark grey circle) binds to RNA/miRNA of target gene (Light grey circle) and regulates its metabolism. B) Shows an illustration of a network representing a dense set of RBP-RNA interactions with different RBPs having diverse targets. The targets of one RBP in the network may be RNA of other genes or miRNA (dark and light circle linked by arrow), the RNA of the RBP itself (loop from dark circle) and the RNA of other RBPs (two dark circles linked by an arrow).

Table 4. Properties defining the structure of the post-transcriptional network of RBPs and their target RNAs in the model eukaryote, *S. cerevisiae*. The dataset employed for characterizing the network structure was obtained from Hogan et al.[19] and all the network properties are calculated using *igraph*, a publicly available R package for analyzing graphs [http://cneurocvs.rmki.kfki.hu/igraph/ and http://www.r-project.org]

Property	Definition	Value*
No. of edges	Each edge corresponds to a single RBP-RNA interaction. Hence, total edges represent all the interactions in the post-transcriptional network.	19396
No. of vertices/ nodes	Total number of nodes, which comprise of both the RBPs as well as the RNAs, encoding for both protein coding and noncoding genes. This network comprises of 41 RBPs which are screened for their RNA targets.	5398
Degree or connectivity	Degree or connectivity refers to the number of interactions a protein or RNA has in this network—the higher the connectivity (*i.e.*, hub nodes) the more the number of targets and/or more the number of RBPs controlling it.	7.18
Clustering coefficient	Clustering coefficient of a node reflects the extent to which the neighbors of a given node are interconnected among themselves to what is expected theoretically and indicates the cohesiveness or local modularity of the network. Average value taken over all nodes reflects the modularity of the network.	0.37
Betweenness	Betweenness centrality of a node measures the number of shortest paths between all pairs of nodes in the network that pass through a node of interest—the higher the number of paths that pass through a node, the more important it is.	43.11
Average path length	Average length of the shortest paths between all pairs of nodes in the network.	2.65
Closeness	Closeness centrality is defined as the inverse of the average length of all the shortest paths from a node of interest to all other nodes in the network—note that closeness centrality defined this way implies that higher the closeness value, the higher the importance (centrality) of a node.	0.38
Diameter	The diameter of a network is the length of the longest path among all the shortest paths defined between two nodes. It gives an estimation of the farthest distance between nodes in the network.	6
Graph density	The density of a network is the ratio of the number of edges to the number of total possible edges.	$1.33x10^{-3}$
Power law fit (exponent-alpha)	Fitting a power-law distribution function to the degree distribution of the network to study whether the network is likely to exhibit a scale-free network structure.	1.77

* Note that average values for the entire network are reported for properties which are defined for specific node or edge.

a combinatorial and multidimensional network of regulation. Indeed, we found that the average connectivity of a node in this network was ~7 indicating that most nodes in this network have more number of targets and/or more number of RBPs controlling them. Other measures of centrality like betweenness and closeness which provide a measure of the importance of a node in a network, shown in this table, also reflect this trend [see[41] and references therein for comprehensive definitions]. For instance, the average length of the shortest path between two nodes in this network which gives an indication of the distance between nodes suggests that most nodes are separated by no more than 3 edges—a measure reflecting the dense networking in this network.[42] Similarly, the diameter of a network which refers to the longest of all the shortest paths between pairs of nodes is about 6, indicating that two nodes in this network are separated by no more than 6 edges. The clustering coefficient which is a proxy for the modularity of the network shows that neighbors of most nodes tend to be highly interconnected among themselves, forming a dense and cohesive network of regulatory linkages at this level of regulation. Finally, although incomplete in size, the scaling exponent of this network is about 1.8 which suggests that the network might obey a scale-free topology with a power-law degree distribution.[42] In simple terms, scale-free topology of a network refers to a degree distribution of nodes, where most nodes in the network have low degrees while few nodes are very highly connected and are referred to as hubs in the network.

EXPRESSION DYNAMICS OF RBPs IN POST-TRANSCRIPTIONAL NETWORKS

Due to their central role in controlling gene expression at the post-transcriptional level, alteration in expression or mutations in either RBPs or their RNA targets (i.e.,the transcripts which physically associate with the RBP) have been reported to be the cause of several human diseases such as muscular atrophies, neurological disorders and cancer.[43-46] In particular, disorders such as myotonic dystrophy (DM) and oculopharyngeal muscular dystrophy (OPMD) have been attributed with RNA's gain-of-function. For instance, CUG repeat expansion in the case of myotonic dystrophy protein kinase (DMPK)[45] and GCG repeat expansion in exon 1 of the RBP, PABPN1 in the case of OPMD[43] are illustrated examples of RNA's gain-of-function. On the other hand, diseases like opsoclonus-myoclonus ataxia (POMA) and spinal muscular atrophy (SMA) have been reported to be due to the RBPs loss of function,[43] suggesting that mutations in either RBP or any of its interacting RNA target sequences can lead to extensive variations in their expression patterns and result in a number of diseases. In addition to the fitness defects that variations in RBPs can bring about in cells, it has been recently shown in yeast that RBPs form an important class of prionogenic proteins.[47] Prions are proteins that convert between structurally and functionally distinct states, which are capable of existing in multiple heritable conformations by associating with distinct phenotypes.

These observations raise the questions: are RBPs finely controlled in terms of their expression patterns and if so, are there constraints on their expression levels depending on the number of distinct RNA targets they control? To address these questions, a recent study analyzed the post-transcriptional network formed by RBPs in yeast at two distinct levels.[48] The first involved asking whether RBPs as a group show distinct dynamic properties in comparison to nonRBPs in the whole genome. The second comprised of understanding the constraints placed on dynamic properties of RBPs in relation to the number of distinct transcripts controlled by them. Analysis at the first level according to this study showed that RBPs, as a functional class, are rapidly turned over (i.e.,less stable) at the transcript level and are tightly controlled at the protein level (Fig. 3). Analysis of the post-transcriptional network formed by RBPs unambiguously revealed that highly connected RBPs are more abundant and ubiquitously present within the cell.[48]

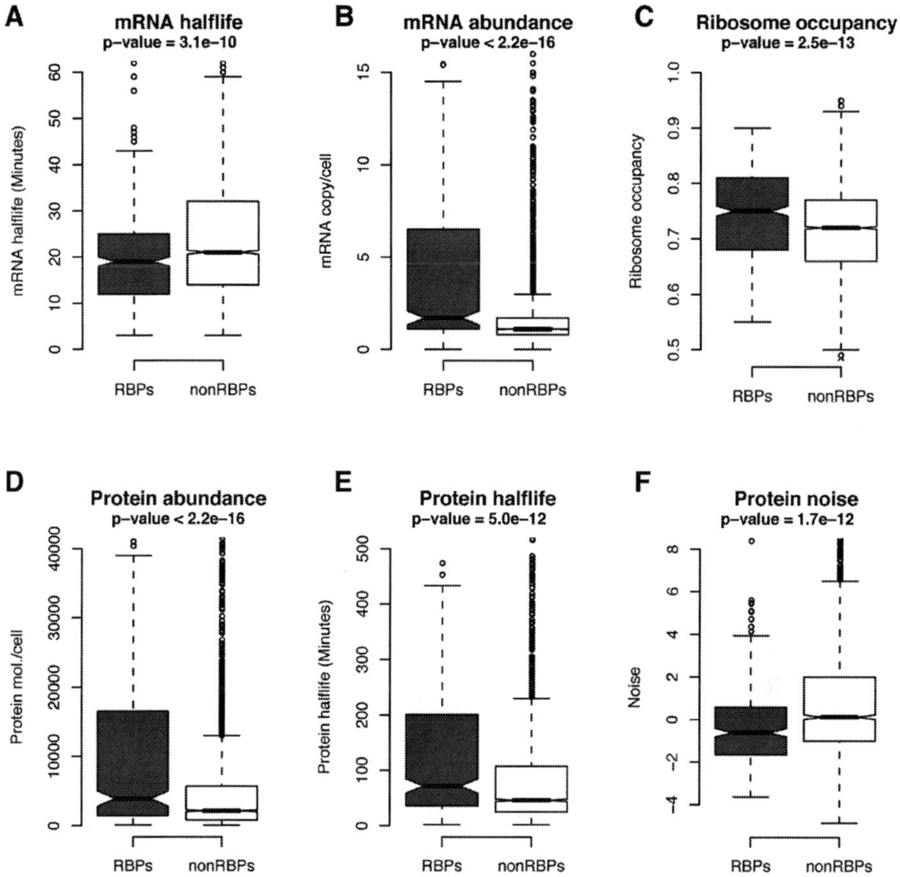

Figure 3. Comparing expression dynamics of RBPs with nonRBPs from yeast as defined by Mittal et al.[48] Box-plots showing the distribution of values for various regulatory properties for the two different groups of proteins (RBPs and nonRBPs) in *S. cerevisiae*. Dark (blue) and light (red) colored bars correspond to RBP and nonRBP populations respectively. A) mRNA half-life B) mRNA copy number C) Ribosome occupancy D) Protein abundance E) Protein half-life F) Protein noise. These datasets were obtained from previously published high-throughput studies.[57-61] In each case, P-values shown correspond to the significance estimated based on Wilcoxon test comparing the RBP and nonRBP group of proteins. RBPs were found to show significantly lower transcript stability, higher mRNA copy number, ribosome occupancy, protein stability and abundance. However protein noise which reflects the extent of cell-to-cell variation in protein levels, was found to be significantly lower for RBPs compared to nonRBPs suggesting that most RBPs are uniformly expressed across a homogenous population of cells. This figure was adapted from Mittal et al.[48] A color version of this figure is available at www.landesbioscience.com/curie.

RBPs form an important class of evolutionarily conserved proteins[49] and are known to be involved in a wide range of cellular processes. In addition to their functional roles in diverse processes (Fig. 1), RBPs are also implicated in a number of disorders due to their mis-expression or mutations in the sequences that are employed to recognize their cognate target RNAs. For instance, in humans, malfunctioning of RBPs like Nova, which is a neuron specific protein responsible for the alternative splicing of a subset of premRNAs and is involved in the pathogenesis of the neurodegenerative syndrome Paraneoplastic Opsoclonus-Myoclonus Ataxia (POMA).[16] In line with this and other observations on the impact of changes in the expression levels of RBPs being associated with diseases and fitness defects,[43,44] this study revealed that RBPs as a functional class show very little variation in their expression across cells suggesting the importance in tightly controlling them (Fig. 3F). In addition, it was found that RBPs regulating multiple transcripts show a significantly reduced expression noise indicating that variations in the expression levels of these key post-transcriptional regulators can have significant impacts on the functioning of the cell thereby leading to a disease phenotype.

CONCLUSION

The fact that RBPs are generally less stable at the transcript level, but exhibit higher stability and abundance at the protein level (Fig. 3), demonstrates that they form a group of proteins which follow the theoretically proposed time averaging effect on noise propagation.[50] According to this theory, if the protein has long half life compared to its mRNA then it averages over the noisy fluctuations in the mRNA decreasing the protein expression noise. These results are compatible with the expression of RBPs being predominantly controlled at the protein level through the use of a number of post-translational modifications (PTMs) such as phosphorylation, arginine methylation and sumoylation, which have been reported to occur in several well-studied RBPs.[51-53] Indeed, a comparison of the number of phosphorylated RBPs and nonRBPs by Mittal et al.[48] revealed the predominance of post-translational control in RBPs. Therefore, it is possible to suggest that a wide variety of these PTMs might be responsible for their ability to spatially and temporally regulate transcripts throughout eukaryotic systems. It is possible to speculate from these observations that the low noise levels in the expression of RBPs together with extensive regulatory flexibility at the protein level might give them an advantage to control gene regulation at a finer and more immediate level compared to transcriptional control by transcription factors. This might thereby provide a quick and extensive framework for controlling gene expression of a wide range of genes. This is also supported based on the observation that RBPs which are central to the cell are not only required in large quantities but are also found to be present for a longer time in the cell (see Fig. 3). All these observations suggest the importance of a post-transcriptional network of interactions in multi-cellular eukaryotes and raise several open questions in the regulation of gene expression beyond transcription. It should be possible to address such questions in the near future as more data from different levels of regulation becomes available.[54-56]

While the postgenomic era has introduced the genomic complement of hundreds of genomes, it has also left us with several unanswered questions regarding the functional relevance of the genes harbored by an organism and of the principles that govern the regulation of such genes. It is noteworthy to mention that even in a model organism like *S. cerevisiae*, regulation of gene expression at the post-transcriptional level is rather poorly understood. Nevertheless with recent improvements and availability of high-throughput approaches to the study of RBPs, such as RNA-sequencing and immunoprecipitation protocols, we can expect to see a wealth of data detailing the dynamic, spatial and tissue-specific nature of the interactions governed by these exciting class of regulatory molecules. Such advances would undoubtedly allow us to gain a deeper understanding of regulation at a level that has been under-appreciated over the past decades. Given the unprecedented detail at which these high-throughput technologies can reveal the link between the regulatory elements on the target RNAs and the RNA-binding proteins specific to environmental conditions, it is possible to use these approaches to interrogate the prevalence of these phenomena in different states and thereby study their relevance to physiology and disease in diverse model systems.

ACKNOWLEDGEMENTS

SCJ acknowledges financial support from MRC Laboratory of Molecular Biology and Cambridge Commonwealth Trust. NM acknowledges research support from the Commonwealth split-site program and National Institute of Pharmaceutical Education and Research (NIPER). We would also like to thank AJ Venkatakrishnan, Guilham Chalancon, Gabriel Moreno-Hagelsieb and Jurg Bahler for providing helpful comments on previous versions of this manuscript.

SUPPLEMENTARY MATERIAL

Additional figures, tables and supporting material related to this work can be obtained from the URL: http://www.mrc-lmb.cam.ac.uk/genomes/sarath/PTN/

REFERENCES

1. Martinez-Antonio A, Janga SC, Salgado H et al. Internal-sensing machinery directs the activity of the regulatory network in Escherichia coli. Trends Microbiol 2006; 14(1):22-27.
2. Babu MM, Janga SC, de Santiago I et al. Eukaryotic gene regulation in three dimensions and its impact on genome evolution. Curr Opin Genet Dev 2008; 18(6):571-582.
3. Janga SC, Collado-Vides J. Structure and evolution of gene regulatory networks in microbial genomes. Res Microbiol 2007; 158(10):787-794.
4. Babu MM, Luscombe NM, Aravind L et al. Structure and evolution of transcriptional regulatory networks. Curr Opin Struct Biol 2004; 14(3):283-291.
5. Ideker T, Thorsson V, Ranish JA et al. Integrated genomic and proteomic analyses of a systematically perturbed metabolic network. Science 2001; 292(5518):929-934.
6. Greenbaum D, Colangelo C, Williams K et al. Comparing protein abundance and mRNA expression levels on a genomic scale. Genome Biol 2003; 4(9):117.
7. Gygi SP, Rochon Y, Franza BR et al. Correlation between protein and mRNA abundance in yeast. Mol Cell Biol 1999; 19(3):1720-1730.
8. Glisovic T, Bachorik JL, Yong J et al. RNA-binding proteins and post-transcriptional gene regulation. FEBS Lett 2008; 582(14):1977-1986.
9. Mata J, Marguerat S, Bahler J. Post-Transcriptional control of gene expression: a genome-wide perspective. Trends Biochem Sci 2005; 30(9):506-514.
10. Keene JD. RNA regulons: coordination of post-transcriptional events. Nat Rev Genet 2007; 8(7):533-543.
11. Collins LJ, Penny D. The RNA infrastructure: dark matter of the eukaryotic cell? Trends Genet 2009; 25(3):120-128.

12. Sanford JR, Gray NK, Beckmann K et al. A novel role for shuttling SR proteins in mRNA translation. Genes Dev 2004; 18(7):755-768.
13. Zhong XY, Wang P, Han J et al. SR proteins in vertical integration of gene expression from transcription to RNA processing to translation. Mol Cell 2009; 35(1):1-10.
14. Gross T, Richert K, Mierke C et al. Identification and characterization of srp1, a gene of fission yeast encoding a RNA binding domain and a RS domain typical of SR splicing factors. Nucleic Acids Res 1998; 26(2):505-511.
15. Pascale A, Amadio M, Quattrone A. Defining a neuron: neuronal ELAV proteins. Cell Mol Life Sci 2008; 65(1):128-140.
16. Ule J, Jensen KB, Ruggiu M et al. CLIP identifies Nova-regulated RNA networks in the brain. Science 2003; 302(5648):1212-1215.
17. Ule J, Stefani G, Mele A et al. An RNA map predicting Nova-dependent splicing regulation. Nature 2006; 444(7119):580-586.
18. Gruter P, Tabernero C, von Kobbe C et al. TAP, the human homolog of Mex67p, mediates CTE-dependent RNA export from the nucleus. Mol Cell 1998; 1(5):649-659.
19. Hogan DJ, Riordan DP, Gerber AP et al. Diverse RNA-binding proteins interact with functionally related sets of RNAs, suggesting an extensive regulatory system. PLoS Biol 2008; 6(10):e255.
20. Maris C, Dominguez C, Allain FH. The RNA recognition motif, a plastic RNA-binding platform to regulate post-transcriptional gene expression. FEBS J 2005; 272(9):2118-2131.
21. Ciesla J. Metabolic enzymes that bind RNA: yet another level of cellular regulatory network? Acta Biochim Pol 2006; 53(1):11-32.
22. Hu S, Xie Z, Onishi A et al. Profiling the human protein-DNA interactome reveals ERK2 as a transcriptional repressor of interferon signaling. Cell 2009; 139(3):610-622.
23. Pinero DJ, Hu J, Connor JR. Alterations in the interaction between iron regulatory proteins and their iron responsive element in normal and Alzheimer's diseased brains. Cell Mol Biol (Noisy-le-grand) 2000; 46(4):761-776.
24. Thomson AM, Rogers JT, Walker CE et al. Optimized RNA gel-shift and UV cross-linking assays for characterization of cytoplasmic RNA-protein interactions. Biotechniques 1999; 27(5):1032-1039, 1042.
25. Wilhelm JE, Vale RD. A one-hybrid system for detecting RNA-protein interactions. Genes Cells 1996; 1(3):317-323.
26. SenGupta DJ, Zhang B, Kraemer B et al. A three-hybrid system to detect RNA-protein interactions in vivo. Proc Natl Acad Sci USA 1996; 93(16):8496-8501.
27. Gerber AP, Herschlag D, Brown PO. Extensive association of functionally and cytotopically related mRNAs with Puf family RNA-binding proteins in yeast. PLoS Biol 2004; 2(3):E79.
28. Hieronymus H, Silver PA. Genome-wide analysis of RNA-protein interactions illustrates specificity of the mRNA export machinery. Nat Genet 2003; 33(2):155-161.
29. Galante PA, Sandhu D, de Sousa Abreu R et al. A comprehensive in silico expression analysis of RNA binding proteins in normal and tumor tissue: Identification of potential players in tumor formation. RNA Biol 2009; 6(4):426-433.
30. Hall DA, Zhu H, Zhu X et al. Regulation of gene expression by a metabolic enzyme. Science 2004; 306(5695):482-484.
31. Fasolo J, Snyder M. Protein microarrays. Methods Mol Biol 2009; 548:209-222.
32. Zhu H, Bilgin M, Bangham R et al. Global analysis of protein activities using proteome chips. Science 2001; 293(5537):2101-2105.
33. Zeng F, Peritz T, Kannanayakal TJ et al. A protocol for PAIR: PNA-assisted identification of RNA binding proteins in living cells. Nat Protoc 2006; 1(2):920-927.
34. Zielinski J, Kilk K, Peritz T et al. In vivo identification of ribonucleoprotein-RNA interactions. Proc Natl Acad Sci USA 2006; 103(5):1557-1562.
35. Ho AM, Johnson MD, Kingsley DM. Role of the mouse ank gene in control of tissue calcification and arthritis. Science 2000; 289(5477):265-270.
36. Eberwine J, Belt B, Kacharmina JE et al. Analysis of subcellularly localized mRNAs using in situ hybridization, mRNA amplification and expression profiling. Neurochem Res 2002; 27(10):1065-1077.
37. Keene JD, Tenenbaum SA. Eukaryotic mRNPs may represent post-transcriptional operons. Mol Cell 2002; 9(6):1161-1167.
38. Townley-Tilson WH, Pendergrass SA, Marzluff WF et al. Genome-wide analysis of mRNAs bound to the histone stem-loop binding protein. RNA 2006; 12(10):1853-1867.
39. Keene JD, Lager PJ. Post-Transcriptional operons and regulons co-ordinating gene expression. Chromosome Res 2005; 13(3):327-337.
40. Pullmann R, Jr., Kim HH, Abdelmohsen K et al. Analysis of turnover and translation regulatory RNA-binding protein expression through binding to cognate mRNAs. Mol Cell Biol 2007; 27(18):6265-6278.

41. Junker BH, Koschutzki D, Schreiber F. Exploration of biological network centralities with CentiBiN. BMC Bioinformatics 2006; 7:219.
42. Barabasi AL, Oltvai ZN. Network biology: understanding the cell's functional organization. Nat Rev Genet 2004; 5(2):101-113.
43. Lukong KE, Chang KW, Khandjian EW et al. RNA-binding proteins in human genetic disease. Trends Genet 2008; 24(8):416-425.
44. Cooper TA, Wan L, Dreyfuss G. RNA and disease. Cell 2009; 136(4):777-793.
45. Musunuru K. Cell-specific RNA-binding proteins in human disease. Trends Cardiovasc Med 2003; 13(5):188-195.
46. Kim MY, Hur J, Jeong S. Emerging roles of RNA and RNA-binding protein network in cancer cells. BMB Rep 2009; 42(3):125-130.
47. Alberti S, Halfmann R, King O et al. A systematic survey identifies prions and illuminates sequence features of prionogenic proteins. Cell 2009; 137(1):146-158.
48. Mittal N, Roy N, Babu MM et al. Dissecting the expression dynamics of RNA-binding proteins in post-transcriptional regulatory networks. Proc Natl Acad Sci USA 2009; 106(48):20300-20305.
49. Anantharaman V, Koonin EV, Aravind L. Comparative genomics and evolution of proteins involved in RNA metabolism. Nucleic Acids Res 2002; 30(7):1427-1464.
50. Paulsson J. Summing up the noise in gene networks. Nature 2004; 427(6973):415-418.
51. Schullery DS, Ostrowski J, Denisenko ON et al. Regulated interaction of protein kinase Cdelta with the heterogeneous nuclear ribonucleoprotein K protein. J Biol Chem 1999; 274(21):15101-15109.
52. Vassileva MT, Matunis MJ. SUMO modification of heterogeneous nuclear ribonucleoproteins. Mol Cell Biol 2004; 24(9):3623-3632.
53. Yu MC, Bachand F, McBride AE et al. Arginine methyltransferase affects interactions and recruitment of mRNA processing and export factors. Genes Dev 2004; 18(16):2024-2035.
54. Halbeisen RE, Galgano A, Scherrer T et al. Post-transcriptional gene regulation: from genome-wide studies to principles. Cell Mol Life Sci 2008; 65(5):798-813.
55. Lackner DH, Beilharz TH, Marguerat S et al. A network of multiple regulatory layers shapes gene expression in fission yeast. Mol Cell 2007; 26(1):145-155.
56. Hieronymus H, Silver PA. A systems view of mRNP biology. Genes Dev 2004; 18(23):2845-2860.
57. Wang Y, Liu CL, Storey JD et al. Precision and functional specificity in mRNA decay. Proc Natl Acad Sci USA 2002; 99(9):5860-5865.
58. Arava Y, Wang Y, Storey JD et al. Genome-wide analysis of mRNA translation profiles in Saccharomyces cerevisiae. Proc Natl Acad Sci USA 2003; 100(7):3889-3894.
59. Belle A, Tanay A, Bitincka L et al. Quantification of protein half-lives in the budding yeast proteome. Proc Natl Acad Sci USA 2006; 103(35):13004-13009.
60. Ghaemmaghami S, Huh WK, Bower K et al. Global analysis of protein expression in yeast. Nature 2003; 425(6959):737-741.
61. Newman JR, Ghaemmaghami S, Ihmels J et al. Single-cell proteomic analysis of S. cerevisiae reveals the architecture of biological noise. Nature 2006; 441(7095):840-846.
62. Finn RD, Mistry J, Tate J et al. The Pfam protein families database. Nucleic Acids Res 38(Database issue):D211-222.
63. Lee MH, Schedl T. RNA-binding proteins. WormBook 2006:1-13.
64. Lasko P. The drosophila melanogaster genome: translation factors and RNA binding proteins. J Cell Biol 2000; 150(2):F51-56.
65. McKee AE, Minet E, Stern C et al. A genome-wide in situ hybridization map of RNA-binding proteins reveals anatomically restricted expression in the developing mouse brain. BMC Dev Biol 2005; 5:14.
66. Sanchez-Diaz P, Penalva LO. Post-Transcription meets postgenomic: the saga of RNA binding proteins in a new era. RNA Biol 2006; 3(3):101-109.
67. Ray D, Kazan H, Chan ET et al. Rapid and systematic analysis of the RNA recognition specificities of RNA-binding proteins. Nat Biotechnol 2009; 27(7):667-670.
68. Tenenbaum SA, Carson CC, Lager PJ et al. Identifying mRNA subsets in messenger ribonucleoprotein complexes by using cDNA arrays. Proc Natl Acad Sci USA 2000; 97(26):14085-14090.
69. Stelzl U, Nierhaus KH. SERF: in vitro election of random RNA fragments to identify protein binding sites within large RNAs. Methods 2001; 25(3):351-357.
70. Paraskeva E, Atzberger A, Hentze MW. A translational repression assay procedure (TRAP) for RNA-protein interactions in vivo. Proc Natl Acad Sci USA 1998; 95(3):951-956.
71. Rodgers ND, Jiao X, Kiledjian M. Identifying mRNAs bound by RNA-binding proteins using affinity purification and differential display. Methods 2002; 26(2):115-122.
72. Butter F, Scheibe M, Morl M et al. Unbiased RNA-protein interaction screen by quantitative proteomics. Proc Natl Acad Sci USA 2009; 106(26):10626-10631.

CHAPTER 8

SAFE KEEPING THE MESSAGE: mRNP Complexes Tweaking after Transcription

Said Hafidh,*[,1] Věra Čapková[1] and David Honys[1,2]

[1]*Laboratory of Pollen Biology, Institute of Experimental Botany AS CR, Rozvojová, Czech Republic.* [2]*Department of Plant Experimental Biology, Faculty of Science, Charles University in Prague, Viničná, Czech Republic*
**Corresponding Author: Said Hafidh—Email: hafidh@ueb.cas.cz*

Abstract: The mRNA-protein complexes (mRNPs, Messenger ribonucleoprotein particles) are the "couriers" of the modern eukaryotes that process, store and deliver messages (transcripts) from the nucleus to the appropriate subcellular compartments and beyond. Presence of mRNPs arbitrates the posttranscriptional control of gene expression by editing the precursor RNA to maturity, postulate its subcellular localization and/or storage and dictate its fate once in the cytoplasm; either to be translated or dispensed through mRNA degradation. Initiation of transcription is coupled with processing of the transcribed message and the immediate association of the transcript with a set of structural and regulatory proteins. *Per se*, mRNP complexes sub-optimize transcription by recruiting RNA-binding proteins which are the core component of the RNP activities that culminate overall distribution and abundance of individual proteins. This asymmetric distribution of the mRNA is the determinant of protein gradient and is known to influence cell polarity, cell fate and overall patterning during development. Embryo patterning in Drosophila, polarization of maternal mRNA to daughter cell in budding yeast and directional growth of mammalian neural cell and pollen tubes of flowering plants, are the most prominent examples of mRNP facilitated posttranscriptional control, influencing cell fates and patterns of development.

This chapter addresses the current knowledge on the mechanisms of posttranscriptional control reinforced by the formation of RNP particles and reviews differences in the underlying mechanisms. The outline of the chapter encompasses step-wise cellular processes leading to the formation of mRNPs and its implication to cellular activities. A dedicated section is also integrated discussing the recent findings on the unique mechanism of RNP formation in the male gametophyte of *Nicotiana tabaccum*. A proposed model outlines the network of posttranscriptional control with a focus on the role of RNPs is also presented aiming to stimulate future research with a perspective of advancing our knowledge on the subject and its plausible application in improving food quality.

RNA Infrastructure and Networks, edited by Lesley J. Collins.

INTRODUCTION

The complexity of the eukaryotic genome presents a challenging task for a cell to selectively "display" precise genetic information at a given time and space during development. Deployment of specific transcription factors and marking of the genome with epigenetic patterns has allowed eukaryotic cells to select and modulate the amount of genetic information in different cell types. Similarly, several classes of noncoding RNAs (ncRNAs) have been well characterized as posttranscriptional modulators of gene expression regulating protein-coding RNAs through transcription, splicing, mRNA turnover, nucleotide modification and translational repression (reviewed in ref. 48). Commitment of the coding RNAs through these multiple fates is dynamic, as a result, a complex network of posttranscriptional processes that interconnect regulatory ncRNAs and the target RNAs exist and is referred as RNA-infrastructure.[48] The building blocks of the RNA-infrastructure are first initiated when noncoding and coding RNAs are assembled into RNA-protein complexes, ribonucleoprotein particles (RNPs), immediately following transcription in the nucleus. Depending on the nature of the bound proteins, RNPs can direct multiple fates of the RNA cargo and thus forms an exclusive pathway of RNA regulation. The networks of RNP mediated-RNA regulation alone represent a major component of the RNA-infrastructure and we referred it as RNP-infrastructure. In this chapter we will focus on the different features of RNP mediated posttranscriptional control of gene expression and how these processes intertwine with different aspects of development.

RNP-INFRASTRUCTURE; POSTTRANSCRIPTIONAL MODULATOR OF GENE EXPRESSION

Developmental architecture in many eukaryotes from embryonic patterning to maturity is preceded by a precise accumulation of specific macromolecules leading to polarity, cell fate establishment, patterning and organ specification. Across all diverse species of living organisms, temporal and spatial control of gene expression is imposed at the transcriptional and posttranscriptional level through chromatin modification, modulated transcriptional initiation and RNA processing, RNA localization and storage and controlled translation and protein turnover. At the centre of these transcriptional events is the assembly of the mRNA-ribonucleoprotein (mRNP) complexes that facilitate all downstream activities to determine the destination of the encoded transcript (Fig. 1). By regulating the quality and destiny of premRNAs, regulation mediated by mRNPs provides qualitative and quantitative assessment of gene expression posttranscriptionally, thereby regulating overall cellular activities. Through this coordinated network of regulation, a foundation for developing fully functional complex structures with distinct biological function is laid and propagated throughout the development of an organism. Thus, it's crucial for mRNAs with specific message to be delivered at the right location, at the right time and at the right dosage.

The exhilarating journey of a transcript following transcription starts in the nucleus where the message is wrapped with a cohort of *trans*-acting factors binding to specific *cis*-elements as it is copied from its DNA template. The message is scanned for its quality and undergoes maturation to its functional form. This complex of mutually interacting mRNA and proteins (termed mRNP for ribonucleoprotein) is fine-tuned by aggregating additional

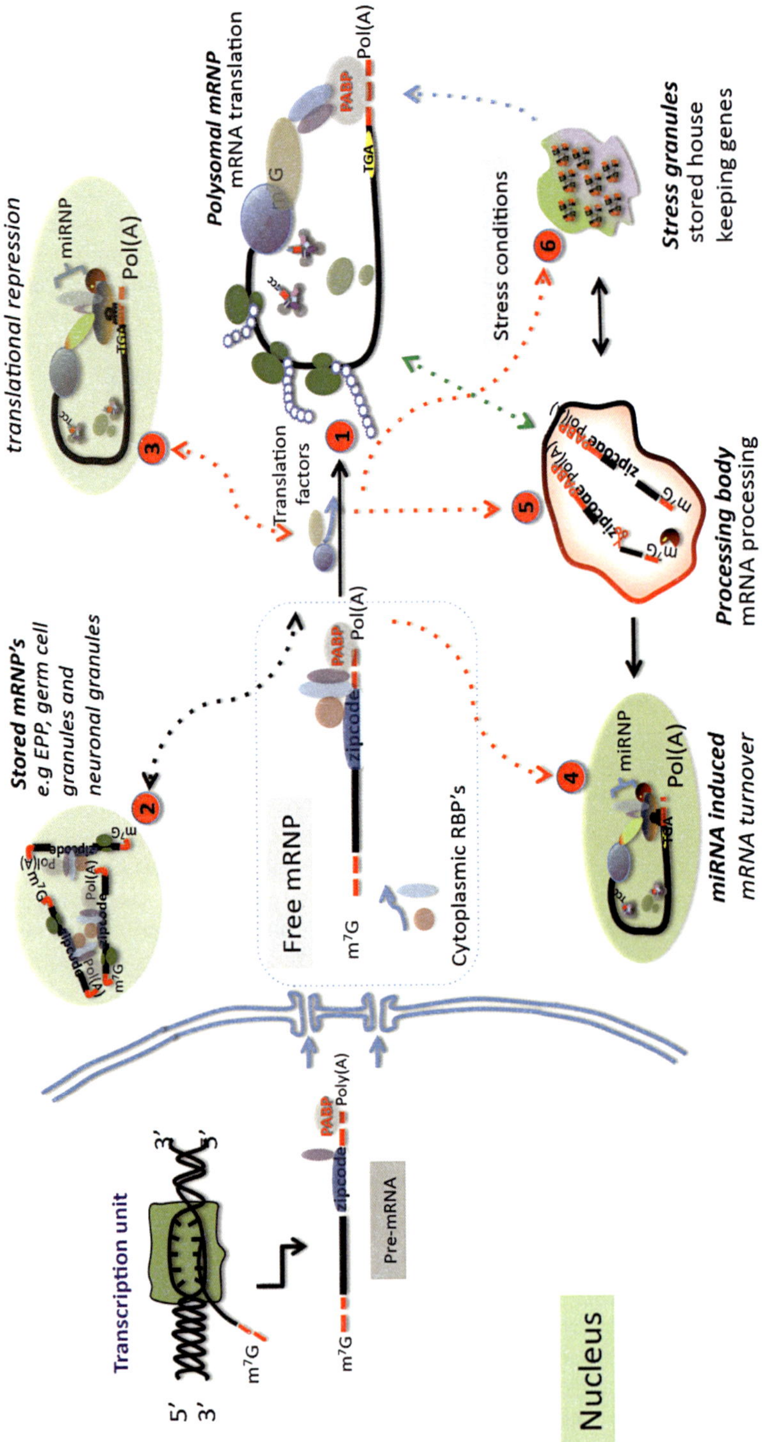

Figure 1. Please see the figure legend on the following page.

Figure 1, viewed on previous page. mRNP complexes mediate mRNA posttranscriptional fate. Transcription in the nucleus is immediately coupled with the binding of RBPs (steel blue and purple) to the appropriate *cis*-elements of the pre-mRNA (coding region in black; 5′ and 3′-*UTR* in red) to facilitate the maturation processes and imminent transport through the nuclear pore. Once cytoplasmic, additional proteins (sky blue and brown) are recruited to the mRNP that determines the fate and proper subcellular localization of the mRNA. Several fates are imminent for the aggregated mRNP; (1) translation factors together with ribosomal subunits (olive green) might be recruited to the complex and remodel the mRNP into a translationally active polysomal mRNP. (2) Emerging mRNP complexes can be aggregated into dormant translationally sequestered granules (stored RNPs) either to be transported or stored to be utilized at later times. (3) Similarly, mRNPs can be aggregated and stored in a miRNA-induced translationally-silent state. (4) For aberrant transcripts or for modulation of gene expression, some mRNAs undergo degradation and are discarded from translation. (5) Sequestered mRNPs can be gathered in processing bodies to be processed for degradation and recycled or to be shuttled back to the cytoplasm for translation. (6) Occasionally under stress conditions, mRNPs are assembled into stress granules and sequestered from translation. Active shuttling of mRNPs between PBs, SGs and polysomal mRNP promote regulation of gene expression linked with cellular development. All features shown are not drawn to scale. Abbreviations: mRNP, mRNA ribonucleoprotein; RBPs, RNA binding proteins; UTR, un-translated region; miRNA, micro-RNA; miRNP, microRNA ribonucleoprotein complex; PBs, processing bodies; SGs, stress granules.

specialized proteins prior to its migration to the cytoplasm. Such a routine is renowned to be one of the established mechanisms of posttranscriptional regulation, necessary for the temporal and spatial localization and distribution of the target transcript within the cell, to ensure localized protein synthesis and delivery to the site of action. Messages that are transcribed but only to be translated at later times, or at later stages of developmental, or in a case of urgency such as stress or tissue damage, are stored in a form of stored RNPs and kept away from translational initiation machinery (Fig. 1).[1,2] Furthermore, their storage has to be "cost-effective" and easily accessible upon the initiation of translation, particularly in response to external stimuli. Aberrant transcripts or those accumulated in excess are directed to processing bodies (PBs) and undergo RNA degradation.[3] As such the mechanism of RNA storage and mobilization in conjunction with RNA turnover provides exquisite control of gene products, advancing the specificity installed in the promoter. The integration of newly encoded transcripts into the mRNP complexes is also critical for the subcellular localization of the mRNA and allows on-site protein synthesis. A considerable number of encoded proteins do not possess a localization signal and thus depend entirely on the localization of their transcript. The importance of correct localization and storage of mRNA during development has been studied intensively in *Drosophila melanogaster* during oogenesis. mRNA encoded by *bicoid* [*bcd*] and *nanos* [*nos*] genes are localized on the interior and posterior cortex of the Drosophila oocyte until the completion of oogenesis.[4,5] The distribution of these two sets of mRNAs is an outcome of opposing protein gradients essential for the initiation of asymmetry along the interior-posterior axis in the early developmental stages of the Drosophila embryo.[4] Localization studies have shown that polarization of the *bcd* mRNA is solely dependent on the specific sequences located within the 3′-UTR of the transcript, together with the intact polarized network of microtubule cytoskeleton.[6] Subcellular localization of mRNA is also a prevailing concept in plants and there is cumulative evidence demonstrating its significance in many aspects of cellular activities. Besides starch, rice accumulates two major classes of proteins, prolamines and globulin-like glutelins.[7] These two proteins are both synthesised on the endoplasmic reticulum (ER) and then translocated to the ER lumen. Whereas prolamines are retained and arranged into protein bodies, glutelins are transported and stored in protein storage vacuoles.[7] This fundamental differential localization has been shown to be the result of an RNA-embedded subcellular RNA localization signal.[7,8]

An exquisite model of mRNP mediated posttranscriptional regulation has also emerged in the development of the male gametophyte of Tobacco (*Nicotiana tabacum*).[1,9] In all flowering plants, the development of the male gametophyte is governed by precisely orchestrated cellular activities leading to developmental maturity accompanied by cellular and morphological changes. The progression through the stages of development is propelled by underlying molecular changes including chromatin remodelling, gene expression profile and posttranscriptional regulation.[10] Once matured, the male gametophyte (pollen) undergoes dramatic progamic changes during the formation of a pollen tube. Initiation of these changes is accomplished through storage and precise localization of specific macromolecules and a `burst` of translation at a local site to promote directional growth of the pollen tube. As such, the architecture of the male gametophyte is ideal to pinpoint and understand mechanisms imposed in the progamic phase switch leading to differential gene expression, protein accumulation and cell signalling with respect to mRNA storage and simulated localization. Transcriptomic studies of the male gametophyte developmental stages have identified a subset of genes which are expressed early during development and those which are detected only at the later stages of pollen development.[11,12] Studies in the developing tobacco pollen have established that some of the early expressed genes accumulate the message at the early stages of pollen development,[13] but, their translation was postponed to the time of pollen tube growth.[10,14] Furthermore, analyses of protein extracts in a time course of pollen development has indicated that the male gametophyte of flowering plants has evolved a more stringent system in which the late-translated messages are stored in aggregated multiple-mRNP particles and sequestered from translation.[1]

PRE-mRNA PROCESSING AND THE INITIATION OF mRNP FORMATION

Activation of gene expression is the first implicated step that determines the accumulation of a protein in a cell. Gene expression starts in the nucleus and the transcripts are ultimately mobilized to the cytoplasm for translation. Local or targeted chromatin modification involves methylation and acetylation exposing *cis*-elements of the gene to be transcribed, resulting in the binding of the transcription-activating complex and transcription of the DNA template. The transcribed RNA is immediately bound with a cohort of noncoding small nuclear RNAs with the appropriate binding proteins (snRNPs) associated with other RNA binding proteins to form a spliceosome complex responsible for splicing of the introns and maturation of the transcript. In addition to intron removal, other nuclear RNA-binding proteins (RBPs) are involved in pre-mRNA 5′- capping and polyadenylation, mRNA export, degradation of some of the transcripts and synthesis of small interfering RNAs (siRNAs) and microRNAs (miRNAs)(see Chapter 7, page 103 for more detail on RNA-binding proteins). Proteomic studies in several plant species have identified numerous RBPs possessing RNA binding domains such as the RNP motif, RNA recognition motif (RRM), Pumilio (PUM), pentatricopeptide (PPR) and other putative motifs that have also been predicted to bind RNA.[3]

The majority of RBPs in eukaryotes have now been characterised and are known to participate in diverse roles of mRNA management including nuclear processes such as the activity of the exon junction complex (EJC) in pre-mRNA splicing and nuclear export, initiation of translation, formation of PBs and SGs, translational repression by miRNAs and siRNAs and mRNA subcellular translocation. The precise role of EJC and the impact the nuclear processes has on correct localization of the subsequent mRNA, was apparent from studies of Drosophila *OSKAR* mRNA localization and investigation of the role of the EJC in mammalian neurons. EJC protein composition between the two organisms remained conserved and the complex was shown to posses four core proteins; RNA helicase eIF4AIII, Barentsz, Mago Nashi and Tsunagi (Y14) (as well as the human orthologs MLN51, Magoh and Y14).[15] In the course of the nuclear processes, the core proteins are anchored at the exon-exon junctions of the *OSKAR* pre-mRNA, actively involved in its splicing and remain bound until the first round of translation is initiated.[16] The correct localization of *OSKAR* mRNA requires the inclusion of the first intron together with the localization motif in the 3′-UTR.[15] The necessity of intron 1 retention in *OSKAR* mRNA localization was demonstrated by sequential deletions of the three introns as well as deletion of more than one intron.[15] Results of these studies further demonstrated that localization of *OSKAR* mRNA was prompted by correct splicing at first exon-exon junction and was independent of the intron sequence identity.[15] As such, precise spatial distribution and binding of EJC between exon-exon junctions is not only critical for correct splicing but also essential for posterior localization of the *OSKAR* mRNA, linking EJC role in splicing and its influence on correct mRNA localization. Identification of many other RBPs in plants associated with mRNP particles and their functional significance within the mRNP complex still remains to be scrutinised.

The characteristics of the RNPs, as determined by the nature and composition of the associated proteins, dictate the fate of the loaded message as the assembled complex exits the nuclear pore. Some of the messages undergo nonsense-mediated decay (NMD),[63] whereas the majority are transported to the cytoplasm for localization, storage or immediate translation (Fig. 1). Thus, the existence of the mRNP complexes is not only essential for intracellular protein distribution and storage, but also provides an opportunity to study a snapshot of the life history of the transcript from the nucleus to its fate in the cytoplasm and subsequently highlight the link with posttranscriptional mode of regulation. Since the formation of the mRNPs is immediately initiated in the nucleus, the fate of the mRNA is predetermined well before its entry into the cytoplasm, directed by the composition and the architecture of the mRNP.

The binding of specific RBPs to the mRNA is solely dependent on *cis*-acting elements, also known as "zipcodes", which are present on the mRNA and are known to depict the fate and subcellular targeting of the message. In most specifically localized mRNAs, the zipcodes are located in the 3′-untranslated region, frequently forming secondary structures and are sometimes present as multiple repeats. *Vg1* mRNA from Xenopus oocyte is one good example.[17] Occasionally, the zipcodes can also be located in the 5′-UTR or even in the coding sequence of the mRNA as in the case of *ASH1* mRNA from yeast, in which three localization repeats have been identified in the coding sequence and an additional repeat embossed in the 3′-UTR.[18] Each *ASH1* sequence motif was confirmed to have a capacity to localize a bound transcript alone, although the presence of multiple repeats enhances the correct localization of the message. Arn et al[19] proposed a functional model that multiple clusters of localization motifs are likely to promote regional concentrations of RBPs, which in turn attract other proteins through protein-protein interactions necessary for RNA localization.[19] One of the best characterized RNA-binding proteins utilizing localization elements is the chicken zipcode binding protein 1 (ZBP1) which binds to a conserved zipcode of the β-actin mRNA in its 3′-UTR and enables the translocation of the message to actin-rich protrusions in primary fibroblasts and neurons.[22] Characterization of the ZBP1 have led to the identification of two RNA recognition motifs (RRM) and four hnRNP K homology (KH)

RNA binding domains.[23] The authors[23] identified distinct roles for each of the found domains in which the KH domains were identified to mediate binding to the zipcode and formation of an RNP, whereas, the RRM motifs were responsible for the subcellular localization of the β-actin RNPs. An equivalent of ZBP1 in plants, OsTudor-SN, is known to be involved in the subcellular trafficking of prolamine and glutelin mRNAs along the actin filaments.[24]

Another important set of proteins associated with mRNP particles are those dedicated for modulating gene expression and mRNA translation, as well as dictating the rate of protein accumulation. Several have been identified including Argonaute proteins (AGO) and Argonaute-like Piwi proteins which together with Piwi interacting RNAs (piRNAs) direct cleavage of the target mRNA, sequestered transcripts from translation and represses transposon translocation in somatic cells and germ cells respectively (Fig. 1).[25,57,58] These proteins are known to interact with miRNAs, siRNAs or piRNAs preloaded in the mRNP to induce transcript cleavage via the RISC complex. Alternatively, some (e.g., human let-7 miRNA)[59] can also impose translational repression via the interaction of AGO protein with the 7-methyl-guanine (m7G) cap of the mRNA outcompeting translation initiation factor elF4E.[26] This alternate role of the RISC complex is a result of partial complementarities between miRNA and the miRNA binding sites, which leads to translational repression or accelerating the degradation of the target mRNA within Processing bodies, PB (see below and refer to Table 1). In numerous studies, miRNA or siRNA-targeted transcripts were demonstrated to be sequestered to PBs and many translationally repressed mRNAs have been localized within these granules.[59-61] Under certain conditions such as stress, some of the repressed transcripts are relieved from inhibition and re-enter the translation machinery. A well-demonstrated example is the human *CAT-1* gene that is expressed in the hepatoma Huh7 cells. The *CAT-1* mRNA is naturally sequestered from translation by miR122 to the PBs, however, this inhibition is lifted during stress and prompt recruitment of *CAT-1* mRNA to polysomes and subsequent translation.[62] Presence of other repressing RBPs also results in the lack of translation through yet unknown mechanisms and their repression is relieved upon posttranslational modification.

MOBILIZATION TO THE SITE OF STORAGE AND mRNA LOCALIZATION

mRNA localization provides an efficient means of establishing a targeted protein synthesis at specific subcellular location and thus generating a gradient of protein accumulation and a subsequent local function. Although several mechanisms of mRNA localization have been proposed that concurrently accomplish the subcellular localization of several transcripts, the most common pathway involves the translocation of mRNPs and anchoring at the site of translation.[6] It is now known that during the mRNP formation in the nucleus, a class of accessory proteins are also bound to the transcript that mark its intracellular localization. Several of these proteins have been identified in mammals and a well described example is the family of heterogeneous nuclear ribonucleoprotein proteins (hnRNP) of which some putative homologs have already been identified in *Arabidopsis thaliana* and rice.[27] It is likely that hnRNPs are recruited to the mRNP particles via zipcodes recognition. Since there is no defined general localization consensus sequence, hnRNPs are unlikely to provide specificity during mRNA translocation, instead other cohorts of transacting factors might be responsible in playing this role in concert with bound hnRNPs. The significance of hnRNPs in mRNA localization was demonstrated in

Table 1. Glossary of the cellular components used in the chapter

Name	Localization	General Description	RNA Processing Role	Refs
Polysomes (Polyribosomes)	Cytoplasm	Cluster of ribosomes bound to mRNA. Exist in free, cytoskeletal or membrane-bound form.	Involved in protein synthesis on the associated mRNA. Can initiate translation at the 5′-UTR or through specific internal motifs.	25,37
mRNA ribonucleoprotein particles (mRNPs)	Cytoplasm	A complex of proteins bound to mRNA associated with ribosomes (polysomal mRNPs) or without (free mRNPs).	Splicing, cytoplasmic translocation, subcellular localization, transcript storage, dictate fate on translation and protein abundance.	9,37
EDTA/ puromycin-resistant particles (EPPs)	Cytoplasm	Aggregated mRNPs with preloaded translation machinery first identified in the tobacco male gametophyte as large RNPs co-sedimenting with polysomes and resistant to polysome-destabilising substances.	Provide robust storage of translationally silenced mRNAs in the early stages of pollen development and promotes immediate translation upon pollen tube growth.	1,9
Stress granules (SGs)	Cytoplasm	Generated in response to stress-induced polysomes disassembly and translational silencing.	Stores silenced aggregated mRNPs and resupply 'preloaded' mRNPs for immediate translation.	3,37,48
Processing bodies (PBs)	Cytoplasm	Sites of mRNA quality control and decay driven by 5′→3′ exoribonuclease	Decapping, degradation and recycling of mRNAs and aberrant RNA's. Interplay with translation machinery.	3,25,37, 48
Exon junction complex (EJC)?	Nucleus	Associate with precursor RNA and possess helicase activities	Intron splicing, mRNA fate and quality control.	16
Cytoskeleton	Cytoplasm	Scaffolding of the cell made of actin fillaments and microtubules.	Intracellular transport of RNA and proteins.	6,20,28, 29,30,49
Endoplasmic reticulum (ER)*	Cytoplasm	Interconnected network of tubular membranes and vesicles with an extension to Golgi apparatus (cisternae).	Rough-ER: site of protein synthesis, packaging and transport.	7,8,20, 27,49

*Three forms of ER; Rough endoplasmic reticulum (protein synthesis and transport), Smooth endoplasmic reticulum (lipids and steroids synthesis, control of cellular metabolism), Sarcoplasmic reticulum (regulate calcium concentration).

neuronal cells whereby hnRNAP A2 was identified to be involved in the localization of the myelin basic protein in oligodendrocytes of the mammalian neurone and in Drosophila in which Hrp48 is required for subcellular localization and a subsequent translational regulation of many RNA molecules.[27]

Once the nuclear mRNPs are translocated into the cytoplasm (cytoplasmic mRNPs), they are packaged into a transported form through the recruitment of cytoplasmic RBPs to the complex, leading to the transformation of the nuclear mRNP into transported granules or particles. At this stage, the mRNP bears several sets of transacting factors involved in mRNA translocation, anchoring and those involved in translational regulation and subsequent protein synthesis. The packed mRNP is sequestered from translation by specific set of bound proteins during translocation until it reaches its destination.

CYTOSKELETON: ACTIN AND MICROTUBULE DYNAMICS AS A FLOATING RAFT FOR mRNPs SUBCELLULAR LOCALIZATION

The cytoplasmic mRNPs are conveyed to their destination via the cytoskeleton, most commonly through microtubules and occasionally through actin microfillaments which provide a basic "road map" interconnecting different parts of the cell, as well as act as a scaffold for translational purposes. The initial understanding that the cytoskeleton is the framework that provides cell shape has expanded to a much wider role particularly in subcellular mRNA localization and control of localized protein synthesis. A combination of biochemical studies together with advances in cellular imaging, has allocated a central role of cytoskeleton in mediating movements of macromolecules. *Per se*, this promotes on-site translational regulation and protein function, which consecutively establishes a local gradient of cellular proteins. This differential protein accumulation influence cellular patterning as well as facilitates mRNA translocation to cellular organelles and the extracellular matrix.[28] Deployment of mRNAs over a short distance along the cytoskeleton network operates through actin filaments, whereas, long distance trafficking such as in oocytes, neural cells and likely in other directionally growing structures utilizes microtubules to move along the mRNP granules.[29] It is not yet known how the mRNPs are translocated along the cytoskeleton in plants, but the involvement of motor proteins has been well demonstrated in other eukaryotes. For instance, the subcellular localization of *ash1* mRNA to the bud tip of the daughter cell in yeast is known to be mediated by myosin, whereas, similar mechanisms involving kinesin and dynein-mediated movement have been proposed in the Drosophila oocyte and human oligodendrocytes.[30] However, a more pinpointed mechanism behind mRNPs trafficking along the cytoskeleton in plants remains to be unravelled. It is tempting to envision and there is already a strong indication pointing towards the involvement of RBPs together with cytoskeleton interacting proteins in tethering and facilitating the movement of the mRNPs along actin and microtubule railings to promote their localization.

INTRA- AND INTERCELLULAR TRANSLOCATION OF mRNPs

The fact that plants are sessile and instead respond to environmental stimuli to induce necessary physiological changes, means that cell-to-cell signalling becomes one of the critical steps to communicate and promote overall changes. Unlike animals, plant cells are gated (made of discontinuous cell wall that forms plasmodesmata openings) suggesting possible exchanges of macromolecules and other cellular components between adjacent cells. Indeed, several examples demonstrate the movement of RNA and other particles through plasmodesmata as a long distance migration of transcripts (signaling molecules) influencing cellular activities of the neighboring cells. One classical example is the cell-to-cell movement of viruses following an infection, which results in the systematic spreading of the infection. The mechanism behind this viral cell-to-cell RNA translocation is the work of the movement proteins, which interact with the plasmodesmata and alters their size-exclusion limit promoting RNA intercellular trafficking.[31] In another example, *KNOTTED1* (*KN1*) and *SUCROSE TRANSPORTER1* (*SUT1*) are two plant genes whereby their encoded transcripts are also translocated between neighbouring cells. The intercellular movement of *KN1* has been shown to mimic the mechanism imposed by viral movement proteins and it is now known to be involved in the initiation and maintenance of meristem in shoot apex as well as between cell layers within the leaf.[32] *In situ* hybridization studies have localized *SUT1* mRNA in companion cells and in nuclear-less sieve elements (SE) of the phloem system. Since SE lack nuclei, *SUT1* mRNA is transcribed in companion cells and then translocated to the SE with an as yet unknown mechanism. The detection of SUT1 protein in SE also signifies a likelihood of intercellular trafficking of core translational components.[32] Similarly, the plant vascular system has also been proposed as a passage for a long distance transport of molecules. The phloem system is responsible for delivering a variety of signalling molecules such as hormones, whereas the xylem system is involved in the transport of water and nutrients. The ability of the small interfering RNAs (siRNAs) to exert a systematic wide spread distribution throughout the plant suggest an existence of an effective system in which these small molecules can be transported. Indeed, Bartel et al[33] and Kidner et al[34] in their analyses of phloem sap identified diverse species of miRNAs, firming up the model of wide spread siRNA induced effect in regulating plant development through the vascular system. A well-demonstrated example of long distance mRNA translocation in plants is that of the *Fowering locus T* (*FT*), which responds to day length and induces flowering. Perceived light signal on the leaf leads to the transcription of the *FT* gene in leaves. However, its effect is imposed in a distantly located floral meristem in the shoot apex.[35] Trafficking of the *FT* transcript along the phloem was later demonstrated by transiently expressing *FT* on a single Arabidopsis leaf using a heat shock promoter. Following induction, expressed transcripts were detected at the shoot apex verifying signal perception in a form of mRNA via phloem system. A similar mode of action has been recently described for the BEL1-like family of transcription factors in potatoes in which *StBEL5* (a gene that regulates tuber formation) is induced in leaf veins and petioles but exerts its effect in stolon tips.[36] Thus, plants have developed sophisticated mechanisms of dispatching the information between cells and between organelles and are able to induce a long distance effect by translocating necessary mobile signals. It is yet to be demonstrated how transcripts are packaged for long distance migration, although future studies are anticipated to uncover stored mRNAs in a form of RNPs.

The co-existence and the symbiotic relationship between plant cells and cellular organelles is also governed by the exchange of macromolecules. The partial transcriptome encoded by the chloroplast and mitochondrial genomes is insufficient for a complete biogenesis and function of these organelles. Nuclear encoded proteins are known to be imported into the chloroplasts and mitochondria (represent >90% of the required proteins), through interaction of peptide motifs (*trans*-peptides) of the imported proteins and the envelope membrane channel import complex (protein translocons). Nonetheless, nuclear encoded RNAs such as *ATM1* from yeast and ChL H and CHL 42 in plants are also targeted to these organelles. The *ATM1* possesses two repeats of mitochondrial zipcodes located in the 3′-UTR and 48-nucleotide sequence in the coding sequence necessary for its import to the mitochondria. Although the precise mechanism of intercellular RNA trafficking still remains elusive, more and more evidence are now emerging demonstrating the movement of RNAs within the cell and beyond, deployed as part of signal transduction or for other cellular functions.

STORAGE OF mRNPs AND mRNA TURNOVER

Although a significant proportion of the mRNAs are translated immediately once they have localized, a subset of the transcripts are redirected to be stored and inactivated from translation in various types of stored RNP particles, stress granules or discarded into processing bodies where they undergo a nonsense-mediated decay and eventually are degraded and recycled (Fig. 1). Likewise, other types of stored RNP granules have also emerged including neuronal granules (found in neurons) and polar or germinal granules that have been identified in germ cells compartments in flies.[37] These types of RNA granules have a similar role to that of SGs, however, they are not produced in response to stress.

Stress granules and processing bodies are active sites of mRNP processing in eukaryotes. SGs are a class of RNA granules that are highly up-regulated during translational repression predominantly in response to stress (Table 1). Stress response induces shuttling of transcripts from translationally active polysomes to mRNPs that aggregates into large cytoplasmic foci and sequesters transcripts from translation. Stored mRNAs are shuttled back to polysomes for rapid translation following recovery from stress. Thus, SGs are the centre of mRNA processing providing emergency storage and protection of functional transcripts and their immediate resupply following recovery from stress. Several proteins have been identified that constitute SGs some of which are in common with PBs and also those that are exclusive to SG's. Some of the identified proteins includes T-cell intracellular antigen 1 (TIA1), TOA1-related (TIA1R), eIF4E, eIF4G, eIF4A, eIF3, PABP, G3BP1 and 40S ribosomal subunits.[3] TIA1 has been characterised as a multitask component serving as a translational silencer, a regulator of alternative splicing and mRNA decay. Plants orthologs of TIA1 have also been identified and a mutation in *Nicotiana plumbaginafolia* TIA1-like oligouridylate-binding protein 1 (UBP1) prevents SGs formation.[3]

Processing bodies on the other hand (Table 1), are the sites for mRNA degradation prompted by the polyA-tail deadenylation, RNA-mediated gene silencing and translational repression. PBs are generally characterised for possessing 5′-m^{7}G decapping enzymes (DCPs), 5′-3′ exonuclease (XRN1), a class of Ago proteins and Ago-like (GW182) in animals cells.[37] Arabidopsis null mutant *xrn4-5* which has reduced cytosolic 5′-3′ exonuclease activities shows increase accumulation of PBs foci demonstrating linkage of PBs with their role in mRNA degradation.[3]

Characterisation of another form of stored RNP particles, germ cell granules, with respect to their protein and RNA constituents in many organisms showed that they resemble both mRNA storage properties of SGs and RNA decay characteristics of PBs. Mammalian germ cell granules (chromatoids) possess several components involved in small RNAs processes and mRNA decay. The Dicer enzyme, Argonaute proteins, GW182, MIWI (a homologue of Drosophila PIWI), several species of miRNAs, DCP1A and XRN1 represent several identified components with known function.[25] In part, it is hypothesised that the mammalian germ cell granules are equipped to modulate gene expression and to dictate the initiation and efficiency of protein translation, although a much wider role has been also implemented. Similar germ cell granules have been also found in the Drosophila germ line linking them with the control of protein translation and miRNA mediated transcriptional control of maternally expressed genes. Distinctively, RNA granules recently identified in the gonadal syncytium (a large cell-like structure consisting of multiple haploid sperm cell nuclei following germ cell division without cytokinesis) of *Caenorhabditis elegans* have been annotated to be compositionally similar to both PBs and SGs.[25] They contain PB markers CGH-1 (conserved germline helicase 1) and CAR-1 (cytokinesis, apoptosis, RNA-associated 1) as well as SG-associated proteins like PAB-1 (poly(A) binding protein 1) and ATX-2 (ataxin-related 2). This suggests their possible dual role in stabilizing maternal mRNAs and simultaneously repressing their translation.[38] Occurrence of similar factors in the germ cell granules of flowering plants remains to be demonstrated, however, their presence is imminent.

The ontogenic development of the male gametophyte in which a small daughter cell (germ cell) is enclosed within a larger daughter cell (vegetative cell), provides a unique model in understanding the mechanism of transcriptional and translational control, in which the two cell types acquires two distinct fates. This distinct specification of the two cell types suggests differences in the underlying molecular pathways leading to cell differentiation. Consistent with this objective, the earlier expression of male gametophytic genes and a need for high rate of translation during pollen tube growth, suggests a demand for a robust storage system that could withstand a long-term storage and yet to deliver the message efficiently. A number of pollen mRNAs were shown to be bound to pollen stored mRNP particles.[9] However, the nature of the isolated mRNPs was "indestructible" in a buffer constituting strong detergents in comparison to other universal forms of RNPs. Furthermore, these two types of mRNPs showed different densities in a sucrose gradient. This novel class of RNP particles were annotated as EDTA/puromycin-resistant particles or EDTA/puromycin-resistant particles (EPPs).[1,9] With NTP303 as a model example,[1,9,13,39] EPP complexes seem to offer that secure storage. The authors hypothesised that the EPP complex is assembled gradually during development by agglomeration of the mRNP monomers following initial assembly of the *ntp303* mRNP particles.

Recent identification of the EPP particles in the tobacco male gametophyte highlights the presence of germ cell-like granules in flowering plants.[1] Just like the role played by Drosophila germinal granules in delivering maternal mRNAs and its role during initial stages of embrogenesis, plants EPP particles demonstrate nicely the significant role of RNA granules (in the form of mRNP) as a developmental "clock" that induces programic and morphological changes in response to activating stimulus (Fig. 2). EPP particles represent preloaded complex machinery devoted to mRNA processing, transport, subcellular localization and protein synthesis. Evidence from the analysis of isolated EPP fractions identified several proteins associated with protein metabolism including; protein synthesis (eIF4A-8, eukaryotic initiation factor 4A-8), protein fate (BiP4, laminal-binding protein 4 precursor) and protein transport (Rab11a, Ras related protein) and those associated with RNA localization and translation including cytoskeletal proteins, protein kinases and phosphotases. The composition of EPP particles also consists of set of mRNAs that are stored and translationally silenced at earlier stages of development and travels with the maturation processes of the pollen grain. Some of the stored messages are massively translated either at the late stages of pollen development and/or transported to the growing

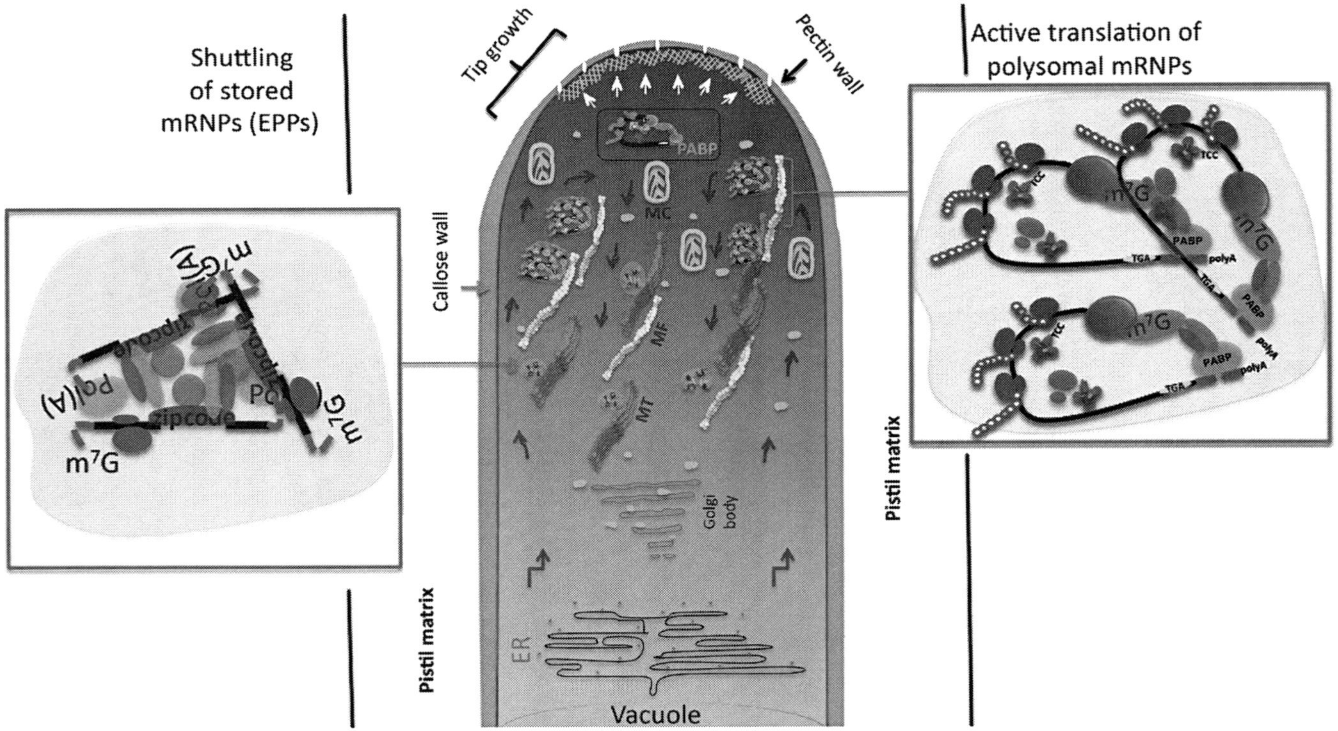

Figure 2. Stored mRNPs facilitate pollen tube tip growth. Directional growth as seen during pollen tube growth as well as in dendritic extension in neuronal cells are preeminent examples of stored mRNA utilization in a form of mRNPs (or EPPs as in tobacco male gametophyte) that provides dynamic and spatial control of gene expression during development. Sequestered stored particles are targeted to the tip region (site of growth) along an array of microtubules (MT) and microfilaments (MF) and delivered to the site of active translation and protein concentration. Neuronal studies have also demonstrated the existence of large RNA granules resembling the characteristics of EPP particles of the tobacco male gametophyte. Features in the diagram are not drawn to scale. ER; endoplasmic reticulum. A color version of this figure is availabe at www.landesbioscience.com/curie.

tip of pollen tube to be translated and deposited at the tip region (Fig. 2). This potential role of EPPs particularly resembles that of the growing dendrites of the neuronal cells in human. The directional growth in neurons is facilitated by the transport of sequestered mRNAs by neuronal granules to the synaptic surfaces for translation. Similar to EPPs, neuronal granules are also preloaded with translational machinery and several regulatory RBPs. Among the identified components of neuronal granules includes silenced mRNAs, RBPs such as; HuD, G3BP, Sam68, SYNCRPI, hnRNP A2, RNG105, FMRP and Staufen, as well as translation initiation factors and small and large ribosomal subunits.[25] As such, the neuronal granules are the mediators of nerve cell networking depositing transcripts to the growing tip and catalyze their efficient translation thereby promoting directional growth. Such a role is already emerging in plant EPPs and pollen tube growth,[1] though further experiments are necessary to validate this role and address other potential functions of EPP's that arbitrates the dynamics of gametophyte development.

Moreover, a recent study by Bayer et al[40] have identified a paternally expressed *SHORT SUSPENSOR* (*SSP*) which promotes elongation of the embryonic suspensor cells by activating *YODA-MAPK* pathway (*YDA*). *SSP* is expressed at the mature pollen stage however the protein can only be detected in the zygote and the endosperm post fertilization suggesting delivery of the transcript by the twin sperm cells. Since the EPP mode of organisation seems to be a male phenomenon so far, the *SSP* transcript is likely to be governed by a similar mechanism as the *NTP303* prior to its delivery in the female gametophyte. In view of the fact that both examples are drawn from genes critical for fertilization and zygotic development, the "indestructible" nature of EPPs points to the notion that EPPs are likely to be a step higher to the role played by regular mRNPs and thus offering maximum storage security and assuring the delivery of the important transcripts.

It remains uncertain how cells control the fate of each transcript, nevertheless, there is an accumulative wealth of evidence that the presence of polyA tails influences mRNA fate either to be stored, decapped and degraded, or returned into the translational machinery for synthesis.[41] An insight into the mechanisms that regulates mRNA shuttling between polysomes and RNA granules has started to emerge and further studies on mRNP characteristics would enlighten the complete network of regulation imposed by mRNPs posttranscriptionally.

SYSTEMATIC TRANSLATION REPRESSION BY SMALL RNAs

The continuous effort that has been focused in the understanding of the mechanisms imposed in posttranscriptional control of gene expression, has lately deciphered a complex network of choreographed interaction between mRNAs, microRNAs, proteins and the surrounding cytosolic structures. Since the discovery of the RNA interference phenomenon in plants, research studies over the last decade have evidently proven that double stranded small RNA molecules are universal and effective regulators of gene expression during development in all eukaryotes. RNA interference (RNAi) particularly the subtype microRNA and small interfering RNA, have long been implicated to modulate expression of their target genes posttranscriptionally by inducing endonucleolytic cleavage of the target transcripts.[42] Similar classes of the small RNAs in animals are now known to be involved in the degradation of the target transcripts as well as in the inhibition of translation initiation.[43] FOG-2 is a transcription factor required for cardiac development in animals and it is regulated by miR-130a.[43] Northern analysis showed steady accumulation of the *FOG-2* RNA transcripts predominantly in the heart and brain, however, immunodetection studies showed the dynamic peak of the protein accumulation at embryonic day 16.5 and its diminution at the neonate stage. This dynamic accumulation profile of the FOG-2 protein corresponds to the alternate accumulation of the miR-130a, pointing to the translational regulation of FOG-2 by miR-130a.[43] Another good example of miRNA translational regulation in animals, is that of the dendritic miRNA-134 required for the outgrowth of hippocampal neurons.[44] The miRNA-134 promotes this outgrowth by inhibiting translation of the translational repressor *PUMILIO-2*. Until recently, the notion of translational repression induced by miRNAs was unique to the animal miRNA pathway. The simplest explanation of this particular feature of miRNA action in animals was the result of imperfect hybridization with their target transcripts, whereas plant miRNAs show perfect complementarity with their target sites. Various studies in plants have now emerged demonstrating the "full-throttle action" of the small RNAs pathways from mere RNA cleavage to the block of translation initiation, hinting that the phenomena of translational repression imposed by miRNAs also exist in plants (Fig. 1). Two independent surveys of flowering abnormalities in Arabidopsis have uncovered a posttranscriptional control of floral organ identity genes named *APETALA2* (*AP2*) and *AP2*-like genes, to be under the regulation of miRNA172.[45,46] It was identified that miRNA172 controls *AP2* activity by modulating *AP2* mRNA levels through cleavage of the mRNA as well as by translational repression of the same transcripts. In respect to these findings, Brodersen et al[47] presented a functional verification of the miRNA role in translation repression in plants, by measuring the abundance of a green fluorescent protein (GFP) integrated with the miRNA171 binding site. Thus, the emerging picture from these findings together with the currently ongoing experiments collectively suggest that miRNAs in plants just like in animals also execute their posttranscriptional regulatory actions at the translational level, independent of directing mRNA cleavage, or even incorporate both levels of posttrancriptional regulation for a more advance control of the target gene expression. Such advance control of gene expression has been hypothesised to be more apparent for genes having a developmental role for the reason that their misregulation can be detrimental during development. Therefore, the dual posttranscriptional control imposed by the formation of mRNP particles together with the action of small RNAs molecules, provide a mechanistic control of gene expression that influence cell fate decisions marking and sustaining the blueprint of development.

REINITIATING TRANSLATION OF THE LOCALIZED, TRANSLATIONALLY SEQUESTERED mRNAs

The innovation of mRNP formation as a posttranscriptional mediator of gene expression and mRNA translation might have occurred primarily to sequester mRNAs from immediate translation. Imminent repression of the mRNP prior to its entry into the cytoplasm is a must, otherwise preloaded translational machinery would initiate premature translation and thus, interfere with the localization "machinery" precluding subcellular localization and preventing delivery to other intended destinations. This particular characteristic of mRNPs has been a key determinant of morphogen gradient and cell fate specification, polarised subcellular activities, in addition to promoting intra- and intercellular mRNA trafficking. Although there is not enough evidence specifically from plants, a firm model depicting numerous mechanisms of translational initiation and reinitiation has been developed generally for prokaryotes and eukaryotes.[48] Once the transcript is anchored to its subcellular location, interaction of cytoskeleton binding proteins with the bound RBPs has been hypothesised to fuel the reinitiation of translation of the sequestered transcript. Recent reports have also provided evidence for the role of posttranslational modification of RBPs in the repression and reinitiation of translation. For instance, a Drosophila ZBP1 protein is required for the localization of β-actin mRNA to the leading edge of a lamellipod (a cytoskeletal actin projections of the moving cell) in fibroblasts and to distal ends in growth cones and dendritic spines in neurons.[49] Binding of ZBP1 to the β-actin zipcodes represses untimed translation during the localization process. This translational inhibition is relieved upon localization of β-actin. On-site phosphorylation of ZBP1 by a member of the family of membrane-associated kinases (*Src*), thereby spatially and temporally separates localization and translational initiation.[50] A similar mechanism has been also implicated for *Ash1* mRNA in growing buds of the yeast *Saccharomyces cerevisiae* to promote mating type by restricting translation of *Ash1* mRNA in the daughter cell nuclei. This repression is attained through the association of the Puf6p protein with the 3′-UTR of *Ash1* mRNA. The deficiency of Puf6p activities results in en-route *Ash1* mRNA translation and subsequent lack of protein symmetry.[51] In instances where the translational repression was an outcome of miRNA binding, no clear studies have emerged to address how the actual repression is lifted and how the mRNP is routed back to the active translational machinery. However, what is known is that in an unrepressed form, the absence of miRNA allows the binding of translation initiation factor eIF4E to the 7-methyl-guanine (m^7G) cap and, through eIF4G, its interaction with poly(A) binding protein (PABP) to form a closed loop necessary for efficient translation. Equally, binding of miRNA to the target mRNA could result in RNA cleavage or create a competition for cap binding between the associated Ago protein and eIF4E, thereby releasing eIF4E/G and halting the initiation of translation.[52] Due to the nature of inhibition, it is tempting to speculate that posttranslational modification most likely of the Ago protein, would be one of the mechanisms that de-represses the inhibition of translation prompted by miRNA binding.

CONCLUSION

Advances in the survey of RNP infrastructure have highlighted the extent through which the role of mRNP complexes becomes indispensable, seemingly orchestrating cellular processes in coordination to developmental cues. Diverse complexes of mRNP particles govern the second level of posttranscriptional control, modulating the abundance of gene expression in response to morphological changes during development. The stepping stones of this type of regulation involve binding of a cohort of RBPs, leading to the titration of the message away from the translation machinery, first localized and then stored at the allocated position for later use or dispensed through several mRNA decay mechanisms. The mRNP-mode of regulation and the repertoire of events thereafter are the major drivers promoting asymmetric distribution of cell fate determinants, mobile signalling simulating developmental and environmental response, as well as concentrating and supplies specific proteins in the localized regions of development. Continued effort in this field is still uncovering a cascade of regulated mRNAs and previously missing components of the network. The sheer scale through which mRNPs operate is well demonstrated in the developing male gametophyte of *Arabidopsis thaliana* in which several independent studies have identified a combined maximum of 48.4% total proteome at the mature pollen stage of the actively expressed late genes.[53-56] These findings suggest a large scale translational repression of the encoded transcripts. From this perspective, comparative studies of the EPP transcriptome would explain "where all the transcripts go" and simultaneously validating translational activation in the growing pollen tube. The rapid growth of the pollen tube, which also resembles that of the human neuronal dentrites, demands a fast delivery of functional proteins to the tip region. Although pollen tubes deliver sperm cells with a negligible amount of cytoplasm, cytoplasmic inheritance through the ovules plays a significant role in transmitting genetic information and reinforcing parental epigenetic patterns. As such, assured storage of the messages becomes vital for inducing programic developmental changes and for stable inheritance of the genetic information. Modern eukaryotes seem to have achieved this by utilizing stored messages in a form of mRNP localized in the proximity of site of action where proteins are synthesised in an instant, guaranteeing fast delivery. Thus, the invention of mRNP particles and its high order arrangement (i.e., EPPs) in the male gametophyte of flowering plants seems to deliver that message without any major pitfalls.

Recent efforts combining genetic, transcriptomic and proteomic studies have led to the identification of more RNA-binding proteins and gave more insight to the mechanism of posttranscriptional control. Similarities and differences to the nature of mRNP-mediated posttranscriptional regulation deserve additional attention to appreciate the significance of the dynamics of the RNPs infrastructure. The application of advanced genomic and proteomic technologies, together with modern histological techniques, is anticipated to shed more insight to this molecular connection of RNPs with patterns of cellular development. Targeted studies of the mechanism of EPP formation and translational repression during male gametophyte development and its significance to the pollen transcriptome and consequently proteome, will be of considerable interest adding the dimension to the mechanisms of posttranscriptional regulation in connection with gametophytic development.

ACKNOWLEDGEMENT

The authors gratefully appreciate the financial support from the Grant Agency of the Czech Republic (grant no. 522/09/0858) and from the Ministry of Education, Youth and Sports of the Czech Republic (projects no. LC06004, OC08011 and OC10054).

REFERENCES

1. Honys D, Reňák D, Feciková J et al. Cytoskeleton-associated large RNP complexes in tobacco male gametophyte (EPPs) are associated with ribosomes and are involved in protein synthesis, processing, and localization. J Proteome Res 2009; 8(4):2015-2031.
2. Wood AJ, Oliver MJ. Translational control in plant stress: the formation of messenger ribonucleoprotein particles (mRNPs) in response to desiccation of Tortula ruralis gametophytes. The Plant Journal 1999; 18(4):359-370.
3. Bailey-Serres J, Sorenson R, Juntawong P. Getting the message across: cytoplasmic ribonucleoprotein complexes. Trends Plant Sci 2009; 14(8):443-453.
4. Ephrussi A, Dickinson LK, Lehmann R. Oskar organizes the germ plasm and directs localization of the posterior determinant nanos. Cell 1991; 66(1):37-50.

5. Gavis ER, Lehmann R. Localization of nanos RNA controls embryonic polarity. CELL-CAMBRIDGE MA- 1992; 71:301-301.
6. Kloc M, Zearfoss NR, Etkin LD. Mechanisms of subcellular mRNA localization. Cell 2002; 108(4):533-544.
7. Yamagata H, Sugimoto T, Tanaka K et al. Biosynthesis of storage proteins in developing rice seeds. Plant Physiol. 1982; 70(4):1094.
8. Choi SB, Wang C, Muench DG et al. Messenger RNA targeting of rice seed storage proteins to specific ER subdomains. Nature 2000; 407(6805):765-767.
9. Honys D, Combe JP, Twell D et al. The translationally repressed pollen-specific ntp303 mRNA is stored in nonpolysomal mRNPs during pollen maturation. Sex Plant Reprod 2000; 13(3):135-144.
10. Twell D. Pollen developmental biology. In: O'Neil SD, ed. Plant Reproduction. Annual Plant Reviews Sheffield, UK: Sheffield Academic Press; 2002. p. 86-153.
11. Honys D, Twell D. Comparative analysis of the Arabidopsis pollen transcriptome. Plant Physiol 2003; 132(2):640.
12. Honys D, Twell D. Transcriptome analysis of haploid male gametophyte development in Arabidopsis. Genome Biol 2004; 5(11):R85.
13. Čapková V, Štorchová H, Tupý J. Protein synthesis in pollen tubes: preferential formation of new species independent of transcription. Sex Plant Reprod 1988; (1):150-155.
14. Štorchová H, Čapková V, Tupý J. A Nicotiana tabacum mRNA encoding a 69-kDa glycoprotein occurring abundantly in pollen tubes is transcribed but not translated during pollen development in the anthers. Planta 1994; 192(3):441-445.
15. Hachet O, Ephrussi A. Splicing of oskar RNA in the nucleus is coupled to its cytoplasmic localization. Nature 2004; 428(6986):959-963.
16. Tange TØ, Nott A, Moore MJ. The ever-increasing complexities of the exon junction complex. Curr Opin Cell Biol 2004; 16(3):279-284.
17. Lewis RA, Kress TL, Cote CA et al. Conserved and clustered RNA recognition sequences are a critical feature of signals directing RNA localization in Xenopus oocytes. Mech Dev 2004; 121(1):101-109.
18. Gonzalez I, Buonomo SBC, Nasmyth K et al. ASH1 mRNA localization in yeast involves multiple secondary structural elementsand Ash1 protein translation. Curr Biol 1999; 9(6):337-340.
19. Arn EA, Cha BJ, Theurkauf WE et al. Recognition of a bicoid mRNA localization signal by a protein complex containing Swallow, Nod and RNA binding proteins. Dev Cell 2003; 4(1):41-51.
20. Lipshitz HD, Smibert CA. Mechanisms of RNA localization and translational regulation. Curr Opin Genet Dev 2000; 10(5):476-488.
21. Martin KC, Ephrussi A. mRNA localization: gene expression in the spatial dimension. Cell 2009; 136(4):719-730.
22. Ross A, Oleynikov Y, Kislauskis E et al. Characterization of a beta-actin mRNA zipcode-binding protein. Mol Cell Biol 1997; 17(4):2158.
23. Farina KL, Huttelmaier S, Musunuru K et al. Two ZBP1 KH domains facilitate {beta}-actin mRNA localization, granule formation and cytoskeletal attachment. J Cell Biol 2003; 160(1):77.
24. Wang C, Washida H, Crofts AJ et al. The cytoplasmic-localized, cytoskeletal-associated RNA binding protein Os Tudor-SN: evidence for an essential role in storage protein RNA transport and localization. The Plant Journal 2008; 55(3):443-454.
25. Anderson P, Kedersha N. RNA granules: posttranscriptional and epigenetic modulators of gene expression. Nat Rev Mol Cell Biol2009; 10(6):430-436.
26. Kiriakidou M, Tan GS, Lamprinaki S et al. An mRNA m^7G cap binding-like motif within human Ago2 represses translation. Cell 2007; 129(6):1141-1151.
27. Crofts AJ, Washida H, Okita TW et al. Targeting of proteins to endoplasmic reticulum-derived compartments in plants. The importance of RNA localization. Plant Physiol 2004; 136(3):3414.
28. Muench DG, Park NI. Messages on the move: the role of the cytoskeleton in mRNA localization and translation in plant cells. Botany 2006; 84(4):572-580.
29. de Heredia ML, Jansen RP. mRNA localization and the cytoskeleton. Curr Opin Cell Biol 2004; 16(1):80-85.
30. St Johnston D. Moving messages: the intracellular localization of mRNAs. Nat Rev Mol Cell Biol 2005; 6:363-375.
31. Lazarowitz SG, Beachy RN. Viral movement proteins as probes for intracellular and intercellular trafficking in plants. The Plant Cell Online 1999;11(4):535.
32. Okita TW, Choi SB. mRNA localization in plants: targeting to the cell's cortical region and beyond. Curr Opin Plant Biol 2002; 5(6):553-559.
33. Bartel DP. MicroRNAs genomics, biogenesis, mechanism and function. Cell 2004; 116(2):281-297.
34. Kidner CA, Martienssen RA. The developmental role of microRNA in plants. Curr Opin Plant Biol 2005; 8(1):38-44.
35. Why cells move messages: the biological functions of mRNA localization. Seminars in Cell and Developmental Biology: Elsevier; 2007.

36. Hannapel DJ. A model system of development regulated by the long-distance transport of mRNA. J Integr Plant Biol 2010; 52(1):40-52.
37. Anderson P, Kedersha N. RNA granules. J Cell Biol 2006; 172(6):803.
38. Boag PR, Atalay A, Robida S et al. Protection of specific maternal messenger RNAs by the P body protein CGH-1 (Dhh1/RCK) during Caenorhabditis elegans oogenesis. J Cell Biol 2008; 182(3):543.
39. Weterings K, Reijnen W, Aarssen R et al. Characterization of a pollen-specific cDNA clone from Nicotiana tabacum expressed during microgametogenesis and germination. Plant Mol Biol 1992; 18(6):1101-1111.
40. Bayer M, Nawy T, Giglione C et al. Paternal control of embryonic patterning in Arabidopsis thaliana. Science's STKE 2009; 323(5920):1485.
41. Brengues M, Parker R. Accumulation of Polyadenylated mRNA, Pab1p, eIF4E and eIF4G with P-Bodies in Saccharomyces cerevisiae. Mol Biol Cell 2007; E06.
42. Baulcombe D. RNA silencing in plants. Nature 2004; 431(7006):356-363.
43. Kim VN, Han J, Siomi MC. Biogenesis of small RNAs in animals. Nat Rev Mol Cell Biol 2009; 10(2):126-139.
44. Fiore R, Siegel G, Schratt G. MicroRNA function in neuronal development, plasticity and disease. Biochim Biophys Acta 2008; 1779(8):471-478.
45. Aukerman MJ, Sakai H. Regulation of flowering time and floral organ identity by a microRNA and its APETALA2-like target genes. The Plant Cell Online 2003; 15(11):2730.
46. Chen X. A microRNA as a translational repressor of APETALA2 in Arabidopsis flower development. Science's STKE 2004; 303(5666):2022.
47. Brodersen P, Sakvarelidze-Achard L, Bruun-Rasmussen M et al. Widespread translational inhibition by plant miRNAs and siRNAs. Science 2008; 320(5880):1185.
48. Collins LJ, Penny D. The RNA infrastructure: dark matter of the eukaryotic cell? Trends Genet 2009; 25(3):120-128.
49. Czaplinski K, Singer RH. Pathways for mRNA localization in the cytoplasm. Trends Biochem Sci 2006; 31(12):687-693.
50. Huttelmaier S, Zenklusen D, Lederer M et al. Spatial regulation of b-actin translation by Src-dependent phosphorylation of ZBP1. Nature 2005; 438:512-515.
51. Gu W, Deng Y, Zenklusen D et al. A new yeast PUF family protein, Puf6p, represses ASH1 mRNA translation and is required for its localization. Genes Dev 2004; 18(12):1452.
52. Meister G. miRNAs get an early start on translational silencing. Cell 2007; 131(1):25-28.
53. Holmes-Davis R, Tanaka CK, Vensel WH et al. Proteome mapping of mature pollen of Arabidopsis thaliana. Proteomics 2005; 5(18):4864-4884.
54. Noir S, Bräutigam A, Colby T et al. A reference map of the Arabidopsis thaliana mature pollen proteome. Biochem Biophys Res Commun 2005; 337(4):1257-1266.
55. Sheoran IS, Sproule KA, Olson DJH et al. Proteome profile and functional classification of proteins in Arabidopsis thaliana (Landsberg erecta) mature pollen. Sex Plant Reprod 2006; 19(4):185-196.
56. Grobei MA, Qeli E, Brunner E et al. Deterministic protein inference for shotgun proteomics data provides new insights into Arabidopsis pollen development and function. Genome Res 2009; 19(10):1786.
57. Forstemann K. Transposon defense in Drosophila somatic cells: A model for distinction of self and nonself in the genome. RNA Biol 2010; 7(2).
58. Malone CD, Brennecke J, Dus M et al. Specialized piRNA pathways act in germline and somatic tissues of the Drosophila ovary. Cell 2009; 137(3):522-535.
59. Pillai RS, Bhattacharyya SN, Artus CG et al. Inhibition of translational initiation by Let-7 MicroRNA in human cells. Science 2005; 309(5740):1573.
60. Sheth U, Parker R. Decapping and decay of messenger RNA occur in cytoplasmic processing bodies. Science 2003; 300(5620):805.
61. Sen GL, Blau HM. Argonaute 2/RISC resides in sites of mammalian mRNA decay known as cytoplasmic bodies. Nat Cell Biol 2005; 7(6):633-636.
62. Bhattacharyya SN, Habermacher R, Martine U et al. Relief of microRNA-mediated translational repression in human cells subjected to stress. Cell 2006; 125(6):1111-1124.
63. Chang Y, Imam JS, Wilkinson MF. The Nonsense-Mediated Decay RNA Surveillance Pathway. Annu Rev Biochem. 2007; 76(1):51-74.

CHAPTER 9

UNEXPECTED FUNCTIONS OF tRNA AND tRNA PROCESSING ENZYMES

Rebecca L. Hurto
Department of Molecular Genetics, The Ohio State University, Columbus, USA
Email: hurto.1@osu.edu

Abstract: tRNA and tRNA processing enzymes impact more than protein production. Studies have uncovered roles for tRNA in the regulation of transcription, translation and protein turnover. Induced by stress or as a programmed part of development, nonrandom tRNA fragments can guide mRNA cleavage, inhibit translation and promote morphological changes. Similarly, tRNA processing enzymes, such as RNaseP and tRNA aminoacyl-synthetases participate in tasks affecting more than tRNA function (i.e., mRNA function and cellular signaling). Unraveling the complexities of their functions will increase our understanding of how mutations associated with disease impact these functions and the downstream consequences. This chapter focuses on how tRNA and tRNA processing enzymes influence cellular function and RNA-infrastructure via pathways beyond the decoding activities that tRNA are known for.

INTRODUCTION

From a protein translation centric viewpoint, the life of mature tRNA is to become aminoacylated by an aminoacyl-synthetase (aaRS), to enter the aminoacyl site of the ribosome, participate in peptide synthesis, to exit the ribosome and then repeat these processes. However, tRNAs perform other functions beyond this cycle. Regulation of transcription, translation and mRNA cleavage, and participating in nonribosomal amino acid transfer are just some ways that tRNAs influence cellular pathways (Fig. 1). Understanding how tRNAs influence these pathways will provide insight as to tRNAs roles in the larger RNA-infrastructure of cells. Recent discoveries indicate that most tRNA fragments found in small RNA extracts are not random and have biological

RNA Infrastructure and Networks, edited by Lesley J. Collins.

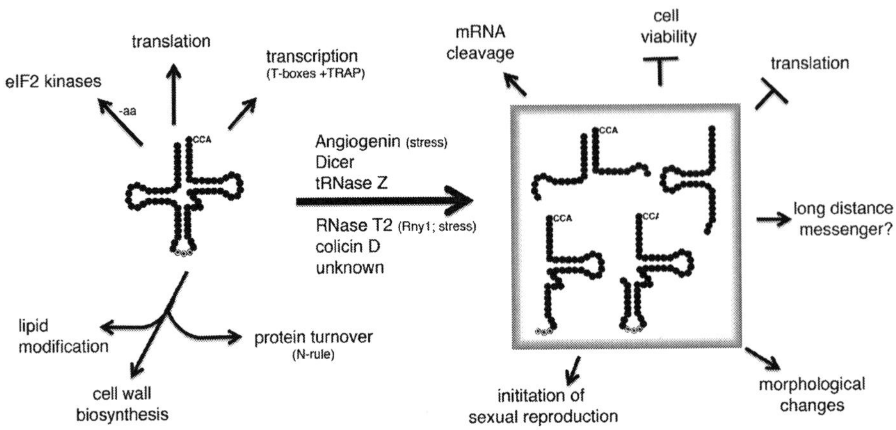

Figure 1. tRNA and tRNA fragments impact a variety of cellular processes.

activities.[1,2] As over-expression of initiator tRNA^{Met} has been shown to be oncogenic,[3] understanding how the various forms of tRNAs function to regulate cellular metabolism will help to unravel the plethora of phenotypes observed in *BRCA1* mutants and other disease related mutations.

tRNAS AND TRANSCRIPTION

Proteins are well known as regulators of cellular transcription. In addition, tRNAs have been shown to influence transcription through their processing and through their indirect and direct interactions. For example, tRNA processing influences the utilization of polycistronic transcripts in several organisms and has led to the tRNA punctuation model (Fig. 2A).[4] tRNAs also have been shown to influence the processing of other RNAs. Uncharged (not aminoacylated) tRNAs can influence transcription by altering translation of specific transcription factors and by shifting the conformational balance of termination/anti-termination switches. These processes are explained in more detail below.

Initial pre-tRNA transcripts usually contain extra sequences at the 5′ and 3′ ends that must be removed during the maturation process. Removal of the 5′ extra sequences is usually catalyzed by a conserved endonuclease called RNase P.[5] The 3′ extra sequences are removed by an assortment of endo- and exonucleases.[6] One of these conserved endonucleases that can catalyze the removal of 3′ extra sequences is tRNase Z.[5] When di- or polycistronic messages contain pre-tRNAs that are interspersed between other mRNAs, rRNAs and ncRNAs (noncoding RNAs), the activities of RNase P and tRNase Z free both the pre-tRNA and the end of the adjoining RNA from the initial transcript. Therefore, the cleavage sites of the pre-tRNA act as "punctuation" marks for initial transcripts. Although first published to describe processing of human mitochondrial transcripts,[4] this method of processing di- and polycistronic RNAs containing pre-tRNA(s) occurs in many organisms. For example, the fruit fly, *Drosophila melanogaster*, has 5 mitochondrial polycistronic primary transcripts containing 11 mRNAs, 22 tRNAs and 2 rRNAs, that are largely processed as predicted.[7] In the plant, *Arabidopsis thaliana*,

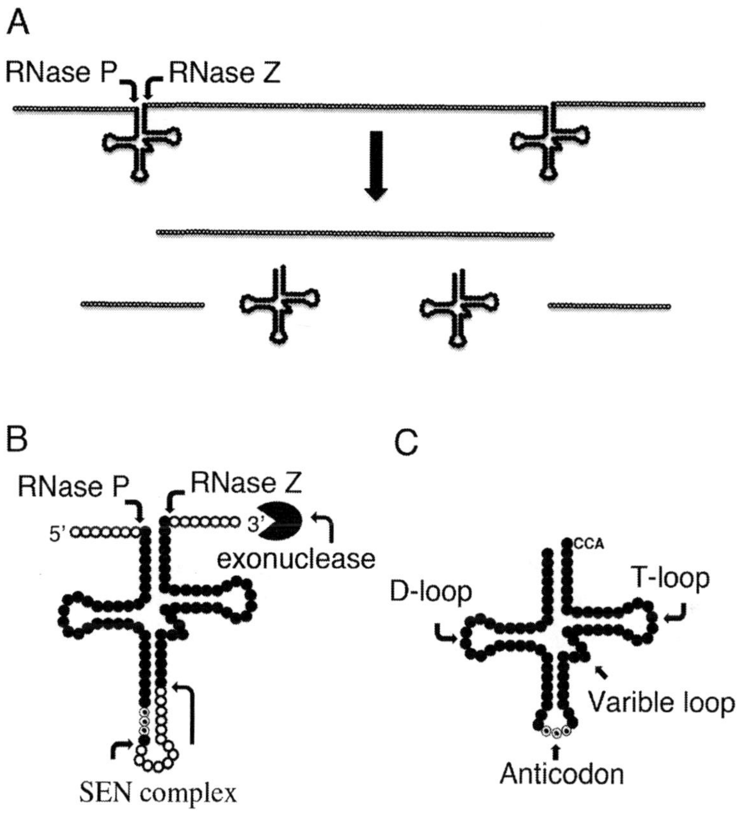

Figure 2. tRNA precursor processing and features. A) The tRNA punctuation model. Polycistronic messages containing pre-tRNAs that are interspersed between other RNAs are cleaved by RNase P and tRNase Z, which recognize the pre-tRNA structure. As a result of these activities, both the pre-tRNA and the adjoining RNAs are freed from the initial transcript. B) Primary pre-tRNAs contain extra sequences (open circles) at the 5′ end (removed by RNaseP), at the 3′ end (removed by tRNaseZ or exonucleases) and sometimes internal to the mature sequence (intron; removed by splicing endonuclease complex [SEN]). C) Structural features of mature tRNA.

which contains 12 nuclear dicistronic tRNA-snoRNA (small nucleolar RNA) precursor RNAs, the $tRNA^{Gly}$-snoR43 precursor was shown to be cleaved by tRNase Z, and snoR43 processing is coupled to snoRNP (small nucleolar ribonucleoprotein complex) formation.[8] As many snoRNAs function as guides for rRNA modifications,[9] the processing of these dicistronic message ties tRNA processing to ribosome modification. In addition, fatty acid synthesis (FAS) is connected to mitochondrial tRNA processing. In the budding yeast, *Saccharomyces cerevisiae*, deletion of FAS genes results in the accumulation of unprocessed pre-tRNAs and unprocessed RPM1, the RNA subunit of mitochondrial RNase P.[10] A second example, in a conserved arrangement found in fish, frogs, mice and humans, the dicistronic RPR14-HTD2 message encodes a RNase P component and FAS protein, providing coregulation of tRNA processing via cotranscription.[10] Transcription of di- or polycistronic messages containing pre-tRNAs can be activated by demand for the mRNA[10] or the tRNA.[11] Processing of the di- or polycistronic messages based on pre-tRNA processing provide additional means for coregulation of mitochondrial tRNA with other cellular processes.

Regulation of transcription via altering translation is a classical theme for tRNAs influence. A well-described eukaryotic example of this regulation is the general amino acid control or amino acid response pathway. Uncharged tRNAs activate translation initiation factor 2 α (eIF2α) kinase, which in turn, phosphorylates eIF2α resulting in decreased translation initiation, thereby allowing translation of the transcription factor Gcn4 (yeast)/Atf4 (human). This transcription factor then activates transcription of many genes including amino acid biosynthetic genes.[12,13]

In prokaryotes, tRNAs can directly or indirectly alter transcription in a number of ways. During the stringent response, uncharged tRNAs enter the ribosomal A site and cause the ribosome to pause which ultimately results in the production of guanosine 3′, 5′-bispyrophosphate (ppGpp). As ppGpp accumulates, it inhibits transcription of rRNAs and tRNAs and inhibits DNA replication.[14] A second system by which tRNA alters transcription of the Trp operon occurs in *E. coli* and takes advantage of cotranscriptional translation. As $tRNA^{Trp}$ becomes increasingly uncharged, ribosomal pausing occurs at a run of Trp codons in the 5′ leader sequence of the Trp operon message. This pausing allows anti-terminator formation to occur resulting in transcription of the entire operon. When $tRNA^{Trp}$ is mostly charged, ribosomal pausing will not occur, the terminator structure forms and causes premature termination of transcription.[15] Another way tRNAs regulate transcription of aaRSs and amino acid biosynthetic genes is by T-box mechanisms. T-boxes are sequences found in the 5′ leader of an mRNA that regulate the balance between termination/anti-termination switches. This regulation can occur during transcription or translation. A recent, detailed review was published on T-box mechanisms;[16] however, a brief overview is presented here. As the nascent transcript is transcribed, T-box sequences encoded in the 5′ leader, have the opportunity to bind specific uncharged tRNAs. Upon binding the uncharged tRNA, the anti-terminator structure is stabilized allowing transcription of the operon. tRNAs with covalently attached amino acids, aminoacylated-tRNAs, act as a non-inducing, binding site competitor. A variation of the T-box mechanism is the TRAP system in the Gram-positive bacteria, *B. subtilis*. In this system, TRAP negatively regulates transcription and translation of the Trp operon. The transcription of the negative regulator anti-TRAP (AT) is regulated by a T-box. T-boxes involved in translation regulation activate translation when uncharged tRNAs bind to sequences that include anti-Shine-Dalgarno sequences, thereby releasing the Shine-Dalgarno sequences to function as the 30S ribosomal subunit binding site.[16]

In summary, eukaryotes and prokaryotes have systems that utilize tRNAs to directly and indirectly regulate transcription. Regulation of the balance of terminator/anti-terminator structures is a common theme in prokaryotes that can be achieved by various mechanisms including second messengers, ribosomal pausing and T-boxes.

tRNA FRAGMENTS

Recent publications on tRNA cleavage in response to stress in ciliates,[17] yeast[18] and mammalian cells[19] have drawn attention to older published data concerning functions of 5′ tRNA halves. This has spurred closer evaluation of the results from "deep" sequencing of cDNA libraries made from small RNAs. The explosion of publications[1,2,18-28] in the last couple of years has provided new insights into the presence, cellular function and mode of action of tRNA fragments. Mature tRNAs can be cleaved as a programmed part of development, in response to stress and in nonstressed cells. The data indicate that tRNA fragments can function as translational inhibitors, long-distance signaling molecules and small guide RNAs. The results of these studies may inspire re-evaluation of existing data sets where tRNA fragments were found but never analyzed. tRNA fragments were previously disregarded as "noise" due to sample degradation in small RNA sequencing; however researchers are realizing that tRNA fragments are a way that cells are singing novel tunes.[2]

Fragments Vary in Source, Size, Abundance and Composition

tRNA fragments are not random, however, the composition, abundance and cleavage site varies by organism and cell type. In order to understand the source of tRNA fragments, it is helpful to highlight the differences between pre-tRNAs and mature tRNAs (Fig. 2B). Initial pre-tRNA transcripts contain extra sequences on the 5′ and 3′ ends and frequently contain extra internal sequences (introns) that must be removed during the maturation process. Mature tRNAs have the extra sequences removed, have additional base modifications that cause predictable mismatches during sequencing (reviewed in ref. 6) and many organisms add the 3′ terminal C, C and A nucleotides posttranscriptionally.[29] Sequencing data provides the most direct evidence that the majority of tRNA fragments are generated from mature tRNAs. Conspicuously absent are reports of clones of tRNA fragments with introns or 5′ sequence extensions. The majority of tRNA fragments cloned with 3′ ends also contain CCA, which is frequently added after removal of extra sequences from the 3′ end of a primary tRNA transcript. There are two noteworthy examples of exceptions. The first is in the ciliate protozoan, *Tetrahymena thermophila*, where the vast majority of 3′ end containing clones lack CCA.[17] There are a several possible explanations for this result. As the cells were undergoing amino acid starvation, uncharged tRNAs may be the preferred substrate for cleavage. An alternative explanation is that the CCA termini may have been lost due to increased expression of tRNase Z, which may cleave off the amino acid and the CCA from mature, aminoacylated tRNAs, a phenomenon observed in *E. coli*.[30] Consistent with the expected mature tRNA cleavage substrate, a number of expected sequence mismatches occurred at sites of predicted nucleoside modification and 5′, 3′ and internal extra sequences were absent. Therefore it is likely that these tRNA fragments were also derived from mature tRNAs in *Tetrahymena thermophila*. A second exception is from a human prostate cancer cell line in which some of the clones contained the 3′ extensions from pre-$tRNA^{Asp}$ and pre-$tRNA^{Ser}$ generated by tRNase Z (ELAC2).[2] Other studies utilized Northern analysis to uncover fragments with sizes and probe hybridization specificities consistent with the source being mature tRNA. The available data indicate that mature tRNAs are the source of the majority of tRNA fragments, though fragments from pre-tRNAs may also be present.

Sequence alignments and Northern analyses have revealed that the tRNA cleavage sites are not random.[1,2,18-28] There are several different types of tRNA fragments that vary by cleavage site, specific species of tRNAs that were cleaved and the abundance of each type of fragments are maintained in that cell type. Cleavage at sites in or near the anticodon generated most of the reported tRNA fragments, usually referred to as tRNA halves, despite the fact that cleavage may generate two unequal pieces.[18,19,22,23] Other tRNA fragments that are generated by cleavage in the 5′ half of the base-paired portion of the tRNA that contains the anticodon,[25,26] the variable loop,[17] the D-loop[22] and the T-loop

(Fig. 2C).[19] Which tRNAs are cleaved and which fragments are retained varies from system to system. For example, in fed and phosphate starved root-tips from *Arabidopsis thaliana*, where tRNA fragments comprise 24 to 34% of the small RNA library, more than 80% of the fragments contained the first 19 nt from the 5′ end of $tRNA^{Gly}$(UCC).[26] In contrast, both 5′ and 3′ halves from all tRNAs tested appeared in approximately equal amounts from *S. cerevisiae* undergoing oxidative stress.[18] Other studies have found 5′ fragments but not 3′ fragments of a particular tRNA and vice versa.[2,19,22] While there are 64 possible codons, wobble base pairing allows most cells to have ~40 different types of mature tRNAs within each cell.[31] As each type of tRNA has its own sequence and many cleavage enzymes have sequence and structural preferences, some species of tRNAs may be more efficiently cleaved by one RNase than another. Additionally, RNAs are less likely to be degraded when bound by protein(s). Analysis of tRNA fragments from different tissues revealed that the abundances of individual tRNA fragments vary from cell-type to cell-type.[2,19,24] While tRNA fragments are not random, there is also organism and tissue specific variability in which tRNAs are targeted and where cleavage occurs. The variability observed in cleavage sites may reflect the variation in functions identified for these fragments.

Fragments and Development

Studies in several organisms have found that tRNA cleavage may have role(s) during development. In many unicellular organisms stress and starvation are triggers for shifting from vegetative growth to sexual reproduction or spore formation. Fragments of tRNAs were observed in stress conditions that induced morphological changes in several species of fungi,[20] bacteria[27] and protists.[17,25] Since starvation often induces transitions from vegetative growth to spore formation, some studies took additional steps to determine whether tRNA fragment appearance is due to starvation or change in morphology. In the fungus, *Aspergillus fumigatus*, tRNA halves were observed during conidiogenesis (spore formation) resulting in reduced levels of mature tRNAs in conidia. tRNA levels recovered during germination.[20] In a second study, Hasier et al[27] noted that the appearance of tRNA halves coincided with initiation of aerial hyphae in the Gram-positive bacteria, *Streptomyces coelicolor*. Some of the mutations in the genes controlling morphology and physiology blocked the appearance of $tRNA^{Met}$ halves and changed the pattern of appearance for other tRNA halves. Another organism for which starvation is important for inducing the mating pathway is *Tetrahymena thermophila*. Starvation and mating pathways can be chemically uncoupled allowing the cells to experience starvation but preventing the cells from becoming mating competent. Under these conditions, tRNA cleavage does not occur, indicating that cleavage is coordinated with life cycle progression.[17] In another study, two different sources of tRNA cleavage products were observed in the flagellated protozoan, *Giardia lamblia*: stress-induced products and differentiation-induced products.[25] The ~46 nt 3′ tRNA fragments observed during differentiation appeared 3 hours after encystation initiation and both the 3′ fragments and mature tRNAs were maintained at constant levels. In contrast, stress caused the appearance of tRNA halves (~37 nt) and the disappearance of mature tRNAs and ~46 nt fragments. Transfer RNA halves may also have a role in human development, as RNA extracts from human fetal hepatic tissue have higher levels of tRNA halves than do cell lines.[24] These data show that the appearance of tRNA fragments can be regulated as part of morphological transitions, indicating that tRNA cleavage may function in development.

Stress-Induced Fragments in Fungi and Plants

Stress-induced tRNA cleavage has been shown to occur in bacteria, fungi, plants and mammalian cells in response to an assortment of stimuli with varied consequences and proposed functions. In view of the crucial role of tRNA as a substrate for translation, the simplest expected function for tRNA cleavage is to halt protein synthesis. While this appears to be the case in some organisms (discussed below), tRNA fragments can also act as long distance messengers and guide RNAs. In *S. cerevisiae*, oxidative stress, heat and high cell density induce accumulation of tRNA halves by inducing the release of Rny1 (RNase T2/ELAC2) from the vacuole. This cleavage resulted in the accumulation of tRNA fragments without strongly influencing the levels of mature tRNAs. Further examination of Rny1 revealed that Rny1 can promote apoptosis independently of its RNase activity.[32] A different study examined the consequences of tRNA cleavage by the exogenous expression of colicin D, which cleaved nearly all of the mature $tRNA^{Arg}$ in *S. cerevisiae*. Colcin D expression induced expression of genes in the mating pathway, induced expression of genes of the pheromone-signaling pathway in mating-type α cells, down-regulated expression of metabolic genes and impaired cell growth. Cumulatively, the data indicated that tRNA cleavage may allow stress survival by providing fast impairment of cell growth and translation (discussed below) in *S. cerevisiae*.[33]

Stress-induced tRNA fragments identified in the phloem sap of the pumpkin, *Cucurbita maxima,* were found to inhibit translation and may function as long distance messengers[22] similar to the micro RNA, miR399.[34] Small RNAs can function as long distance messengers by traveling through phloem from the source tissue producing it to the "sink" tissues which receive the message though the alteration of mRNA expression.[22] Analogous to the high phosphate-induced transport of miR399 from shoots to roots via the phloem where it acts to suppress phosphate uptake,[34] tRNA cleavage products found in the phloem may serve to alter transcription/translation of unknown target genes in response to nutrient availability. Cloning and sequencing of small RNAs from fed and phosphate starved root-tips of *A. thaliana* revealed that the first 19 nt from the 5′ ends of $tRNA^{Gly}$(UCC) comprised 18% and 28% of the total sequence reads. While abundant in root tips, low levels of tRNA fragments were observed in shoots.[26] In contrast, oxidative stress produced high levels of $tRNA^{Arg}$ halves in flowers.[18] These fragments may serve the same functions: inhibiting translation and acting as long distance messengers by providing distant cells information on metabolic status.

Fragments in Mammalian Cells

Small RNA extracts from mammalian cells contain tRNA fragments. A study targeting 17 to 26 nt long RNAs from prostate cancer cells found that ~40% of sequences that appeared at least 5 times mapped to loci containing tRNA and the flanking 25nt. Of the sequences, 77% had precise sites at 5′ and 3′ ends providing evidence that most of the tRNA fragments are not random degradation products caused by cell lysis.[2] Sequencing of a small RNA library from liver carcinoma cells (HepG2) revealed two different populations of small tRNA fragments with sizes of ~22 nt and of 31 to 39 nt, respectively.[19] However, tRNA fragments may be under represented in these data due to limitations of this technique. Reverse transcription, required for these sequencing studies, is inhibited by a subset of nucleoside modifications that are prevalent in mature tRNAs.

Data from other studies (described in the following sections) provide evidence that the two different populations result from the activities of three different enzymes and serve different cellular functions. Although previously unappreciated, basal and stress-induced tRNA fragments are likely to have important roles in cancer.

Dicer Cleaves Mature tRNAs in Human Cells

One of the main functions of Dicer is the cleavage of double stranded RNAs to produce siRNAs; siRNAs function to regulate mRNA stability, alter translation and silence histone modifications.[35] Deep sequencing of small RNAs from HeLa cell extracts uncovered a population of ~19 nt tRNA fragments. The vast majority of the tRNA fragment sequences aligned at the 5′ end with the 5′ end of mature tRNA sequences and the 3′ end aligned with the D-loop of the tRNA. Knockdown of Dicer caused a reduction of ~19 nt tRNA fragments.[1] Dicer-dependent tRNA fragments from a single tRNA were also observed in mouse ES cells.[28] The RNA-induced silencing complex (RISC) incorporates small RNAs that guide its cleavage activities. These tRNA fragments were found to associate with Ago2, a core RISC component, by fractionation and immunoprecipitation in HeLa cells.[1] Association of $tRNA^{Glu}$ fragments with Ago1, a core RISC component, was observed in the fission yeast, *Schizosaccharomyces pombe,* TRAMP mutants. This would not usually occur as the TRAMP complex, a polyA polymerase complex that targets aberrant RNAs for degradation via the nuclear exosome, prevents tRNA association with RNAi machinery in wild type cells.[36] This data led to the proposal that either high-levels of tRNA fragments permit tRNA fragments to bind RISC (competition) or that miRNA processing machinery is processing misfolded tRNAs.[1,36] By extension, a likely conclusion is that tRNA fragments alter Dicer-mediated mRNA cleavage.

tRNase Z is Involved in Pre-tRNA Fragment Production

This highly conserved endonuclease removes the extra sequences from the 3′ end of tRNA and occurs in short (tRNase Z^S) and long (tRNase Z^L) forms.[37] The human tRNase Z^L form, encoded by ELAC2, has been identified as a prostate cancer susceptibility gene.[38] When Lee et al[2] found 5′ and 3′ tRNA halves in small RNA clones, they also identified fragments from pre-tRNAs that start precisely at tRNase Z sites, the 3′ extra sequences. To determine if one of the cloned 3′ extra sequences derived from $tRNA^{Ser}$ (TGA), tRF-1001, had a biological function, tRF-1001 was knocked down in HCT116 cells. The reduction of tRF-1001 caused reduced cell viability, G2 cell cycle block, but no change in pre or mature serine-$tRNA^{TGA}$ levels were observed. Knock down of tRNase Z^L caused a reduction in tRF-1001. While the authors were unable to determine its mode of action, tRF-1001 does not act via a siRNA or miRNA as evaluated by a luciferase reporter.[2] Surprisingly, tRF-1001 and pre-$tRNA^{Ser}$ (TGA) were found to be in the cytoplasm by cell fractionation experiments. It was proposed that pre-$tRNA^{Ser}$ (TGA) has two possible ends, either processed to become a mature tRNA or be exported to become a small RNA.[2] This is the first time that pre-tRNA derived fragments have been shown to have biological activity. While this is an exciting and novel result, perhaps one should not be surprised as previously published data has provided clues that tRNase Z has functions beyond pre-tRNA processing.[39]

Fragment Guided tRNase Z Cleaves mRNA

In a separate parallel function, tRNase Z^L degrades mRNAs in a process that is guided by tRNA halves. An early study recognized that 3′ truncated tRNA could be used to target RNAs, such as HIV, for cleavage by mouse tRNase Z (3′ tRNase) by forming a pre-tRNA-like complex. The sequence and structure of the T-loop were found to be important, but not essential for cleavage.[40] The studies with pig tRNase Z demonstrated that simple double stranded RNAs are not substrates and that DNA is not cleaved. However, the RNA-RNA complex of *lin-4* and *lin-14* from the nematode, *C. elegans,* is a substrate for tRNase Z.[41] Recently, human tRNase Z^L was found to possess the same 5′ tRNA half guided mRNA cleavage activity. tRNA halves copurified with human tRNase Z^L. Importantly, PPM1F was identified as a naturally occurring, endogenous target of a 5′-half-$tRNA^{Glu}$/tRNase Z^L complex. Other small RNA/tRNase Z^L complexes were evaluated.[39] Artificial tRNA-like constructs have been used to target endogenous mRNAs, such as vascular endothelial growth factor (VEGF), for tRNase Z^L mediated cleavage, which demonstrates its potential as a therapeutic approach.[42] This alternative activity may help to explain why some of the ELAC2 mutations associated with prostate cancer do not inhibit its pre-tRNA 3′ end-processing activity.[43] Perhaps these mutations inhibit tRNase Z^L mRNA cleavage activities or alter its specificity resulting in failure to regulate target mRNA stability.

Angiogenin Cleaves Mature tRNAs

Oxidative stress,[18,23] heat shock, UV exposure,[23] PBS treatment, nonserum starvation, hypoxia and hypothermia[24] caused increased accumulation of tRNA halves (~33nt). While many of these stimuli cause apoptosis, other apoptosis inducing agents, such as stauroporine, α-Fas, gamma-irradiation[24] and caffeine,[23,24] did not induce tRNA half accumulation indicating that increased tRNA cleavage is a stress response, not a result of the apoptotic cell death. Mature tRNAs are cut near the anticodon loop (Fig. 2C) by stress-activated angiogenin.[23,24] Angiogenin is a ribonuclease that is secreted, endocytosed by endothelial cells and then transported to the nucleolus. It functions to cleave tRNA, promote rRNA transcription and promote new blood vessel growth, an important factor for tumor formation and growth. Entry into the nucleus was not required for angiogenin-mediated tRNA cleavage as neomycin, a drug that blocks entry of angiogenin into the nucleus, did not affect cleavage.[23] RNAi against RNase T2 (homologue of *S. cerevisiae* Rny1), RNase L, or tRNase Z (ELAC2) did not prevent stress-induced tRNA cleavage, however RNAi against RNH1, a negative regulator of angiogenin, did result in increased tRNA cleavage.[23] In summary, a select variety of stresses can activate angiogenin-mediated tRNA cleavage.

tRNA Halves Can Inhibit Translation

Translation is repressed during stress in several model systems. As the ratio of cleaved tRNAs to mature tRNAs is less than 0.1 in arsenite treated U2OS (human bone osteosarcoma epithelial) cells, the lack of tRNA is unlikely to cause the decrease in translation observed during oxidative stress. Similarly, levels of mature tRNAs were maintained in *S. cerevisiae* during oxidative stress[44] and in *A. thaliana* during phosphate starvation,[26] supporting the model that amount of mature tRNA is not limiting translation.

In contrast, *E. coli* has a two-part response to T4 phage infection. First, the infection induces expression of tRNase Z, which stops translation by cleaving mRNAs[45] and the 3′ ends of mature tRNAs regardless of aminoacylation status.[30] Second, the infection induces expression of colicins, which cleave all or nearly all of the target tRNAs in the anticodon loop,[33] further impairing translation. If lack of tRNA did not cause the observed reduction in translation, then how is translation being repressed in this U2OS system? Transient transfection of 5′ tRNA halves into U2OS and MEF (mouse embryonic fibroblast) cells inhibited translation.[23] This data indicates that tRNA halves can cause translation repression and are not a merely side product. Additionally, tRNA halves have been shown to inhibit translation in wheat germ lysate.[22]

This leaves the very important question of how tRNA halves inhibit overall translation. Translation is frequently down regulated by eIF2α kinases in response to a variety of stimuli. The eIF2α kinase activity can be induced directly through interactions with uncharged tRNAs, or indirectly by phosphorylation by different kinases that were activated by various cellular stresses. Phosphorylation of eIF2α at selected sites results in a reduction in translation. The transfected tRNA halves significantly inhibited translation in MEFs with nonphosphorylatible eIF2α (S51A). This indicates that eIF2α phosphorylation is not necessary for tRNA half induced translational repression. Knockdown of HR1, the 1 of 4 mammalian eIF2α kinases induced by arsenite (oxidative stressor), caused increased tRNA cleavage, suggesting that inhibition of translation may suppress tRNA cleavage.[23] This is consistent with data from *T. thermophila*, where inhibition of translation by addition of cyclohexamide prevented accumulation of tRNA halves regardless of media composition.[17] In *S. cerevisiae,* regulation of translation repression during oxidative stress depends on the oxidant. While the role of tRNA fragments in translation inhibition has yet to be addressed in yeast, the translational apparatus has been evaluated under conditions of oxidative stress. Gcn2, the sole eIF2α kinase in *S. cerevisiae* and Eap1, a regulator of the cap-binding protein, eIF4E, are required for translation repression in response to oxidative stress caused by peroxide, or cadmium and diamide, respectively.[46] Therefore, it is possible that proteins may recognize the tRNA halves and alter global translation. Alternatively, the stress-induced tRNA halves may directly interfere with translation. Determining the consequences of accumulating tRNA halves will provide insight in the anti-tumor and anti-HIV activity of onconase, a tRNA anticodon endonuclease.[47] Stress-induced tRNA halves function to regulate translation by an undefined mechanism.

The discovery of nonrandom tRNA fragments has led to the investigation of new modes of post-transcriptional mRNA regulation. As tRNA fragments are observed in a variety of organisms, from *Tetrahymena* to humans, the mechanisms by which they influence cellular function may be conserved. Most of the fragments are derived from mature tRNAs by the activities of angiogenin, Dicer, tRNase Z^L and yet to be identified endonucleases. While the activities of angiogenin are clearly regulated by stress, production of other tRNA fragments could provide different venues of control and coordinate regulation. The evidence in hand indicates that tRNA fragments influence mRNA stability and global translation and act as signaling molecules. High levels of tRNAs are observed in several types of cancer, including ovarian and cervical.[48] Understanding the role(s) of tRNA fragments may help to explain how over-expression of initiator $tRNA^{Met}$ promotes cell growth and tumor formation in immortalized mouse embryonic fibroblasts.[3] There are a large number of studies that sequenced small RNAs that dismissed tRNA fragments as degradation products, never analyzing them. It is exciting to think of the information that these data sets may hold.

AMINOACYL-tRNA FUNCTIONS BEYOND TRANSLATION

At least twenty different amino acids are carried by specific tRNAs to their next destination. The formation of an aminoacyl bond between a specific amino acid and the specific tRNA is usually catalyzed by a specific aaRS with the aaRS being responsible for ensuring fidelity. However, many unicellular organisms do not possess a full complement of aaRSs needed to add canonical amino acids to tRNA. The most commonly absent aaRSs catalyze the addition of glutamine, asparagine, or cysteine. Also, an aaRS for selenocysteine has yet to be found.[49] To circumvent this issue, some aaRSs will mis-acylate the tRNA and then a second enzyme will complete the process by converting it to the cognate amino acid/tRNA pair. For example, tRNA-dependent amidotransferase (AdT) converts a Glu-tRNA^{Gln} to Gln-tRNA^{Gln} (Fig. 3; reviewed in ref. 50). The presence of AdTs are not limited to unicellular organisms, as homologues of AdT subunits have been identified in flies and humans.[50] Mischarged tRNAs serve as a substrate for amino acid converting enzymes acting as a specialized amino acid biosynthesis pathway. As glutamine is an ammonia storage form and amide donor within the cell, requiring AdT activity ties translation to nitrogen availability via aminoaclyation. These indirect pathways of aminoacylation may help the cells to link translation to metabolism.[49]

Aminoacyl-tRNAs (aa-tRNAs) can transfer amino acids to substrates outside of the ribosome. Through these alternative pathways aa-tRNA contribute to the biosynthesis of cell walls,[51] cell membranes[52] and antibiotics[53] and participate in protein degradation.[54] Factors essential for methicillin (Fem) family of transferases utilize specific aa-tRNAs to build peptide bridges that cross-link peptidoglycan, the major structural component of bacterial cell wall. Selection for tRNA^{Ala}, tRNA^{Ser}, or tRNA^{Gly} is very important, as incorporation of other amino acids inhibits the final step in cross-linking, thereby acting as a chain terminator. Loss of the cross-linking activity leads to increased antibiotic sensitivity,[51] which makes it an attractive therapeutic target. A second group of activities were identified by their role in antibiotic resistance. Aminoacyl-phosphatidylglycerol synthetases utilize aa-tRNAs to attach amino acids, such as lysine, alanine and proline, to the polar head groups of phosphatidylglycerol. These modifications help to neutralize the charge on the cellular membranes and increase the number of intermolecular ionic interactions that confer resistance to cationic antibiotics. These modifications provide a growth advantage during acid or osmotic stress conditions.[52] While aa-tRNAs are sometimes used to evade antibiotics, they also are used as substrates for the production of valanimycin and cyclodipeptides, which have antibacterial and antifungal activity. Additionally, Glu-tRNA^{Glu} is used for the production of delta-aminolevulinic acid, the precursor for the production of hemes, chlorophylls and billins in plants.[53] Thus, cells use aa-tRNAs as amino acid donors in a variety of biosynthetic processes.

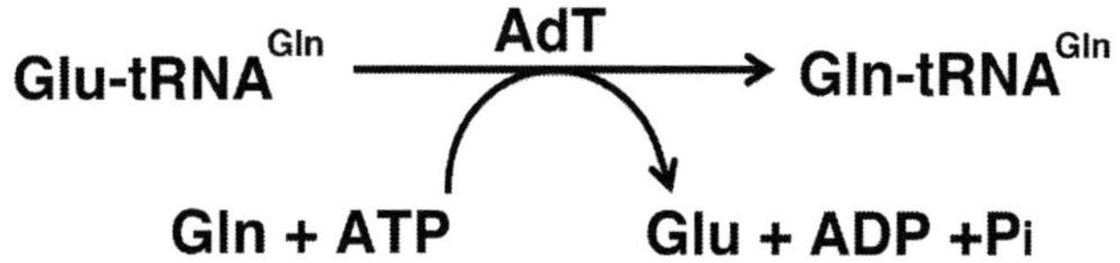

Figure 3. The tRNA-dependent amidotransferase (AdT) reaction. AdT converts a Glu-tRNA^{Gln} to Gln-tRNA^{Gln} by activating glutamic acid (attached to tRNA) using ATP, hydrolysis of the amide group from glutamine and transfer of the amide group to the glutamic acid.

At the other end of the spectrum, tRNAs participate in protein turnover through the N-end rule pathway. The N-end rule is a cellular pathway that reduces the half-life of targeted proteins by adding amino acids to the N-termini, thereby marking targeted proteins for polyubiquitination (eukaryotes) or AAA+ ClpA (prokaryotes) mediated destruction. The addition of amino acids is catalyzed by aminoacyl-transferases that utilize aa-tRNAs as amino acid donors.[53] The N-end rule pathway not only contributes to general turnover, but also is known to mediate various cellular and developmental processes through regulation of these aminoacyl-transferases. For example, *Arabidopsis* shoot and leaf development is regulated in part by *AtATE1* and *AtATE2*, which transfer arginine from Arg-$tRNA^{Arg}$ to BREVIPEDICELLUS, a meristem promoting protein.[54] The requirement of aa-tRNAs as an amino acid donor for nonribosomal biosynthesis, protein degradation and translation has put these processes in competition for the same pool of amino acids. When amino acids are readily available, all of these processes can proceed at maximal rates. However, when amino acids are in low supply, substrate binding affinities and other posttranslational modifications determine how well these processes compete for aa-tRNAs, providing cells with another venue for coordinating regulation.

tRNA PROCESSING ENZYMES AFFECT MORE THAN tRNA

The enzymes involved in tRNA processing and aminoacylation have been shown to moonlight by processing substrates other than pre and mature tRNAs. As mutations of these enzymes have been implicated in disease, determining the affect of each mutation upon each activity carried out by that enzyme will be important for understanding the disease state. For example, the recently identified subunits of the human mitochondrial RNase P complex have other previously established functions (for more on RNase P, please see Chapter 1, Figure 2 page 5 and Chapter 14, Figure 1, page 222).[55] Intriguingly, this complex does not require a RNA component for its pre-tRNA processing function providing evidence to support the controversial protein-only mitochondrial enzyme. This is in sharp contrast to the mitochondrial RNase P in *S. cerevisiae* and other organisms where RNA subunits have been identified.[56] One of the human mitochondrial RNase P proteins, MRPP2, is also known as HSD17B10 (hydroxysteroid [17-β] dehydrogenase 10) due to its 3-hydroxyacyl-CoA dehydrogenase type 2 activity and known as ABAD (amyloid-β-binding protein) demonstrating its association with Alzheimer's disease, a neurodegenerative disease that causes protein aggregate formation and dementia.[57] Additionally, mutations of *HSD17B10* have been linked to mental retardation.[58] However, it is not clear which of the functions of *HSD17B10* are altered in the disease states. As mitochondrial RNase P is required for maturation of polycistronic RNAs containing pre-tRNAs in several organisms,[4,7,10] the mutation of mitochondrial RNase P subunits affects not only pre-tRNA processing but also processing of adjacent RNAs.

Nuclear encoded RNase P is required for processing of several RNAs that are not tRNAs. Cleavage of *MALAT1* (metastasis-associated lung adenocarcinoma transcript-1) by RNase P produces a nuclear long noncoding RNA and a tRNA-like cytoplasmic RNA. *Men*-β, also cleaved by RNase P, is a noncoding RNA that is differentially expressed during myoblast differentiation.[59] In *S.cerevisiae*, RNase P is also involved in the processing of intron-encoded C/D box small nucleolar RNAs.[60] Some RNase P targets contain tRNA-like structures, while other targets do not.[59,60] Similarly, tRNase Z targets have been identified in humans.[39] In addition, nuclear RNase P functions as a transcription factor for Pol III in yeast and in humans [61,62] thereby influencing the production a variety of small RNAs associated with cancer.[48] The highly conserved tRNA splicing endonuclease (SEN) complex removes introns from pre-tRNAs. The human SEN forms two different complexes; both efficiently cleave pre-tRNAs and associate with pre-mRNA 3′ end processing.[63] Mutations of some of the human SEN components were found to have causative roles in pontocerebellar hypoplasia, a developmental neural disorder resulting in small, underdeveloped cerebellum and pons leading to severe retrandation, spinal muscular atrophy and early death. As a mutation of the mitochondrial arginyl-synthetase (ArgRS) was found to have a causative role in a subtype of pontocerebellar hypoplasia, it seems possible that loss or reduction of functional aa-tRNA may contribute to this disease.[64]

AMINOACYL-tRNA SYNTHETASES, tRNA AND DISEASE

In addition to their roles in tRNA aminoacylation, aaRSs participate in a variety of other cellular activities. Some of these additional activities include splicing mRNAs,[65] cytokine activity,[66] and immune responses.[67] For example, yeast and *Neurospora* mitochondrial aaRSs can act to splice group I introns.[65] Several aaRSs have shown to possess cytokine activity. The short form of tyrosyl-synthetase (TyrRS), for example, has cytokine activities that allow it to participate in cell migration and angiogenesis by interacting with the CXCR1 receptor.[66] Lysyl-sythetase (LysRS) is another example of an aaRS that participates in the immune response. When phosphorylated by MAPK/ERK kinase (MEK) in response to IgE, LysRS dissociates from the MSC and increases production of a second messenger diadenosine tetraphosphate. After leaving the MSC, is LysRS imported into the nucleus where it binds the transcription factor MITF to activate transcription of MITF target genes.[67] Therefore, aminoacyl-tRNA synthetases perform functions beyond the expected catalysis of tRNA aminoacylation.

Understanding the alternative functions of each aaRS is important for determining how each of the aaRS functions contribute to disease. For example, mutations of tyrosyl-synthetase (TyrRS)[66] and glycyl-synthetase (GlyRS)[68] cause neuronal dysfunctions that lead to diseases such as Charcot-Marie-Tooth disease, that involves a progressive loss of peripheral nerve function. In a different example, cerebellar ataxia, where muscle control is lost due to cerebellar dysfunction, is caused by a mutation of alanyl-tRNA synthetase that is predicted to block editing, the process that removes misacylated amino acids.[69] Additionally, mutations of the mitochondria aspartyl- and arginyl-synthetases cause leukoencephalopathy (degeneration of myelin sheaths insulating neurons) and pontocerebellar hypoplasia, respectively.[70] The most common approaches to determine if the aminoacylation of tRNAs or alternative activities of aaRSs contribute to the observed phenotypes are to evaluate each known aaRS activity of the mutants and to create mutations that block each known aaRS activity for evaluation. For example, the aminoacylation activity of a mutant aaRS can be evaluated in vitro. Only careful evaluation of mutations that affect each of the individual proteins functions will provide a complete picture of its role in pathogenesis.

Surprisingly, some of the pathogenic mutations of TyrRS and GlyRS do not affect their aminoacylation activities in vitro.[66,71] Careful compilation and analysis of available data by Antonellis and Green[70] led to the proposal that the pathogenic mutations of TysRS and GlyRS cause disease by one of three mechanisms: loss of aminoacylation activity, toxic gain of function, or impaired undiscovered secondary functions. Several lines of data support these possibilities. All but two characterized mutations of TysRS and GlyRS, which do not affect in vitro aaRS activity, have been associated with loss of function characteristics such as mislocalization. One aspect of the aminoacylation activity that has yet to be evaluated is the effect of these mutations on the rate of misincorporation of amino acids. As increased amino acid misincorporation is the predicted effect of the pathogenic gain-of-function mutation of AlaRS, this aspect of aaRS function may have significant impact on neuronal translation and contribute to a disease phenotype. The fact that aaRSs secondary activities are often regulated and are in competition with their aminoacylation activity, has piqued curiosity of this author. For example, when LysRS is activated by MEK, LysRS increases its second messenger production and dissociates from the multisynthetase complex (MSC).[67] What happens to the in vivo Lys-tRNA^{Lys} levels? Is the aaRS activity of other multisynthetase complex components affected? Similar questions can be asked of TyrRS, GlyRS and their mutants. As the absence of one amino acid can cause loss of aminoacylation for several different, noncognate tRNA species in *S. cerevisiae,*[72] it is possible that altering the level of one aa-tRNA may cause the cell to alter levels of other aa-tRNAs. Changing the availability of aa-tRNAs is likely to impact translation and contribute to the disease state.

The identified pathogenic mutations of aaRS genes cause diseases of the nervous system indicating that aaRSs may also be important for neuronal function. Many of these mutations cause loss-of-function phenotypes by impairing aminoacylation or causing mislocalization. However, toxic gain-of-function mutations are also predicted to cause disease. Further evaluation of TyrRS, GlyRS and their mutants will be needed to determine which altered functions are important for the disease phenotypes. Clarification of the contributions of aaRSs to normal neuronal function and determining how the mutations alter these functions will increase our understanding of metabolic regulation.

The functionality of aa-tRNAs is important for normal cellular function. Mutations that result in a loss of aa-tRNA function would be expected to be pathogenic. For example, some mutations of mitochondrial tRNAs that result in reduced mitochondrial translation cause mitochondrial myopathy, encephalopathy, lactic acidosis and stroke-like episodes (MELAS) or myoclonus epilepsy associated with ragged-red fibers (MERRF).[73] Additionally, loss of functional nuclear-cytoplasmic tRNA^{Sec} by conditional knockout in macrophages or liver caused oxidative stress resulting in transcription of stress genes mediated by Nrf2. Although they are known for inhibiting the sterol pathway, 3-hydroxy-3-mehtylglutaryl-coenzyme A inhibitors, "statins", inhibit modification of tRNA^{Sec} at a base necessary for effective incorporation of selenocysteine into proteins. Known side effects of statins include muscle-related complaints, myopathies and oxidative stress.[74] The co-occurrence of oxidative stress and the appearance of myopathies suggest that alteration of aa-tRNA functionality may contribute to disease phenotypes. Another drug used for cancer treatment, 5-flurouracil (5-FU), also affects the function of tRNA. Studies in *S. cerevisiae* revealed that mutation of tRNA modification genes causes increased sensitivity to 5-FU and that 5-FU sensitivity increases with temperature.[75] Additionally, tRNA modification enzymes, psuedouridine synthease and 5-methyluridine tRNA methyltransferase, which contribute to the stability of tRNA tertiary structure,[76] are inhibited by 5-FU.[77,78] The loss of nucleoside modifications and increased temperature likely destabilized the tRNAs, contributing to the cytotoxic effects of 5-FU.[75] These data are consistent with the observation that 5-FU is more cytotoxic in hyperthermic conditions, increasing the effectiveness of 5-FU in cancer therapy.[79] Cumulatively, the data indicate

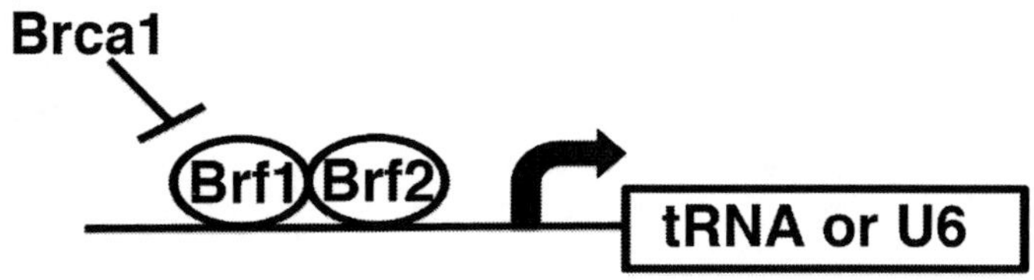

Figure 4. Brca1 regulation of Pol III transcripts. Brca1 negatively regulates the activity of Brf1 and Brf2, TFIIIB subunits, which recruit PolIII to promoters of tRNA and U6 genes.

that impairing tRNA function negatively impacts cellular function, suggesting that loss of tRNA function can contribute to disease states.

Understanding all of the functions that tRNAs perform in the cell is important for understanding disease states. For example, individuals with *BRCA1* mutations are more likely to develop breast cancer.[80] How are cells affected by loss of *BRCA1* function? As Brca1 is a negative regulator of the transcription factors Brf1 and Brf2, part of TFIIIB complex that recruits Pol III to promoters (Fig. 4), there is increased expression of Brf1 and Brf2 target genes. These target genes include tRNAs and snRNAs.[81] Brf1 selectively induces transcription of tRNAs and 5S rRNAs, but not snRNAs, mRNAs or other rRNAs. Over-expression of Brf1 or initiator $tRNA^{Met}$, but not elongator $tRNA^{Met}$, caused increased cell proliferation and promoted transformation in cultured MEF cells. When injected into mice, MEF cells over-expressing either Brf1 or initiator $tRNA^{Met}$, but not elongator $tRNA^{Met}$, caused tumor formation.[3] Additionally, the mutations of *BRCA1* associated with disease frequently occur in the domains of *BRCA1* required for repression of tRNA transcription.[82] These data indicate that over-expression of tRNA likely contributes to the disease state. However, these data provide novel questions. How are cells affected by the over-expression of tRNA(s)? Why does over-expression of initiator $tRNA^{Met}$, but not elongator $tRNA^{Met}$, promote transformation? One can imagine several different possibilities. Perhaps initiator $tRNA^{Met}$ is somehow limiting or regulated and its over-expression alters translation. Another possibility is that fragments generated from initiator $tRNA^{Met}$ have different target RNAs than those generated from elongator $tRNA^{Met}$. Over-expression of initiator $tRNA^{Met}$ may promote degradation or translational repression of mRNAs that would normally produce negative metabolic regulators or tumor suppressor proteins. Clarifying the contribution(s) of tRNAs to breast and other cancers will lead to a better understanding of disease initiation and progression. It will also help to resolve the ambiguity for diseases that are associated with mutations of tRNA processing enzymes and aaRSs. This will aid in the design of new drugs and provide insight into current therapies where the mechanisms of action and downstream consequences are not well described.

CONCLUSION

From their well-known role in translation to their more recently discovered roles in transcription and mRNA cleavage, tRNAs actively participate in numerous cellular activities (Fig. 1) and thus provide numerous opportunities for cells to coordinate metabolism. tRNAs are also involved in translational regulation, often in response to amino acid availability. In eukaryotes, uncharged tRNAs regulate translational repressors such as Gcn2[12] and Atf4,[13] indirectly causing translation of transcription factors leading to the synthesis of proteins involved in amino acid biosynthesis and import. In prokaryotes, uncharged tRNAs can act directly or indirectly to activate production of amino acid biosynthetic proteins and aaRS proteins. An example of an indirect mechanism is ribosomal pausing caused by uncharged tRNAs entering the A-site. Ribosomal pausing leads to production of ppGpp[14] and/or, in the case of the Trp operon, formation of a transcriptional anti-terminator.[15] The T-box mechanism is an example of how uncharged tRNAs can directly affect the balance of terminator/anti-terminator formation allowing tRNAs to regulate transcription and translation in response to the availability of amino acids.[16] Another way that cells use tRNAs to regulate and coordinate metabolism is through the production of tRNA fragments. The majority of fragments are derived from cleavage of mature tRNAs. These fragments are not random, but do vary in size, abundance and tRNA species cleaved depending on organism, tissue type and current cellular environment. Fragments are produced under stress and nonstress conditions and as part of normal development. Some of the demonstrated and proposed functions for tRNA fragments include inhibitors of translation inhibition, long-distance messengers, mRNA turnover machinery and end-processing guides. Other important functions of tRNA include acting as an amino acid donor for nonribosomal aminoacyl transfer and a substrate for amino acid transformation.[49] Through nonribosomal aminoacyl transfer processes tRNAs participate in biosynthesis of prokaryotic cell walls,[51] cell membranes,[52] and antibiotics,[53] and in prokaryotic and eukaryotic protein turnover pathways.[54,82] tRNAs impact cellular function through a wide variety of pathways providing numerous avenues for cross talk and competition that cells use to coordinate their cellular activities. Discovery and elucidation of these functions will aid in the understanding of normal cellular function and provide insight into diseases caused by mutations of aaRSs and tRNA processing enzymes.

ACKNOWLEDGEMENTS

Thanks to the members of the A. K. Hopper Lab, especially A. K. Hopper, E. Kramer and N. Dhungel for comments on the manuscript and to National Institutes of Health for support from grant GM-27930 (to A. K. H.).

REFERENCES

1. Cole C, Sobala A, Lu C et al. Filtering of deep sequencing data reveals the existence of abundant Dicer-dependent small RNAs derived from tRNAs. RNA 2009; 15(12):2147-2160.
2. Lee YS, Shibata Y, Malhotra A et al. A novel class of small RNAs: tRNA-derived RNA fragments (tRFs). Genes Dev 2009; 23(22):2639-2649.
3. Marshall L, Kenneth NS, White RJ. Elevated tRNA(iMet) synthesis can drive cell proliferation and oncogenic transformation. Cell 2008; 133(1):78-89.
4. Ojala D, Montoya J, Attardi G. tRNA punctuation model of RNA processing in human mitochondria. Nature 1981; 290(5806):470-474.
5. Hartmann RK, Gossringer M, Spath B et al. The making of tRNAs and more—RNase P and tRNase Z. Prog Mol Biol Transl Sci 2009; 85:319-368.
6. Hopper AK, Phizicky EM. tRNA transfers to the limelight. Genes Dev 2003; 17(2):162-180.
7. Stewart JB, Beckenbach AT. Characterization of mature mitochondrial transcripts in Drosophila and the implications for the tRNA punctuation model in arthropods. Gene 2009; 445(1-2):49-57.

8. Barbezier N, Canino G, Rodor J et al. Processing of a dicistronic tRNA-snoRNA precursor: combined analysis in vitro and in vivo reveals alternate pathways and coupling to assembly of snoRNP. Plant Physiol 2009; 150(3):1598-1610.
9. Dieci G, Preti M, Montanini B. Eukaryotic snoRNAs: a paradigm for gene expression flexibility. Genomics 2009; 94(2):83-88.
10. Hiltunen JK, Schonauer MS, Autio KJ et al. Mitochondrial fatty acid synthesis type II: more than just fatty acids. J. Biol. Chem 2009; 284(14):9011-9015.
11. Kruszka K, Barneche F, Guyot R et al. Plant dicistronic tRNA-snoRNA genes: a new mode of expression of the small nucleolar RNAs processed by RNase Z. EMBO J 2003; 22(3):621-632.
12. Hinnebusch AG. Gene-specific translational control of the yeast GCN4 gene by phosphorylation of eukaryotic initiation factor 2. Mol Microbiol 1993; 10(2):215-223.
13. Kilberg MS, Shan J, Su N. ATF4-dependent transcription mediates signaling of amino acid limitation. Trends Endocrinol Metab 2009; 20(9):436-443.
14. Potrykus K, Cashel M. (p)ppGpp: still magical? Annu Rev Microbiol 2008; 62:35-51.
15. Yanofsky C. The different roles of tryptophan transfer RNA in regulating trp operon expression in E. coli versus B. subtilis. Trends Genet 2004; 20(8):367-374.
16. Green NJ, Grundy FJ, Henkin TM. The T box mechanism: tRNA as a regulatory molecule. FEBS Lett;584(2):318-324.
17. Lee SR, Collins K. Starvation-induced cleavage of the tRNA anticodon loop in Tetrahymena thermophila. J Biol Chem 2005; 280(52):42744-42749.
18. Thompson DM, Lu C, Green PJ et al. tRNA cleavage is a conserved response to oxidative stress in eukaryotes. RNA 2008; 14(10):2095-2103.
19. Kawaji H, Nakamura M, Takahashi Y et al. Hidden layers of human small RNAs. BMC Genomics 2008; 9:157.
20. Jochl C, Rederstorff M, Hertel J et al. Small ncRNA transcriptome analysis from Aspergillus fumigatus suggests a novel mechanism for regulation of protein synthesis. Nucleic Acids Res 2008; 36(8):2677-2689.
21. Kawaji H, Hayashizaki Y. Exploration of small RNAs. PLoS Genet 2008; 4(1):e22.
22. Zhang S, Sun L, Kragler F. The phloem-delivered RNA pool contains small noncoding RNAs and interferes with translation. Plant Physiol 2009; 150(1):378-387.
23. Yamasaki S, Ivanov P, Hu GF et al. Angiogenin cleaves tRNA and promotes stress-induced translational repression. J Cell Biol 2009; 185(1):35-42.
24. Fu H, Feng J, Liu Q et al. Stress induces tRNA cleavage by angiogenin in mammalian cells. FEBS Lett 2009; 583(2):437-442.
25. Li Y, Luo J, Zhou H et al. Stress-induced tRNA-derived RNAs: a novel class of small RNAs in the primitive eukaryote Giardia lamblia. Nucleic Acids Res 2008; 36(19):6048-6055.
26. Hsieh LC, Lin SI, Shih AC et al. Uncovering small RNA-mediated responses to phosphate deficiency in Arabidopsis by deep sequencing. Plant Physiol 2009; 151(4):2120-2132.
27. Haiser HJ, Karginov FV, Hannon GJ et al. Developmentally regulated cleavage of tRNAs in the bacterium Streptomyces coelicolor. Nucleic Acids Res 2008; 36(3):732-741.
28. Babiarz JE, Ruby JG, Wang Y et al. Mouse ES cells express endogenous shRNAs, siRNAs and other Microprocessor-independent, Dicer-dependent small RNAs. Genes Dev 2008; 22(20):2773-2785.
29. Xiong Y, Steitz TA. A story with a good ending: tRNA 3′-end maturation by CCA-adding enzymes. Curr Opin Struct Biol 2006; 16(1):12-17.
30. Takaku H, Nashimoto M. Escherichia coli tRNase Z can shut down growth probably by removing amino acids from aminoacyl-tRNAs. Genes Cells 2008; 13(11):1087-1097.
31. Agris PF, Vendeix FA, Graham WD. tRNAs wobble decoding of the genome: 40 years of modification. J Mol Biol 2007; 366(1):1-13.
32. Thompson DM, Parker R. The RNase Rny1p cleaves tRNAs and promotes cell death during oxidative stress in Saccharomyces cerevisiae. J Cell Biol 2009; 185(1):43-50.
33. Shigematsu M, Ogawa T, Kido A et al. Cellular and transcriptional responses of yeast to the cleavage of cytosolic tRNAs induced by colicin D. Yeast 2009; 26(12):663-673.
34. Lin SI, Chiou TJ. Long-distance movement and differential targeting of microRNA399s. Plant Signal Behav 2008; 3(9):730-732.
35. Giles KE, Ghirlando R, Felsenfeld G. Maintenance of a constitutive heterochromatin domain in vertebrates by a Dicer-dependent mechanism. Nat Cell Biol;12(1):94-99; sup pp 91-96.
36. Buhler M, Spies N, Bartel DP et al. TRAMP-mediated RNA surveillance prevents spurious entry of RNAs into the Schizosaccharomyces pombe siRNA pathway. Nat Struct Mol Biol 2008; 15(10):1015-1023.
37. Zhao Z, Su W, Yuan S et al. Functional conservation of tRNase ZL among Saccharomyces cerevisiae, Schizosaccharomyces pombe and humans. Biochem J 2009; 422(3):483-492.

38. Tavtigian SV, Simard J, Teng DH et al. A candidate prostate cancer susceptibility gene at chromosome 17p. Nat Genet 2001; 27(2):172-180.
39. Elbarbary RA, Takaku H, Uchiumi N et al. Modulation of gene expression by human cytosolic tRNase Z(L) through 5′-half-tRNA. PLoS One 2009; 4(6):e5908.
40. Nashimoto M. Specific cleavage of target RNAs from HIV-1 with 5′ half tRNA by mammalian tRNA 3′ processing endoribonuclease. RNA 1996; 2(6):523-524.
41. Nashimoto M. Anomalous RNA substrates for mammalian tRNA 3′ processing endoribonuclease. FEBS Lett 2000; 472(2-3):179-186.
42. Elbarbary RA, Takaku H, Tamura M et al. Inhibition of vascular endothelial growth factor expression by TRUE gene silencing. Biochem Biophys Res Commun 2009; 379(4):924-927.
43. Takaku H, Minagawa A, Takagi M et al. A candidate prostate cancer susceptibility gene encodes tRNA 3′ processing endoribonuclease. Nucleic Acids Res 2003; 31(9):2272-2278.
44. Thompson HJ, Heimendinger J, Gillette C et al. In vivo investigation of changes in biomarkers of oxidative stress induced by plant food rich diets. J Agric Food Chem 2005; 53(15):6126-6132.
45. Perwez T, Kushner SR. RNase Z in Escherichia coli plays a significant role in mRNA decay. Mol Microbiol 2006; 60(3):723-737.
46. Mascarenhas C, Edwards-Ingram LC, Zeef L et al. Gcn4 is required for the response to peroxide stress in the yeast Saccharomyces cerevisiae. Mol Biol Cell 2008; 19(7):2995-3007.
47. Saxena SK, Sirdeshmukh R, Ardelt W et al. Entry into cells and selective degradation of tRNAs by a cytotoxic member of the RNase A family. J Biol Chem 2002; 277(17):15142-15146.
48. Marshall L, White RJ. Non-coding RNA production by RNA polymerase III is implicated in cancer. Nat Rev Cancer 2008; 8(12):911-914.
49. Sheppard K, Yuan J, Hohn MJ et al. From one amino acid to another: tRNA-dependent amino acid biosynthesis. Nucleic Acids Res 2008; 36(6):1813-1825.
50. Sheppard K, Soll D. On the evolution of the tRNA-dependent amidotransferases, GatCAB and GatDE. J Mol Biol 2008; 377(3):831-844.
51. Barreteau H, Kovac A, Boniface A et al. Cytoplasmic steps of peptidoglycan biosynthesis. FEMS Microbiol Rev 2008; 32(2):168-207.
52. Roy H. Tuning the properties of the bacterial membrane with aminoacylated phosphatidylglycerol. IUBMB Life 2009; 61(10):940-953.
53. Francklyn CS, Minajigi A. tRNA as an active chemical scaffold for diverse chemical transformations. FEBS Lett 584(2):366-375.
54. Graciet E, Walter F, Maoileidigh DO et al. The N-end rule pathway controls multiple functions during Arabidopsis shoot and leaf development. Proc Natl Acad Sci USA 2009; 106(32):13618-13623.
55. Holzmann J, Frank P, Loffler E et al. RNase P without RNA: identification and functional reconstitution of the human mitochondrial tRNA processing enzyme. Cell 2008; 135(3):462-474.
56. Ellis JC, Brown JW. The RNase P family. RNA Biol 2009; 6(4):362-369.
57. Holzmann J, Rossmanith W. tRNA recognition, processing and disease: hypotheses around an unorthodox type of RNase P in human mitochondria. Mitochondrion 2009; 9(4):284-288.
58. Yang SY, He XY, Olpin SE et al. Mental retardation linked to mutations in the HSD17B10 gene interfering with neurosteroid and isoleucine metabolism. Proc Natl Acad Sci USA 2009; 106(35):14820-14824.
59. Wilusz JE, Spector DL. An unexpected ending: Noncanonical 3′ end processing mechanisms. RNA 2009.
60. Coughlin DJ, Pleiss JA, Walker SC et al. Genome-wide search for yeast RNase P substrates reveals role in maturation of intron-encoded box C/D small nucleolar RNAs. Proc Natl Acad Sci USA 2008; 105(34):12218-12223.
61. Ishiguro A, Kassavetis GA, Geiduschek EP. Essential roles of Bdp1, a subunit of RNA polymerase III initiation factor TFIIIB, in transcription and tRNA processing. Mol Cell Biol 2002; 22(10):3264-3275.
62. Jarrous N, Reiner R. Human RNase P: a tRNA-processing enzyme and transcription factor. Nucleic Acids Res 2007; 35(11):3519-3524.
63. Paushkin SV, Patel M, Furia BS et al. Identification of a human endonuclease complex reveals a link between tRNA splicing and pre-mRNA 3′ end formation. Cell 2004; 117(3):311-321.
64. Budde BS, Namavar Y, Barth PG et al. tRNA splicing endonuclease mutations cause pontocerebellar hypoplasia. Nat Genet 2008; 40(9):1113-1118.
65. Lambowitz AM, Perlman PS. Involvement of aminoacyl-tRNA synthetases and other proteins in group I and group II intron splicing. Trends Biochem Sci 1990; 15(11):440-444.
66. Kapoor M, Otero FJ, Slike BM et al. Mutational separation of aminoacylation and cytokine activities of human tyrosyl-tRNA synthetase. Chem Biol 2009; 16(5):531-539.
67. Yannay-Cohen N, Carmi-Levy I, Kay G et al. LysRS serves as a key signaling molecule in the immune response by regulating gene expression. Mol Cell 2009; 34(5):603-611.

68. Banks GT, Bros-Facer V, Williams HP et al. Mutant glycyl-tRNA synthetase (Gars) ameliorates SOD1(G93A) motor neuron degeneration phenotype but has little affect on Loa dynein heavy chain mutant mice. PLoS One 2009; 4(7):e6218.
69. Lee KW, Briggs JM. Molecular modeling study of the editing active site of Escherichia coli leucyl-tRNA synthetase: two amino acid binding sites in the editing domain. Proteins 2004; 54(4):693-704.
70. Antonellis A, Green ED. The role of aminoacyl-tRNA synthetases in genetic diseases. Annu Rev Genomics Hum Genet 2008; 9:87-107.
71. Antonellis A, Lee-Lin SQ, Wasterlain A et al. Functional analyses of glycyl-tRNA synthetase mutations suggest a key role for tRNA-charging enzymes in peripheral axons. J Neurosci 2006; 26(41):10397-10406.
72. Zaborske JM, Narasimhan J, Jiang L et al. Genome-wide analysis of tRNA charging and activation of the eIF2 kinase Gcn2p. J Biol Chem 2009; 284(37):25254-25267.
73. Kirino Y, Suzuki T. Human mitochondrial diseases associated with tRNA wobble modification deficiency. RNA Biol 2005; 2(2):41-44.
74. Suzuki T, Kelly VP, Motohashi H et al. Deletion of the selenocysteine tRNA gene in macrophages and liver results in compensatory gene induction of cytoprotective enzymes by Nrf2. J Biol Chem 2008; 283(4):2021-2030.
75. Gustavsson M, Ronne H. Evidence that tRNA modifying enzymes are important in vivo targets for 5-fluorouracil in yeast. RNA 2008; 14(4):666-674.
76. Chernyakov I, Baker MA, Grayhack EJ et al. Chapter 11. Identification and analysis of tRNAs that are degraded in Saccharomyces cerevisiae due to lack of modifications. Methods Enzymol 2008; 449:221-237.
77. Huang L, Pookanjanatavip M, Gu X et al. A conserved aspartate of tRNA pseudouridine synthase is essential for activity and a probable nucleophilic catalyst. Biochemistry (Mosc) 1998; 37(1):344-351.
78. Frendewey DA, Kladianos DM, Moore VG et al. Loss of tRNA 5-methyluridine methyltransferase and pseudouridine synthetase activities in 5-fluorouracil and 1-(tetrahydro-2-furanyl)-5-fluorouracil (ftorafur)-treated Escherichia coli. Biochim Biophys Acta 1982; 697(1):31-40.
79. Kouloulias V, Plataniotis G, Kouvaris J et al. Chemoradiotherapy combined with intracavitary hyperthermia for anal cancer: feasibility and long-term results from a phase II randomized trial. Am J Clin Oncol 2005; 28(1):91-99.
80. Hayes DF. Tumor markers for breast cancer. Ann Oncol 1993; 4(10):807-819.
81. Veras I, Rosen EM, Schramm L. Inhibition of RNA polymerase III transcription by BRCA1. J Mol Biol 2009; 387(3):523-531.
82. Rosen EM, Fan S, Ma Y. BRCA1 regulation of transcription. Cancer Lett 2006; 236(2):175-185.

CHAPTER 10

PROGRAMMED DNA ELIMINATION IN *TETRAHYMENA*: A Small RNA-Mediated Genome Surveillance Mechanism

Kensuke Kataoka and Kazufumi Mochizuki*

Institute of Molecular Biotechnology of the Austrian Academy of Sciences, Vienna, Austria
**Cooresponding Author: Kazufumi Mochizuki—kazufumi.mochizuki@imba.oeaw.ac.at*

Abstract RNA interference (RNAi) was initially discovered as a post-transcriptional gene silencing mechanism in which short RNAs are used to target complementary RNAs for degradation. During the past several years, it has been demonstrated that RNAi-related processes are also involved in transcriptional gene silencing by directing formation of heterochromatin. The dynamic DNA rearrangement during sexual reproduction of the ciliated protozoan *Tetrahymena* provides an extreme example of RNAi-directed heterochromatin formation. In this process, small RNAs of ~28-29 nt, which are processed by the Dicer-like protein Dcl1p and are associated with the Argonaute family protein Twi1p, induce heterochromatin formation at complementary genomic sequences by recruiting the histone H3 lysine 9/27 methyltransferase Ezl1p and chromodomain proteins. Eventually these heterochromatinized regions are targeted for DNA elimination. In many eukaryotes, one of the major roles for RNAi-related mechanisms is silencing transposons, and DNA elimination in *Tetrahymena* is also believed to have evolved as a transposon defense by removing transposon-related sequences from the somatic genome. Because DNA elimination is achieved by the coordinated actions of noncoding RNA transcription, RNA processing, RNA transport, RNA-RNA and RNA-protein interactions, RNA degradation and RNA-directed chromatin modifications, DNA elimination in *Tetrahymena* is a useful model to study 'RNA infrastructure'.

RNA Infrastructure and Networks, edited by Lesley J. Collins.

INTRODUCTION

Transposable elements and viruses are genomic 'parasites' that are capable of moving between cells and from one genomic position to another. It is imperative that they should be inactivated because they are otherwise potentially detrimental to genome stability. Therefore host cells need to distinguish these molecular parasites from their own genomes to selectively silence the former.

Prokaryotic bacteria and archaea employ a restriction-modification (R/M) system, which is composed of restriction enzymes and modification methylases, for selective digestion of nonmodified invading DNAs. This system serves as a defense against invading viruses (aka bacteriophages).[1] Recently, the CRISPR (Clustered Regularly Interspaced Short Palindromic Repeat) system, another bacteriophage defense mechanism, has been discovered (Fig. 1A).[2-4] CRISPR loci consist of multiple short nucleotide repeats (typically 25-40 nt) separated by similarly-sized unique spacer sequences that are homologous to bacteriophage genomes and plasmid sequences. The entire CRISPR loci are transcribed as primary transcripts containing a full set of repeats and spacer sequences. The transcripts are processed into small RNAs (~25-40 nt) by a set of Cas (CRISPR-associated) proteins encoded adjacent to the CRISPR clusters. Each small RNA includes a unique spacer sequence. The small RNA associates with an effector Cas protein complex and directly promotes degradation of bacteriophage DNA as well as possibly the RNA product that is complementary to the small RNA.[5-7] In this system, the unique spacer sequences serve as a catalog of bacteriophages to be silenced. Moreover, new spacer sequences are acquired from novel phage genomes.[8] Therefore, the RNA-based CRISPR system provides a heritable acquired immunity against bacteriophages (For more information on the CRISPR system, please see Chapter 13, pages 213-218).

Eukaryotes also utilize small RNAs for gene silencing. However, the mechanism of small RNA production is most likely evolutionarily unrelated to that employed by the CRISPR system.[4] In eukaryotes, RNAi-related pathways are responsible for generating and employing small RNAs. RNAi was first discovered as a post-transcriptional gene silencing mechanism in which double-stranded (ds) RNAs trigger degradation of complementary mRNAs.[9] In this pathway, dsRNA is processed by a Dicer enzyme into small (20-30 nt) RNAs that subsequently form effector complexes with Argonaute proteins, which induce small RNA-directed RNA degradation.[10] An RNAi-related mechanism can also induce transcriptional gene silencing at the chromatin level in organisms ranging from unicellular eukaryotes to plants and humans.[11] Although RNAi has intrigued biologists as a useful genetic tool, it may have evolved as a guardian of cellular genomic DNA from molecular parasites.[12]

The RNAi-related mechanism plays important roles in the defense against viruses.[13] Some viruses possess a dsRNA genome, and single-stranded RNA viruses also have a dsRNA phase of their replication cycles. These dsRNAs are recognized by the RNAi machinery and processed into small RNAs called small interfering (si) RNAs, which direct degradation of viral genomes and their transcripts. RNAi-related pathways also act in silencing repeated elements (Fig. 1B). Because of the ability to amplify their own genomes, successfully integrated retrotransposons such as LINEs and SINEs are repeated in their host genome. Read-through transcripts from adjacent genes result in an accumulation of dsRNAs for these elements.[14] These dsRNAs are processed to siRNAs by RNAi-related pathways, which then induce degradation of transcripts from molecular parasites as well as transcriptional silencing through alteration of chromatin structure. By these mechanisms, any dsRNA is recognized as a signature of invaders.

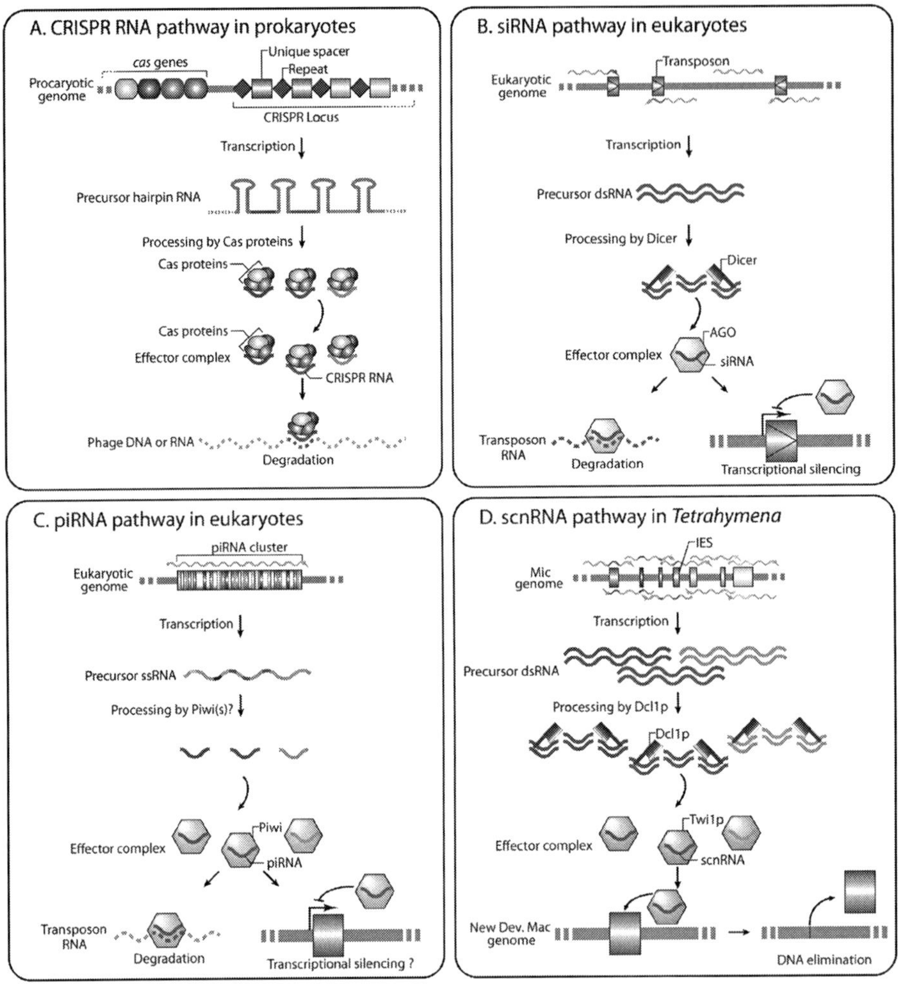

Figure 1. Examples of small RNA-directed defense systems against molecular parasites in prokaryotes and eukaryotes. A) The CRISPR pathway, a prokaryotic defense system. CRISPR RNAs are processed by a set of Cas proteins and the complex targets phage DNA or RNA for degradation. B) siRNA pathway in eukaryotes. Read-through transcripts form adjacent genes produces dsRNA containing a transposon sequence. The dsRNA is processed to siRNA by Dicer. Argonaute protein of the AGO subfamily associates with siRNA and induces degradation of transcripts from transposons as well as transcriptional silencing through alteration of chromatin structure. C) The piRNA pathway in the metazoan germline. piRNA cluster produces single-stranded RNA that is processed to piRNAs, probably by Piwi protein(s). Argonaute proteins of the Piwi subfamily associate with piRNAs and induce silencing of transposons complementary to the piRNAs at the post-transcriptional and transcriptional levels. D) The scnRNA pathway in *Tetrahymena*. dsRNA produced from the micronuclear (Mic) genome is processed to scnRNA by the Dicer-like enzyme Dcl1p. scnRNA interacts with the Piwi subfamily protein Twi1p and promotes heterochromatin formation on IES in the newly developing macronucleus (New Dev. Mac). Eventually, IES is eliminated from the gnome. Gray lines and waved lines represent genomes and transcripts, respectively.

Recent studies in mammals and flies have revealed that these metazoans also silence transposons with small RNAs derived from loci concentrated in degenerated transposon sequences (Fig. 1C).[15,16] These loci, called piRNA clusters, produce single-stranded RNA that is processed to small Piwi-associated (pi) RNAs, probably by sequential actions of endoribonucleolytic activities of two Piwi proteins.[16,17] Argonaute proteins of the Piwi subfamily associate with piRNAs and induce silencing of transposons complementary to piRNAs at the post-transcriptional and transcriptional levels. Although this is reminiscent of the CRISPR system in prokaryotes, the machinery employed in piRNA-mediated transposon silencing in metazoans differs from that in the prokaryotic system.

The ciliated protozoan *Tetrahymena thermophila* employs a unique strategy to silence transposable elements, utilizing an evolutionarily conserved RNAi-related mechanism (Fig. 1D). Its nuclear dimorphism allows *Tetrahymena* to identify transposable elements by comparing the whole genome of vegetative and germline nuclei and to completely eliminate the transposons from the transcriptionally active vegetative nucleus. Small RNAs produced in the germline nucleus are selected for transposable elements specific to the vegetative nucleus; these RNAs then induce heterochromatin formation followed by DNA elimination in the newly developing vegetative nucleus. This small RNA-directed programmed DNA elimination is orchestrated by amazingly complex RNA infrastructure[18], which is composed of spatiotemporally regulated processes including noncoding RNA transcription, RNA processing, RNA transport, RNA-RNA and RNA-protein interactions, RNA degradation and RNA-directed heterochomatin formation. This chapter reviews roles of these processes in DNA elimination in *Tetrahymena* and discusses evolutionary relationship between RNA-directed DNA elimination and transposon-silencing mechanisms in other eukaryotes.

NUCLEAR DIMORPHISM IN *TETRAHYMENA*

Tetrahymena thermophila is a ciliated protozoan, a free-living unicellular eukaryote. *Tetrahymena* and most other ciliates have two structurally and functionally different nuclei within a single cell: a diploid micronucleus and a polyploid macronucleus (Fig. 2).

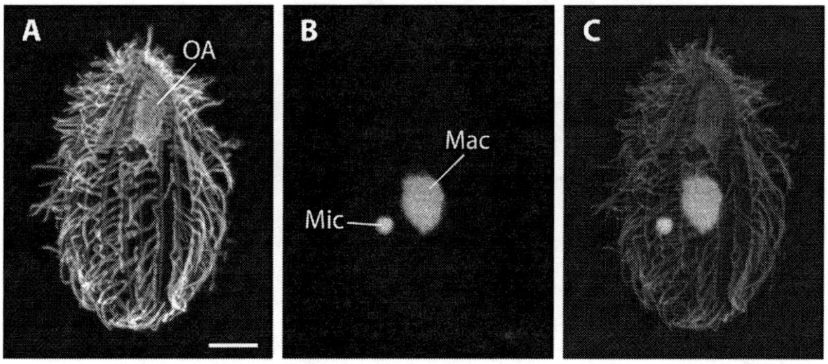

Figure 2. Nuclear dimorphism in *Tetrahymena thermophila*. *Tetrahymena* possesses two distinct nuclei within a single cell. A confocal microscopic view of *Tetrahymena* shows the larger macronucleus (Mac) containing the somatic genome and the smaller micronucleus (Mic) containing the germline genome. A cilium stained with anti-alpha tubulin antibody is shown in (A), a nucleus stained with DAPI in (B) and the merged image in (C). OA: oral apparatus. Scale bar, 5 μm.

The micronucleus is mostly inert transcriptionally (an exception to this generalization is described below), but it has the ability to make both the macro- and micronuclei during sexual reproduction. In contrast, the macronucleus is responsible for all gene expression but is degraded after sexual reproduction. Elimination of 'junk' DNA (see below) and a ~50-fold endoreplication of its chromosomes make the macronucleus a highly specialized machine for gene expression. This division of labor between micro- and macronuclei is reminiscent of the germline-soma separation in metazoans.[19]

When enough nutrients are available, *Tetrahymena* vegetatively (asexually) proliferates by binary fission, and the diploid micronucleus divides mitotically while the polyploid macronucleus divides by amitosis. In the absence of plentiful nutrients, *Tetrahymena* undergoes the sexually reproductive process of conjugation (Fig. 3). During

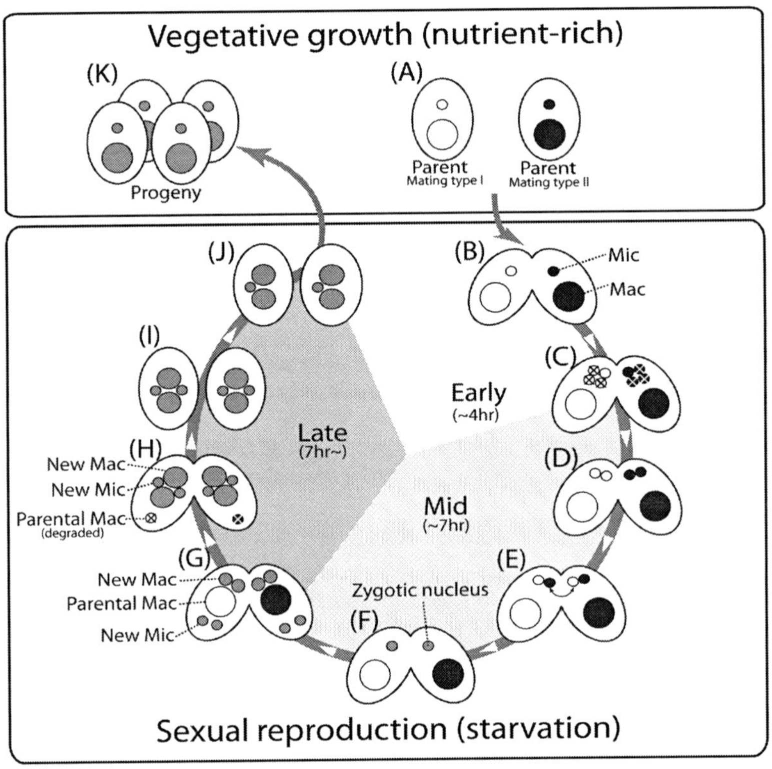

Figure 3. Life cycle of *Tetrahymena*. A) In nutrient-rich conditions *Tetrahymena* grows vegetatively by binary fission. B) Starvation induces sexual reproduction by conjugation. Conjugation begins with pairing of complementary mating types; the parent expressing mating type I has nuclei in white and nuclei from mating type II are in black. C) The micronucleus undergoes meiosis to produce four haploid nuclei, three of which degrade. D) The remaining haploid micronucleus divides mitotically. E-F) The haploid nuclei are exchanged reciprocally and fuse to make a zygotic diploid nucleus shown in gray. G) The zygotic nucleus, undergoes two mitotic divisions. From the four daughter nuclei of these divisions, the two located in the anterior cytoplasm differentiate into macronuclei and the two located posteriorly become micronuclei. H) The parental macronucleus is destroyed by an apoptosis-like DNA degradation process. I) The cells separate and one of the two micronuclei is degraded. J-K) The remaining micronucleus divides mitotically, and the subsequent cell division yields four progeny that are vegetative cells. The anterior to posterior axis of the cells is shown from top to bottom, respectively. Mac, macronucleus; Mic, micronucleus.

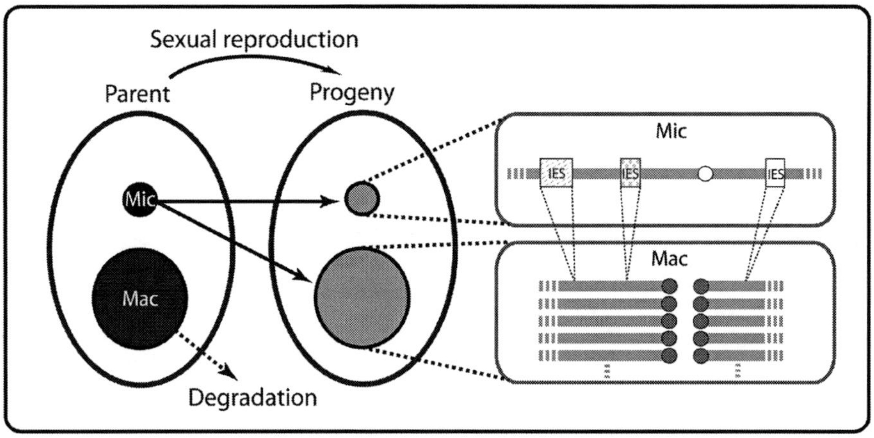

Figure 4. The micro- and macronuclei differentiate from the same zygotic nucleus that is generated after meiosis of the parental micronucleus and subsequent karyogamy during sexual reproduction. The diploid micronucleus always maintains the entire genome. The chromosome contains both macronucleus-destined sequences (gray line) and micronucleus-limited sequences (IES; boxes). Approximately 6000 IESs (~0.5-20 kb in size) are excised from the developing macronuclear chromosomes during differentiation. Simultaneously, the chromosomes are fragmented at approximately 250 sites (white circles) by chromosome breakage coupled with addition of new telomeres (dark gray circles) and are endoreplicated ~50 times. In parallel, the parental macronucleus is degraded at the end of sexual reproduction. Mac, macronucleus; Mic, micronucleus.

conjugation, two cells of complementary mating types partially fuse to make a pair and their micronuclei undergo meiosis (Fig. 3B). One of the four meiotic products in each cell divides once mitotically to produce two haploid pronuclei while the other three meiotic products are degraded (Fig. 3C,D). The paired cells exchange one of their pronuclei, then the stationary and exchanged pronuclei fuse to make a zygotic nucleus (Fig. 3E,F). The zygotic nucleus divides twice. In the second post-zygotic nuclear division, the spindles are parallel to the anteroposterior axis of the cells, and the two daughter nuclei in the anterior cytoplasm differentiate into macronuclei while the other two in the posterior cytoplasm become micronuclei (Fig. 3G,H). In parallel, the parental macronucleus is eliminated by an apoptosis-like process (Fig. 3H).

The micronucleus possesses five chromosomes per haploid set. During differentiation of the new macronucleus, these germline chromosomes are fragmented into approximately 250 pieces by chromosome breakage, which is coupled to the addition of new telomeres.[20-22] Moreover, 15-20 Mbp (~15%) of DNA in the micronuclear genome is removed from the new macronuclear chromosomes (Fig. 4).[23] DNA destined to be eliminated is called an Internal Eliminated Sequence (IES). IESs are ~0.5-20 kb in size and it has been estimated that there are approximately 6000 IES loci in the micronuclear genome. IESs have not been found in gene coding sequences in *Tetrahymena*, although some are located in introns. The majority of known IESs are transposon-like repetitive DNAs and other repeated sequences, which are often categorized as 'junk' DNA in eukaryotic genomes.[24-26] Elimination of IESs from the somatic macronucleus is thought to be essential for cell viability, because all known mutant strains with defective DNA elimination produce nonviable progeny.[27-36] This may be because transposons in IESs are activated in the macronucleus and compromise essential cellular functions.

EPIGENETIC REGULATION OF DNA ELIMINATION BY SMALL RNAS

IES elimination in *Tetrahymena* occurs very precisely such that all given IESs are eliminated in each sexual reproductive event, and their boundaries are conserved to within several base pairs (Fig. 5A). However, no common sequence signature that

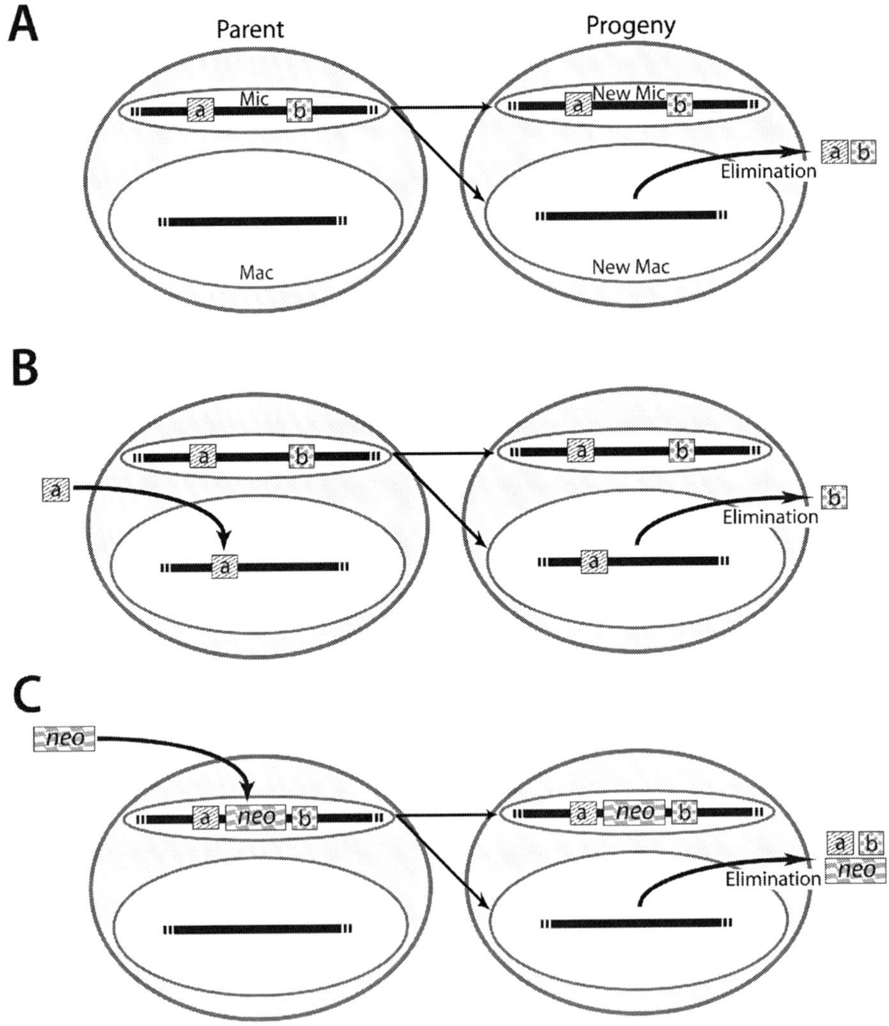

Figure 5. Epigenetic regulation of DNA elimination. A) Genome rearrangement in a wild type strain. IES-(a) and IES-(b) are precisely eliminated from the chromosome during new macronucleus development. During this process, the parental macronucleus is degraded. B) Effect on an IES in the parental macronucleus. When IES-(a) is artificially introduced into the parental macronucleus, elimination of the endogenous IES-(a) is inhibited in the new macronucleus.[37] C) Effect of an exogenous sequence in the parental micronucleus. When a bacteriophage *neo* sequence is artificially introduced into the parental micronucleus, this sequence is eliminated from the new macronucleus.[38,39] Mac, macronucleus; Mic, micronucleus.

might support precise IES elimination has been identified in or around IESs. Chalker and Yao[37] demonstrated that IES elimination in *Tetrahymena* is epigenetically controlled by a mechanism that prevents sequences within the parental macronucleus from being eliminated in the newly developing macronucleus. They first introduced an ectopic IES element, which is normally absent from the mature macronucleus, into the macronuclear genome of the parental strains. Conjugation was then induced using these IES-introduced parental strains, and DNA elimination in their progeny was observed. They found that endogenous IES elements complementary to the artificially introduced IESs failed to be eliminated from the new macronuclear genome (Fig. 5B). Conversely, it has also been reported that a transgene introduced only into the micronucleus of the parental strain is deleted from the developing new macronucleus irrespective of where the transgene is inserted (Fig. 5C).[38,39]

These observations indicate that the new macronucleus epigenetically inherits the pattern of DNA elimination from the parental macronucleus. In other words, *Tetrahymena* can compare micronuclear and parental macronuclear DNA sequences to remove micronuclear-specific DNAs (i.e., IESs) from the newly developing macronucleus. Because there are ~6000 different IESs, it had been expected that DNA or RNA molecules, but not sequence-specific DNA binding proteins, would be used to identify IESs for epigenetic inheritance.[26]

The requirement for the Argonaute protein Twi1p in DNA elimination and accumulation of conjugation-specific small (~28-29 nt) RNAs immediately suggested that these small RNAs, produced by an RNAi-related mechanism, direct DNA elimination.[32] Indeed, later studies have revealed that these small RNAs (named scan RNAs or scnRNAs) are produced by the Dicer-like protein Dcl1p, which is also essential for DNA elimination,[31,33] and form a complex with the Argonaute protein Twi1p.[40] Moreover, it has been demonstrated that injection of dsRNA into conjugating cells can induce ectopic elimination of DNA that is complementary to the injected RNA.[39] Injected dsRNAs are probably processed to small RNAs which then induce IESs elimination. These results strongly argue that scnRNAs are the primary factors that determine which sequences will be eliminated. In the following sections, we describe our current understandings about how scnRNAs epigenetically direct DNA elimination event.

BIOGENESIS OF scnRNA

It has been known for several decades that transcription in the micronucleus can be detected only at early stages of conjugation during prophase meiosis.[41,42] The micronuclear transcripts are ~0.2-1.0 kb in length and are produced from both strands of the genome (Early Stage in Fig. 6).[43] Although the exact mechanism that produces these transcripts is not clear, temporal micronuclear localization of RNA polymerase II (RNAPII) during prophase meiosis[44] indicates that they are transcribed by RNAPII. Knockout strains for *DCL1* do not produce scnRNA and result in over-accumulation of micronuclear RNA,[31,33] indicating that micronuclear transcripts are precursors of scnRNAs. Since all IESs and macronuclear-destined sequences examined so far are transcribed during meiotic prophase, it was proposed [27,43] that the entire micronuclear genome is transcribed in this stage. However, a large-scale genome-wide analysis of micronuclear transcripts will be necessary to confirm this assumption.

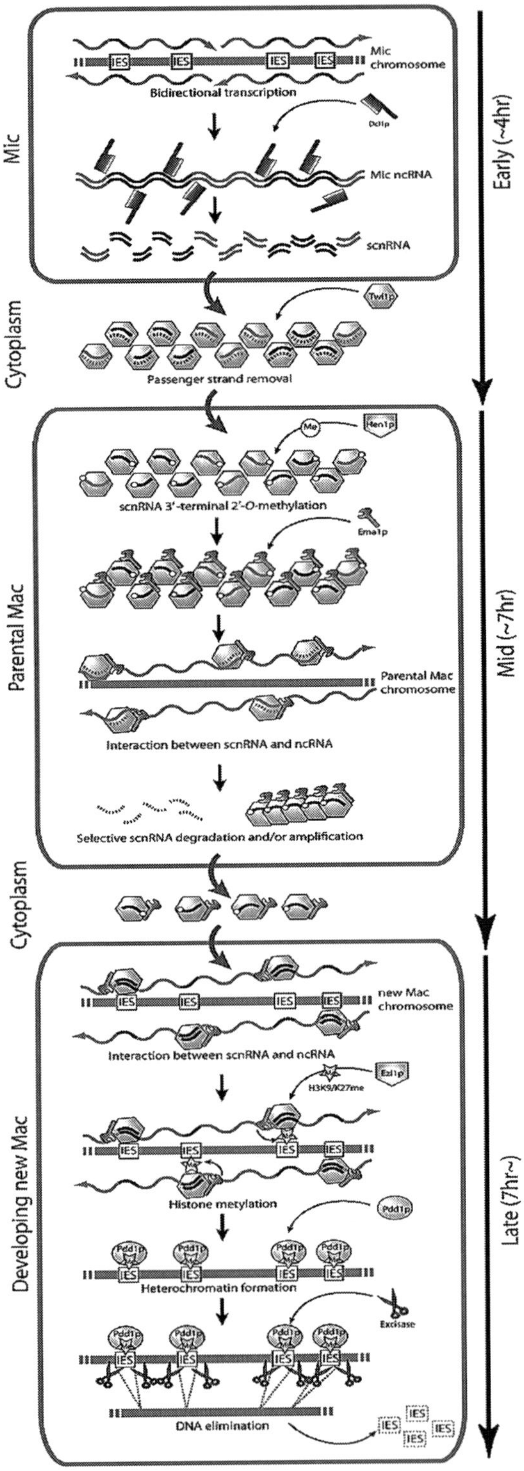

Figure 6. Scan RNA model. Events occurring sequentially are shown from top to bottom. The approximate stages when events occur are indicated on the right by arrows. Stages correspond to those of Figure 3. See text for details. Mic: micronucleus, Mac: Macronucleus.

The scnRNAs produced by Dcl1p in the micronucleus are thought to be transported to the cytoplasm where they form a complex with the Argonaute protein Twi1p (Early Stage in Fig. 6)[40], since Dcl1p exclusively localizes to the micronucleus while Twi1p first appears in the cytoplasm.[31-33] In the absence of Twi1p, scnRNAs disappear rapidly.[40] Therefore, complex formation between Twi1p and scnRNAs must stabilize scnRNAs. The mechanisms by which scnRNAs are exported to the cytoplasm and loaded into Twi1p are not yet well understood.

Although the scnRNAs processed by Dcl1p are double-stranded, one of the strands (the passenger strand) is removed from the scnRNA-Twi1p complex.[35] Like other Argonaute proteins,[45-50] Twi1p possesses endoribonuclease (Slicer) activity that cleaves the passenger strand of scnRNA. This endoribonucleolytic cleavage is necessary to remove the passenger strand from the complex and to eliminate DNA.[35]

It has been shown that scnRNAs are modified by 2'-O-methylation at their 3' ends (Mid Stage in Fig. 6).[51] The RNA methyltransferase Hen1p interacts with Twi1p and is responsible for modification and stable accumulation of scnRNAs.[51] Therefore, 2'-O-methylation at their 3' ends probably protects scnRNAs from exonuclease attack. Since *HEN1* knockout strains show a partial defect in DNA elimination, stabilization of scnRNAs is important to induce proper DNA elimination.[51] Hen1p modifies only single-stranded scnRNAs, both in vitro and in vivo,[51] indicating that Hen1p modifies scnRNA after the passenger strand of scnRNA has been removed from the scnRNA-Twi1p complex.

scnRNA SELECTION

Twi1p first localizes to the cytoplasm in early conjugation and is then exclusively detected in the parental macronucleus at mid-stages of conjugation.[32] In *DCL1* KO strains, which lack scnRNAs, and in Slicer-defective *TWI1* mutant strains, where Twi1p cannot remove the passenger strand of scnRNA, Twi1p does not localize to the parental macronucleus and remains in the cytoplasm.[35] According to these observations, it has been suggested that formation of the scnRNA-Twi1p complex and removal of the passenger strand of scnRNA occur in the cytoplasm (Early Stage in Fig. 6). In the late stages of conjugation, the scnRNA-Twi1p complex translocates from the parental to the new macronucleus where it induces DNA elimination (Late Stage in Fig. 6).

DNA elimination occurs in the new macronucleus. Why then does the scnRNA-Twi1p complex localize first to the parental macronucleus? A key experiment was conducted by Mochizuki and Gorovsky.[40] scnRNAs extracted at different stages of conjugation were radiolabeled and hybridized to genomic DNAs extracted from isolated macro- and micronuclei on a Southern blot. scnRNAs extracted from cells at early stages of conjugation, when Twi1p was predominantly in the cytoplasm, hybridized to both macro- and micronuclear DNA. This is consistent with the prediction that the entire micronuclear genome is transcribed to produce scnRNAs. In contrast, scnRNAs complementary to micronuclear-restricted DNA were gradually enriched during mid-stages of conjugation when the scnRNA-Twi1p was localized to the parental macronucleus, and those extracted from late stages of conjugation, when Twi1p localizes to the new macronucleus, preferentially hybridized to micronuclear DNAs. These observations indicated that, although scnRNAs complementary to both macronuclear-destined sequences and IESs were produced, those complementary to IESs are specifically selected in the parental macronucleus. Thus, the scnRNA-Twi1p complex must localize first to the parental macronucleus in order to interact with the macronuclear chromosome to induce scnRNA selection (Mid Stage in Fig. 6).

The exact molecular mechanism for this scnRNA selection process remains unknown; however, two possible mechanisms have been proposed. The first is the selective degradation of scnRNAs complementary to macronuclear sequences. Since scnRNA is highly unstable in the absence of Twi1p,[32,40] selective degradation of scnRNAs can be achieved if interaction of scnRNA to the macronuclear genome induces dissociation of the scnRNA-Twi1p complex. An alternative possibility is the selective amplification of scnRNAs complementary to the micronuclear-limited sequence. In the yeast *Schizosaccharomyces pombe* and the plant *Arabidopsis thaliana*, siRNAs are amplified by an RNA-dependent RNA polymerase complex, and in the fly *Drosophila* and mouse germlines, piRNAs are amplified probably by 'ping-pong' cycles mediated by Argonaute proteins.[52] Therefore, small RNA amplification mechanisms are widespread among eukaryotes, and a system to selectively amplify scnRNAs that are complementary to IESs may exist in *Tetrahymena*.

Whatever the molecular mechanism for scnRNA selection, scnRNAs complementary to the macronuclear genome must interact by base pairing either directly to macronuclear DNA, or indirectly to RNA transcribed from the parental macronucleus. Our recent study supports the latter possibility (Mid Stage in Fig. 6). ncRNAs transcribed from the parental macronucleus can be co-immunoprecipitated with Twi1p[27] and the interaction between Twi1p and chromatin is RNA-dependent (KM, unpublished results). Moreover, the RNA helicase Ema1p has been identified as an essential factor for the Twi1p-ncRNA interaction, the Twi1p-chromatin interaction, scnRNA selection and DNA elimination.[27] All of these results suggest that the scnRNA-Twi1p complex interacts with macronuclear chromatin via base-pairing between scnRNA and nascent macronuclear noncoding transcripts, and that this interaction either induces degradation of scnRNA or inhibits scnRNA amplification. Ema1p possibly modulates the structures of nascent macronuclear noncoding transcripts to facilitate interactions of the scnRNA-Twi1p complex with chromatin.

scnRNA-INDUCED HETEROCHROMATIN FORMATION AND DNA ELIMINATION

When the new macronucleus forms, the scnRNA-Twi1p complex together with Ema1p translocates from the parental macronucleus to the new macronucleus (Late Stage in Fig. 6).[27,32] Like scnRNA selection in the parental macronucleus, it has been demonstrated that ncRNAs transcribed from the new macronucleus can be co-immunoprecipitated with Twi1p[27] and Twi1p interacts with new macronuclear chromatin in an RNA-dependent manner (KM, unpublished results). Disruption of Ema1p abolishes this interaction and hinders DNA elimination.[27] These observations strongly suggest that the interaction between scnRNA and chromatin via nascent noncoding transcripts is required for targeting DNA for elimination in the newly developing macronucleus.

Methylation of histone H3 at lysines 9 (H3K9me) and 27 (H3K27me) occurs in the newly developing macronucleus.[30,53] These modifications are the hallmarks of heterochromatin in diverse eukaryotes from plants, yeasts to humans. Chromatin immunoprecipitation analyses using anti-H3K9me and H3K27me antibodies reveal that these methylated histones accumulate on IESs but not on macronuclear-destined regions.[30,53] Disruption of the histone methyltransferase Ezl1p, which catalyzes methylation of H3K9 and H3K27 in *Tetrahymena*, as well as amino acid substitutions at H3K9 or H3K27, inhibit DNA elimination.[30,54] Therefore, these histone H3 modifications are required for DNA elimination. H3K9me and H3K27me bind to the chromodomain protein Pdd1p.[30,53] Pdd1p is also localized to heterochromatin of the newly developing macronucleus and is essential for DNA elimination.[29] Elimination of parental *PDD1* expression greatly reduces H3K9me level,[53] indicating that there may be a positive amplification loop through which association of Pdd1p with H3K9/27me recruits Ezl1p to induce greater accumulation of H3K9/27me and Pdd1p. All of these observations strongly suggest that IES elimination is mediated by heterochromatin formation, which includes histone H3 modification (H3K9/27me) and accumulation of the chromodomain protein Pdd1p (Late Stage in Fig. 6). Since disruption of the Dicer gene *DCL1* and the Argonaute gene *TWI1* greatly inhibit H3K9me accumulation or its IES targeting[31,33,54] the RNAi-related pathway is upstream of heterochromatin formation. On the other hand, Ezl1p and Pdd1p are not required for scnRNA accumulation.[30] There may be no feedback mechanism that heterochromatin regulates the RNAi-related pathway in *Tetrahymena* although such mechanism has been reported in the fission yeast.[55]

How heterochromatin structure induces precise DNA elimination is largely unknown. Since artificial tethering of the heterochromatin component Pdd1p to a locus is sufficient to induce its ectopic DNA elimination,[53] Pdd1p can recruit all downstream proteins required for DNA elimination. Cytological observations have shown that heterochromatinized IESs accumulate into nuclear peripheral foci within the developing new macronucleus when DNA elimination occurs.[56,57] It has been suggested that the foci include a putative protein complex called 'Excisase', which catalyzes double-stranded DNA breaking and ligation for each fragmented DNA (Late Stage in Fig 6). Recently it was reported that a PiggyBac transposase-like protein is required for DNA elimination in both *Tetrahymena* and the other ciliate *Paramecium*.[58,59] Therefore, a PiggyBac transposase-like protein is a likely component of Excisase and may directly recognize heterochromatin structures to excise IESs.

EVOLUTIONARY CONSIDERATIONS

The process of DNA elimination in *Tetrahymena* bears a striking similarity to piRNA-directed transposon silencing in metazoans. Although several different RNAi-related pathways collaborate with siRNA, micro(mi)RNA and piRNA in eukaryotes, they all associate with core effector Argonaute family proteins.[12] The Argonaute family proteins can be divided into two subfamilies: AGO and Piwi. siRNAs and miRNAs are associated with AGO proteins while piRNAs bind Piwi proteins. piRNAs (~24-30 nt) are slightly, but significantly, longer than si- and miRNAs (~21-24 nt). Most identified piRNAs are 2'-O-methylated at their 3' ends for all metazoans studied.[60-65] In *Tetrahymena*, scnRNAs specifically interact with the Piwi protein Twi1p.[40] scnRNAs are about 28 to 29 nt in length[32] and are modified by 2'-O-methylation at their 3' ends.[51] Therefore, the biochemical features of scnRNAs in *Tetrahymena* are similar to those of piRNAs in metazoans.

In the germlines of flies, fish and mice, Piwi proteins play pivotal roles in transposon silencing at post-transcriptional as well as at transcriptional levels.[52,66,67] In *Tetrahymena*, the scnRNA-Twi1p complex is essential for IES elimination. Since many IESs are similar to transposons, this process might have evolved as a transposon silencing mechanism for discarding transposons from the transcriptionally active macronucleus. Therefore, transposon silencing might be a task common to the scnRNA-Twi1p complex in *Tetrahymena* and piRNA-Piwi protein complexes in metazoans.

Despite biochemical and functional similarities between scnRNAs and piRNAs, their biogenesis pathways are notably different. scnRNAs are generated in *Tetrahymena* from long double-stranded precursor RNAs by the Dicer protein Dcl1p. Therefore, in terms of biogenesis, scnRNAs must be classified as endogenous siRNAs. In contrast, piRNAs are produced from single-stranded RNAs by a Dicer-independent mechanism.[14] It is believed that piRNAs are produced by the sequential action of two Slicer activities of Piwi proteins described as a 'ping-pong' mechanism.[16,17] It is not yet clear how Dicer-dependent production of Piwi-associated small RNAs (scnRNAs) in *Tetrahymena* and Slicer-dependent production of Piwi-associated small RNAs (piRNAs) in metazoans have evolved. One process could represent the ancestral form of Piwi-associated small RNA biogenesis, or both could have evolved from a different ancestral RNAi mechanism. In this context, it would be interesting to know how Piwi-associated small RNAs are produced in another group of eukaryotes. Since some amoeba species have Piwi proteins,[68,69] investigation of their associated small RNAs is eagerly awaited. Also, because *Tetrahymena* has eleven Piwi proteins besides Twi1p,[70] studying their functions and the biogenesis of their interacting small RNAs would help to understand how Piwi proteins have evolved in eukaryotes.

The process of DNA elimination in *Tetrahymena* is similar to RNAi-directed heterochromatin formation in other eukaryotes such as yeasts, plants, flies and mammals.[11,71-73] The best characterized RNAi-directed heterochromatin formation process is centromeric heterochromatin formation in the fission yeast *S. pombe*.[11] In this process, siRNAs (20-22 nt) complementary to centromeric repeat sequences induce heterochromatin formation. The siRNAs are processed from long dsRNA by the Dicer protein Dcr1 and interact with the Argonaute protein Ago1. The siRNA-Ago1 complex associates with centromeric repeat sequences through RNAPII-transcribed nascent transcripts. Subsequently, the complex recruits the H3K9 methyltransferase Clr4 to the target locus and induces H3K9me. This methylated histone recruits the chromodomain protein Swi6 to establish heterochromatin. The siRNA-Ago1 complex also recruits an RNA-directed RNA polymerase complex and Dcr1 to the target locus, leading to amplification of siRNA signals. As described in this chapter, the *Tetrahymena* scnRNA-Twi1p complex interacts with nascent ncRNAs and is required for Ezl1p-dependent accumulation of H3K9/K27me, which leads to heterochromatin formation. Therefore, RNAi-directed DNA elimination in *Tetrahymena* may be an evolutionary cousin of RNAi-directed heterochromatin formation in other eukaryotes. Since DNA elimination in *Tetrahymena* employs the E(z) ortholog Ezl1p for H3K9/K27me, it might be also related to the Polycomb silencing mechanism. In this context, it is interesting to note that the RNAi pathway may be involved in Polycomb silencing in flies and mice.[74,75]

Developmentally programmed DNA elimination, also called chromatin or chromosome diminution, has long been observed not only in ciliates[76] but also in several taxonomically diverged metazoans such as nematodes,[77,78] sciarid flies,[79] copepods,[80,81] lampreys[82] and hagfish.[83,84] As in *Tetrahymena*, these DNA elimination events take place in the somatic lineage during early embryogenesis, and the eliminated DNAs are often transposon-like repetitive sequences. Interestingly, most eliminated DNA in these programmed elimination processes is at some point embedded into heterochromatin before it is eliminated. Moreover, as described earlier, the molecular mechanism of DNA elimination in *Tetrahymena* is related to RNA-directed heterochromatin formation in other eukaryotes. Therefore, programmed DNA elimination in general might have evolved from a conserved RNA-directed heterochromatin formation mechanism that removes harmful genetic elements from somatic lineages. Since DNA elimination occurs in several phylogenetically separated species, heterochromatin-mediated DNA excision processes might have independently arisen in each organism or conversely, have been preserved from the common ancestor in a few organisms which are currently exist. PiggyBac-like proteins, which may be domesticated transposases, play an essential role in removing heterochromatinized IESs in the ciliates *Paramecium*[58] and *Tetrahymena*.[59] Investigating how these PiggyBac-like proteins have been domesticated and how they interact with heterochromatin will help to understand how ciliates have acquired a DNA excision mechanism during their evolution. Likewise, future studies on the mechanisms of DNA elimination processes in other organism may give us a broader framework for how these processes have evolved in multiple different taxa.

FUTURE PROSPECTS

DNA elimination occurs precisely and reproducibly, and only minor variations of fewer than 10 bp from the elimination boundary have been observed. Although heterochromatin formation, including accumulations of H3K9/27me and the chromodomain protein Pdd1p, is necessary and sufficient to mark DNA sequences for elimination, it is not likely that mere changes in the nucleosome, which contains ~150 bp DNA, are sufficient to support this precision. We speculate that the enzyme (hypothetically called Excisase) that removes IESs may have some sequence preference that limits elimination boundaries to a small range. Future studies of the recently identified PiggyBac-like protein, which is a potential Excisase component, will help to clarify this issue.

Besides transposon silencing, DNA elimination is thought to be involved in mating-type determination. An individual *Tetrahymena* cell expresses one of seven mating types, and mating can occur between cells of any two different types. Since mating-type switching does not occur during vegetative growth, mating types are thought to be determined in the developing new macronucleus by alternative DNA rearrangements. Although a potential mating-type locus named *mat* has been identified,[85] the gene encoded by the *mat* locus has not yet been cloned. Identification of the *mat* locus-encoding gene, as well as a genome-wide comparison between micro- and macronuclear genome sequences from strains expressing different mating types, may enable us to identify the molecule(s) involved in mating-type determination and to understand how DNA rearrangements regulate gene expression.

Another area of interest for future research is the directed transportation of scnRNAs. scnRNAs are produced in the parental micronucleus and exported to the cytoplasm to form complexes with the Argonaute protein Twi1p, after which the complex migrates to the parental macronucleus and finally to the newly developing macronucleus (Fig. 6). Since this sequential localization of scnRNAs is essential for proper DNA elimination, the timing and direction of scnRNA-Twi1p translocations must be precisely regulated. Recently, micro- and macronucleus-specific importin alpha subunits and nucleoporins have been identified.[86,87] Selective interactions between the scnRNA-Twi1p complex and macronucleus-dedicated transporting systems may regulate the spatiotemporal behavior of scnRNAs. Further research into these nucleus-specific transportation machineries might elucidate how selective nuclear transport of scnRNA-Twi1p complexes is achieved.

In terms of the specific import of the scnRNA-Twi1p complex into the newly developing macronucleus, a more fundamental question arises. How are the fates of the somatic macronucleus and germline micronucleus determined? Because macronuclear differentiation (including expansion of the nucleus) starts before Twi1p localizes to the new macronucleus (KM, unpublished results), the scnRNA-Twi1p complex is not likely the fate determinant. In many metazoans, segregation of germline and somatic lineages depends on cytoplasmic determinants that are maternally preserved in the egg.[88] Although the molecular mechanism of nuclear fate determination in *Tetrahymena* has not been elucidated, it has long been known that the developmental fates of nuclei are related to their locations within the cell at a critical time.[89-91] Two consecutive mitoses produce four nuclei from a fertilized zygotic nucleus. Nuclei located in the anterior cytoplasm develop into macronuclei while others that are posteriorly localized become micronuclei (Fig. 3G). Thus, as in many metazoans, an asymmetrically localized molecule(s) (protein and/or RNA) along the anteroposterior axis may act as a determinant for macro- and/or micronuclear fates in *Tetrahymena*. Identification and characterization of this determinant would provide us not only with a mechanistic understanding of how nuclear fates and nucleus-specific transportation systems are established in *Tetrahymena*, but also yield unique insights into how germline-soma segregation has evolved in nonmetazoan eukaryotes.

CONCLUSION

An RNA infrastructure, a network of processes including small RNA biogenesis, post-transcriptional selection of small RNAs, two different nuclear transport processes for small RNAs and small RNA-induced heterochromatin formation, corporately conduct DNA elimination in *Tetrahymena*. Because the process leading to DNA elimination can readily be induced synchronously in several billion *Tetrahymena* cells[92] and we are able to discriminately analyze the RNA infrastructure in different developmental stages in this organism, DNA elimination in *Tetrahymena* is a useful model to study how RNA infrastructure regulates chromatin organization in eukaryotes.

ACKNOWLEDGEMENTS

Research in our laboratory is supported by the Naito Foundation to KK, by the European Research Council (ERC) Starting Grant (204986) under the European Community's Seventh Framework Programme, by the Austrian Science Fund (FWF) Doktoratskolleg RNA Biology, and by the Austrian Academy of Sciences to KM.

REFERENCES

1. Bickle TA, Kruger DH. Biology of DNA restriction. Microbiol Rev 1993; 57(2):434-450.
2. Horvath P, Barrangou R. CRISPR/Cas, the immune system of bacteria and archaea. Science 2010; 327(5962):167-170.
3. Sorek R, Kunin V, Hugenholtz P. CRISPR--a widespread system that provides acquired resistance against phages in bacteria and archaea. Nat Rev 2008; 6(3):181-186.
4. van der Oost J, Jore MM, Westra ER et al. CRISPR-based adaptive and heritable immunity in prokaryotes. Trends Biochem Sci 2009; 34(8):401-407.

5. Brouns SJ, Jore MM, Lundgren M et al. Small CRISPR RNAs guide antiviral defense in prokaryotes. Science 2008; 321(5891):960-964.
6. Hale CR, Zhao P, Olson S et al. RNA-guided RNA cleavage by a CRISPR RNA-Cas protein complex. Cell 2009; 139(5):945-956.
7. Marraffini LA, Sontheimer EJ. CRISPR interference limits horizontal gene transfer in staphylococci by targeting DNA. Science 2008; 322(5909):1843-1845.
8. Barrangou R, Fremaux C, Deveau H et al. CRISPR provides acquired resistance against viruses in prokaryotes. Science 2007; 315(5819):1709-1712.
9. Fire A, Xu S, Montgomery MK et al. Potent and specific genetic interference by double-stranded RNA in Caenorhabditis elegans. Nature 1998; 391(6669):806-811.
10. Matranga C, Zamore PD. Small silencing RNAs. Curr Biol 2007; 17(18):R789-793.
11. Huisinga KL, Elgin SC. Small RNA-directed heterochromatin formation in the context of development: what flies might learn from fission yeast. Biochim Biophys Acta 2009; 1789(1):3-16.
12. Obbard DJ, Gordon KHJ, Buck AH, Jiggins FM. The evolution of RNAi as a defence against viruses and transposable elements. Phil Trans R Soc B 2009; 364:99-115.
13. Ding SW, Voinnet O. Antiviral immunity directed by small RNAs. Cell 2007; 130(3):413-426.
14. Kim VN, Han J, Siomi MC. Biogenesis of small RNAs in animals. Nat Rev Mol Cell Biol 2009; 10(2):126-139.
15. Aravin AA, Sachidanandam R, Girard A et al. Developmentally regulated piRNA clusters implicate MILI in transposon control. Science 2007; 316(5825):744-747.
16. Brennecke J, Aravin AA, Stark A et al. Discrete small RNA-generating loci as master regulators of transposon activity in Drosophila. Cell 2007; 128(6):1089-1103.
17. Gunawardane LS, Saito K, Nishida KM et al. A slicer-mediated mechanism for repeat-associated siRNA 5' end formation in Drosophila. Science 2007; 315(5818):1587-1590.
18. Collins LJ, Penny D. The RNA infrastructure: dark matter of the eukaryotic cell? Trends Genet 2009; 25(3):120-128.
19. Karrer KM. Tetrahymena genetics: two nuclei are better than one. Methods Cell Biol 2000; 62:127-186.
20. Fan Q, Yao M. New telomere formation coupled with site-specific chromosome breakage in Tetrahymena thermophila. Mol Cell Biol 1996; 16(3):1267-1274.
21. Hamilton EP, Williamson S, Dunn S et al. The highly conserved family of Tetrahymena thermophila chromosome breakage elements contains an invariant 10-base-pair core. Eukaryot Cell 2006; 5(4):771-780.
22. Yao MC, Zheng K, Yao CH. A conserved nucleotide sequence at the sites of developmentally regulated chromosomal breakage in Tetrahymena. Cell 1987; 48(5):779-788.
23. Yao MC, Gorovsky MA. Comparison of the sequences of macro- and micronuclear DNA of Tetrahymena pyriformis. Chromosoma 1974; 48(1):1-18.
24. Fillingham JS, Thing TA, Vythilingum N et al. A non-long terminal repeat retrotransposon family is restricted to the germ line micronucleus of the ciliated protozoan Tetrahymena thermophila. Eukaryot Cell 2004; 3(1):157-169.
25. Wuitschick JD, Gershan JA, Lochowicz AJ, Li S, Karrer KM. A novel family of mobile genetic elements is limited to the germline genome in Tetrahymena thermophila. Nucleic Acids Res 2002; 30(11):2524-2537.
26. Yao MC, Duharcourt DL, Chalker DL. Genome-wide rearrangements of DNA in ciliates. In: Craig NL, Craigie R, Gellert M, Lambowitz AM, eds. Mobile DNA II. Washington, DC: ASM Press, 2002:730–758.
27. Aronica L, Bednenko J, Noto T et al. Study of an RNA helicase implicates small RNA-noncoding RNA interactions in programmed DNA elimination in Tetrahymena. Genes Dev 2008; 22(16):2228-2241.
28. Bednenko J, Noto T, DeSouza LV et al. Two GW repeat proteins interact with Tetrahymena thermophila argonaute and promote genome rearrangement. Mol Cell Biol 2009; 29(18):5020-5030.
29. Coyne RS, Nikiforov MA, Smothers JF et al. Parental expression of the chromodomain protein Pdd1p is required for completion of programmed DNA elimination and nuclear differentiation. Mol Cell 1999; 4(5):865-872.
30. Liu Y, Taverna SD, Muratore TL et al. RNAi-dependent H3K27 methylation is required for heterochromatin formation and DNA elimination in Tetrahymena. Genes Dev 2007; 21(12):1530-1545.
31. Malone CD, Anderson AM, Motl JA et al. Germ line transcripts are processed by a Dicer-like protein that is essential for developmentally programmed genome rearrangements of Tetrahymena thermophila. Mol Cell Biol 2005; 25(20):9151-9164.
32. Mochizuki K, Fine NA, Fujisawa T, Gorovsky MA. Analysis of a piwi-related gene implicates small RNAs in genome rearrangement in Tetrahymena. Cell 2002; 110(6):689-699.
33. Mochizuki K, Gorovsky MA. A Dicer-like protein in Tetrahymena has distinct functions in genome rearrangement, chromosome segregation, and meiotic prophase. Genes Dev 2005; 19(1):77-89.
34. Nikiforov MA, Smothers JF, Gorovsky MA, Allis CD. Excision of micronuclear-specific DNA requires parental expression of pdd2p and occurs independently from DNA replication in Tetrahymena thermophila. Genes Dev 1999; 13(21):2852-2862.

35. Noto T, Kurth HM, Kataoka K et al. The Tetrahymena Argonaute-binding protein Giw1p directs a mature Argonaute-siRNA complex to the nucleus. Cell 2010; 140(5):692-703
36. Rexer CH, Chalker DL. Lia1p, a novel protein required during nuclear differentiation for genome-wide DNA rearrangements in Tetrahymena thermophila. Eukaryot Cell 2007; 6(8):1320-1329.
37. Chalker DL, Yao MC. Non-Mendelian, heritable blocks to DNA rearrangement are induced by loading the somatic nucleus of Tetrahymena thermophila with germ line-limited DNA. Mol Cell Biol 1996; 16(7):3658-3667.
38. Liu Y, Song X, Gorovsky MA, Karrer KM. Elimination of foreign DNA during somatic differentiation in Tetrahymena thermophila shows position effect and is dosage dependent. Eukaryot Cell 2005; 4(2):421-431.
39. Yao MC, Fuller P, Xi X. Programmed DNA deletion as an RNA-guided system of genome defense. Science 2003; 300(5625):1581-1584.
40. Mochizuki K, Gorovsky MA. Conjugation-specific small RNAs in Tetrahymena have predicted properties of scan (scn) RNAs involved in genome rearrangement. Genes Dev 2004; 18(17):2068-2073.
41. Martindale DW, Allis CD, Bruns PJ. RNA and protein synthesis during meiotic prophase in Tetrahymena thermophila. J Protozool 1985; 32(4):644-649.
42. Sugai T, Hiwatashi K. Cytologic and autoradiographic studies of the micronucleus at meiotic prophase in Tetrahymena pyriformis. J Protozool 1974; 21(4):542-548.
43. Chalker DL, Yao MC. Nongenic, bidirectional transcription precedes and may promote developmental DNA deletion in Tetrahymena thermophila. Genes Dev 2001;15(10):1287-1298.
44. Mochizuki K, Gorovsky MA. RNA polymerase II localizes in Tetrahymena thermophila meiotic micronuclei when micronuclear transcription associated with genome rearrangement occurs. Eukaryot Cell 2004;3(5):1233-1240.
45. Leuschner PJ, Ameres SL, Kueng S, Martinez J. Cleavage of the siRNA passenger strand during RISC assembly in human cells. EMBO reports 2006;7(3):314-320.
46. Maiti M, Lee HC, Liu Y. QIP, a putative exonuclease, interacts with the Neurospora Argonaute protein and facilitates conversion of duplex siRNA into single strands. Genes Dev 2007;21(5):590-600.
47. Matranga C, Tomari Y, Shin C et al. Passenger-strand cleavage facilitates assembly of siRNA into Ago2-containing RNAi enzyme complexes. Cell 2005;123(4):607-620.
48. Miyoshi K, Tsukumo H, Nagami T et al. Slicer function of Drosophila Argonautes and its involvement in RISC formation. Genes Dev 2005; 19(23):2837-2848.
49. Rand TA, Petersen S, Du F, Wang X. Argonaute2 cleaves the anti-guide strand of siRNA during RISC activation. Cell 2005; 123(4):621-629.
50. Steiner FA, Okihara KL, Hoogstrate SW et al. RDE-1 slicer activity is required only for passenger-strand cleavage during RNAi in Caenorhabditis elegans. Nat Struct Mol Biol 2009; 16(2):207-211.
51. Kurth HM, Mochizuki K. 2'-O-methylation stabilizes Piwi-associated small RNAs and ensures DNA elimination in Tetrahymena. RNA 2009;15(4):675-685.
52. Siomi MC, Kuramochi-Miyagawa S. RNA silencing in germlines--exquisite collaboration of Argonaute proteins with small RNAs for germline survival. Curr Opin Cell Biol 2009; 21(3):426-434.
53. Taverna SD, Coyne RS, Allis CD. Methylation ofhistone h3 at lysine 9 targets programmed DNA elimination in tetrahymena. Cell 2002; 110(6):701-711.
54. Liu Y, Mochizuki K, Gorovsky MA. Histone H3 lysine 9 methylation is required for DNA elimination in developing macronuclei in Tetrahymena. Proc Natl Acad Sci USA 2004; 101(6):1679-1684.
55. Grewal SI. RNAi-dependent formation of heterochromatin and its diverse functions. Curr Opin Genet Dev. 2010 Apr; 20(2):134-41.
56. Madireddi MT, Coyne RS, Smothers JF et al. Pdd1p, a novel chromodomain-containing protein, links heterochromatin assembly and DNA elimination in Tetrahymena. Cell 1996; 87(1):75-84.
57. Yokoyama RW, Yao MC. Elimination of DNA sequences during macronuclear differentiation in Tetrahymena thermophila, as detected by in situ hybridization. Chromosoma 1982; 85(1):11-22.
58. Baudry C, Malinsky S, Restituito M et al. PiggyMac, a domesticated piggyBac transposase involved in programmed genome rearrangements in the ciliate Paramecium tetraurelia. Genes Dev 2009; 23(21):2478-2483.
59. Cheng CY, Vogt A, Mochizuki K, Yao MC. A Domesticated piggyBac transposase plays a key role in heterochromatin dynamics and DNA cleavage during programmed DNA deletion in Tetrahymena thermophila. Mol Biol Cell 2010; 21(10):1753-62.
60. Horwich MD, Li C, Matranga C et al. The Drosophila RNA methyltransferase, DmHen1, modifies germline piRNAs and single-stranded siRNAs in RISC. Curr Biol 2007;17(14):1265-1272.
61. Houwing S, Kamminga LM, Berezikov E et al. A role for Piwi and piRNAs in germ cell maintenance and transposon silencing in Zebrafish. Cell 2007;129(1):69-82.
62. Kirino Y, Mourelatos Z. Mouse Piwi-interacting RNAs are 2'-O-methylated at their 3' termini. Nat Struct Mol Biol 2007;14(4):347-348.

63. Ohara T, Sakaguchi Y, Suzuki T et al. The 3' termini of mouse Piwi-interacting RNAs are 2'-O-methylated. Nat Struct Mol Biol 2007; 14(4):349-350.
64. Saito K, Sakaguchi Y, Suzuki T et al. Pimet, the Drosophila homolog of HEN1, mediates 2'-O-methylation of Piwi- interacting RNAs at their 3' ends. Genes Dev 2007; 21(13):1603-1608.
65. Vagin VV, Sigova A, Li C et al. A distinct small RNA pathway silences selfish genetic elements in the germline. Science 2006; 313(5785):320-324.
66. Aravin AA, Hannon GJ, Brennecke J. The Piwi-piRNA pathway provides an adaptive defense in the transposon arms race. Science 2007; 318(5851):761-764.
67. Malone CD, Hannon GJ. Small RNAs as guardians of the genome. Cell 2009; 136(4):656-668.
68. Cerutti H, Casas-Mollano JA. On the origin and functions of RNA-mediated silencing: from protists to man.Curr Genet 2006; 50(2):81-99.
69. Seto AG, Kingston RE, Lau NC. The coming of age for Piwi proteins. Mol Cell 2007; 26(5):603-609.
70. Couvillion MT, Lee SR, Hogstad B et al. Sequence, biogenesis, and function of diverse small RNA classes bound to the Piwi family proteins of Tetrahymena thermophila. Genes Dev 2009; 23(17):2016-2032.
71. Buhler M, Moazed D. Transcription and RNAi in heterochromatic gene silencing. Nat Struct Mol Biol 2007; 14(11):1041-1048.
72. Matzke M, Kanno T, Daxinger L et al. RNA-mediated chromatin-based silencing in plants. Curr Opin Cell Biol 2009; 21(3):367-376.
73. Matzke MA, Birchler JA. RNAi-mediated pathways in the nucleus. Nat Rev Genet 2005; 6(1):24-35.
74. Grimaud C, Bantignies F, Pal-Bhadra M et al. RNAi components are required for nuclear clustering of Polycomb group response elements. Cell 2006; 124(5):957-971.
75. Ogawa Y, Sun BK, Lee JT. Intersection of the RNA interference and X-inactivation pathways. Science 2008; 320(5881):1336-1341.
76. Prescott DM. The DNA of ciliated protozoa. Microbiol Rev 1994; 58(2):233-267.
77. Boveri T. Uber Differenzierung der Zellkerne während der Furchung des Eies von Ascaris megalocephala. Anat Anz 1887; 2:688-693.
78. Muller F, Bernard V, Tobler H. Chromatin diminution in nematodes. Bioessays 1996; 18(2):133-138.
79. Goday C, Esteban MR. Chromosome elimination in sciarid flies. Bioessays 2001; 23(3):242-250.
80. Degtyarev S, Boykova T, Grishanin A et al. The molecular structure of the DNA fragments eliminated during chromatin diminution in Cyclops kolensis. Genome Res 2004; 14(11):2287-2294.
81. Drouin G. Chromatin diminution in the copepod Mesocyclops edax: diminution of tandemly repeated DNA families from somatic cells. Genome 2006; 49(6):657-65.
82. Smith JJ, Antonacci F, Eichler EE, Amemiya CT. Programmed loss of millions of base pairs from a vertebrate genome. Proc Natl Acad Sci USA 2009; 106(27):11212-11217.
83. Kubota S, Kuro-o M, Mizuno S, Kohno S. Germ line-restricted, highly repeated DNA sequences and their chromosomal localization in a Japanese hagfish (Eptatretus okinoseanus). Chromosoma 1993;102(3):163-173.
84. Kubota S, Takano J, Tsuneishi R et al. Highly repetitive DNA families restricted to germ cells in a Japanese hagfish (Eptatretus burgeri): a hierarchical and mosaic structure in eliminated chromosomes. Genetica 2001; 111(1-3):319-328.
85. Bleyman LK, Baum MP, Bruns PJ, Orias E. Mapping the mating type locus of Tetrahymena thermophila: meiotic linkage of mat to the ribosomal RNA gene. Dev Genet 1992; 13(1):34-40.
86. Iwamoto M, Mori C, Kojidani T et al. Two distinct repeat sequences of Nup98 nucleoporins characterize dual nuclei in the binucleated ciliate Tetrahymena. Curr Biol 2009; 19(10):843-847.
87. Malone CD, Falkowska KA, Li AY et al. Nucleus-specific importin alpha proteins and nucleoporins regulate protein import and nuclear division in the binucleate Tetrahymena thermophila. Eukaryot Cell 2008; 7(9):1487-1499.
88. Extavour CG, Akam M. Mechanisms of germ cell specification across the metazoans: epigenesis and preformation. Development 2003; 130(24):5869-5884.
89. Gaertig J, Fleury A. Spatio-temporal reorganization of intracytoplasmic microtubules is associated with the nuclear selection and differentiation during developmental process in the ciliate Tetrahymena thermophila. Protoplasma 1992; 167:74-87.
90. Kaczanowski A, Ramel M, Kaczanowska J, Wheatley D. Macronuclear differentiation in conjugating pairs of Tetrahymena treated with the antitubulin drug nocodazole. Exp Cell Res 1991;195(2):330-337.
91. Nanney DL. Nucleo-cytoplasmic interactions during conjugation in Tetrahymena. Biol bull 1953; 105:133-148.
92. Collins K, Gorovsky MA. Tetrahymena thermophila. Curr Biol 2005;15(9):R317-318.

CHAPTER 11

LONG NONCODING RNA AND EPIGENOMICS

Chandrasekhar Kanduri
Department of Genetics and Pathology, Uppsala University, Uppsala, Sweden
Email: kanduri.chandrasekhar@genpat.uu.se

Abstract: Accumulating evidence over the last decade has presented us with the intriguing observation that the majority of eukaryotic genomes are pervasively transcribed to encode a complex network of small and long noncoding RNAs. Long noncoding RNAs are of particular interest, as they were once thought to be restricted to housekeeping functions and are now linked to a wide variety of biological functions related to physiology, embryology and development. Emerging evidence indicates that a subset of long noncoding RNAs mediate their biological functions by using chromatin as a substrate, to index the genetic information encoded in the genome. This chapter will discuss how noncoding RNAs and the processes underlying their transcription mediate transcriptional regulation, by epigenetically regulating the structure of chromatin in various biological contexts.

INTRODUCTION

Until recently the noncoding portion of the genome in higher eukaryotes was considered "junk" and was thought to have evolved as a hot spot for mutations that permitted the coding portion of the genome to be highly conserved across the evolutionary spectrum. This view led to the belief that the protein-coding portion of the genome acted as the main architect of organismal development by controlling genetic programming. However, the advent of new high-throughput technologies such as high resolution tiling arrays and RNA sequencing (which can probe the transcriptional landscape of the entire genome with relative ease), has led to the realization that the protein-coding portion of the genome remains relatively constant across different eukaryotic species and that it is only the noncoding portion of the genome that has evolved, thus linking the noncoding

RNA Infrastructure and Networks, edited by Lesley J. Collins.

portion of the genome to eukaryotic complexity.[1] Recent evidence indicates that although more than 70% of the eukaryotic genome is transcribed, only approximately 1 to 2 % of the transcriptome contributes to protein-coding RNA, suggesting that transcription is not just limited to the protein-coding portion of the eukaryotic genome but is pervasive throughout it.[2,3] Several recent analyses have documented that widely transcribed noncoding RNAs (ncRNAs) play a critical role in various biological functions linked to development and differentiation.[4-9] Importantly, the majority of ncRNAs are expressed in a spatio-temporal manner and often exhibit precise sub-cellular localization. These observations lend support to the contention that transcription of the noncoding portion of the genome contributes to the evolution of complex organisms. However, the functional role of alternative splice forms of protein-coding mRNAs and posttranslational modification of proteins, which increase the diversity of their functions, have not been ruled out as also playing a role in the evolution of complex organisms.

Based on their size, ncRNAs can be classified as small, medium and long. Small ncRNAs range from 18-31 nucleotides in length, whereas long ncRNAs range in size from 200 bp to over several hundred kb. Medium length ncRNAs, measure between 31 and 200 nucleotides, contains of mainly snRNAs and snoRNAs. Unlike small and medium sized ncRNAs, long ncRNAs are not highly conserved at the primary sequence level. However, a subset of long intergenic ncRNAs (linc RNAs) identified using chromatin signatures, have been shown to be highly conserved at the sequence level.[10] Small ncRNAs such as small interfering RNAs (siRNAs), microRNAs (miRNAs) and Piwi-interacting RNAs (piRNAs) regulate gene activity at the transcriptional and/or posttranscriptional level through distinct mechanisms via the RNA interference (RNAi) pathway. miRNAs are 21-23-bp single-stranded RNA molecules that serve as posttranscriptional regulators of gene expression in plants and animals. They act by binding to complementary sites on target RNAs to induce cleavage or repression of productive translation. In contrast, siRNAs are double-stranded RNA molecules 20-25 nucleotides in length which have been shown to participate in numerous biological functions such as antiviral mechanisms and chromosome segregation by maintaining the integrity of centromeric heterochromatin.[11] Piwi-interacting RNAs or piRNAs are poorly conserved at the primary sequence level and are longer than both siRNA and miRNA, measuring in size between 26-31 nucleotides. piRNAs are implicated in transcriptional gene silencing of repeat elements such as of retrotransposons in germ line cells via methylating their promoters.[12,13]

Long ncRNAs were once thought to be restricted to housekeeping functions such as DNA replication, posttranscriptional processing and protein synthesis. However, increased focus on the functions of long ncRNA in eukaryotes resulted in the identification of a broad functional repertoire of ncRNAs that take part in important biological functions beyond the realm of housekeeping functions, such as X chromosome inactivation,[14,15] genomic imprinting,[16] sub-cellular structural organization,[17,18] telomere[19] and centromere organization[20,21] and nuclear trafficking.[22] Interestingly, accumulated evidence over the last few years suggests that the majority of functionally characterized ncRNAs act at the level of transcription as *cis* or *trans* elements by modulating chromatin structure or transcriptional programs.[6,23] In certain instances ncRNAs themselves have been observed to act as regulators, whereas in other instances their transcription and/or the RNA itself plays a critical role, indicating that no common theme applies to ncRNA-mediated functions.[16,24,25]

Relatively few ncRNAs have been studied so far, but since more than 70% (70% in nematode and 93% in humans) of the eukaryotic genome is transcribed, thousands of ncRNAs are likely to interact in a complex network. Although many of the functionally implicated ncRNAs, as well as their target genes, are poorly conserved at the primary sequence level, there seem to be striking parallels in their mechanisms of action. For example, despite lacking homology at the sequence level, a subset of long ncRNAs mediates gene regulation of chromosomal domains[16,26] or chromosomes[14] at the transcriptional level by targeting chromatin remodeling complexes as *cis* or *trans* elements. This mechanism has been described in relatively well-characterized biological phenomena such as genomic imprinting and X-chromosome inactivation.

The act of ncRNA transcription (sense or antisense) is also sufficient to regulate the expression of neighboring genes. Contrary to widespread belief, ncRNA transcription does not always result in the silencing of flanking genes. In certain instances, it activates transcription.[27] How does ncRNA transcription control two distinctly regulated processes such as gene activation and silencing? The prevailing view has been that ncRNA transcription across the promoter region of a downstream protein-coding gene directly interferes with the transcription initiation complex, thus silencing the protein-coding gene. Interestingly, ncRNA transcription through the promoter regions of certain tumor suppressor genes in human disease conditions[28] and *GAL1-10* clusters in yeast[29] results in heterochromatinization and gene silencing. This gene silencing does not depend on the polarity of transcription, as it is induced by both sense[30] and antisense[31] ncRNA transcription, indicating that the act of transcription epigenetically regulates gene silencing by interfering with chromatin structure. In contrast, continuous ncRNA transcription through homeobox gene clusters in flies has been linked to homeobox gene activation by interfering with the binding of polycomb (PcG) proteins.[32] Similarly, ncRNA transcription upstream of the fbp1$^+$ gene in yeast induces chromatin remodeling and gene activation,[33] indicating that ncRNA transcription mediates transcriptional gene silencing or activation through chromatin-level regulation in a context-dependent manner. It remains unknown how the act of transcription specifically modifies the chromatin structure around the promoter region. Though the act of ncRNA transcription elicits transcriptional regulation through multiple pathways, transcriptional interference could be a common theme in these pathways.

Here, I focus on a network of RNA infrastructural ncRNAs that mediate gene regulation by epigenetically regulating chromatin structure in various biological contexts from fission yeast *S. pombe*, *Drosophila melanogaster*, mouse and human.

LONG ncRNAs AND TUMOR SUPPRESSOR GENES

DNA methylation of tumor suppressor gene promoters in the context of global hypomethylation is a common feature of several cancers. The mechanisms underlying the epigenetic silencing of tumor suppressor genes by CpG methylation are unknown, as loss of function mutations affecting the CpG methylation machinery have not been detected in the majority of cancers. Emerging evidence[28,34] from recent investigations implicates a functional role for long ncRNAs in epigenetic silencing of the tumor suppressor genes *p15* and *p21*. Both *p15* and *p21* genes are overlapped by antisense ncRNAs and ncRNA transcription through the *p15* and *p21* promoters is correlated with DNA methylation in *cis*. Although both the promoters become methylated in response to antisense ncRNA transcription, the available evidence to date suggests that the underlying mechanisms leading to their methylation differ.

In the case of the *p15* tumor suppressor gene, increased antisense ncRNA transcription was specifically detected in patients with acute myeloid leukemia and acute lymphatic leukemia but not from normal individuals, indicating that it is aberrantly transcribed only in disease conditions.[28] Aberrant antisense transcription is associated with *p15* promoter methylation and gene silencing, indicating that the antisense transcription process and/or its RNA product epigenetically silences the *p15* promoter. By employing in vitro strategies, the antisense RNA was found to silence the *p15* promoter in *cis* or in *trans* and to be involved in the establishment but not in the maintenance of silencing, which persisted even in the absence of the antisense RNA. Interestingly, the dsRNA-mediated RNAi pathway was found not to play a functional role in the *p15* antisense RNA-mediated epigenetic silencing, indicating that the antisense RNA uses currently unknown mechanisms in its epigenetic silencing pathway. Since the in vitro system utilizes only a portion of the antisense RNA sequence, it is not clear whether the in vitro synthesized RNA is fully representative of the functions of the native antisense RNA.

In the case of *p21*, balanced bidirectional transcription from the *p21* sense and antisense promoters seems to play an important role in maintaining the activity of the *p21* promoter, as the silencing of either the *p21* sense or antisense RNA transcript affects the transcription level of the other.[34] For example, down regulation of the sense RNA leads to increased expression of its antisense RNA, followed by silencing of the *p21* promoter by epigenetic modification of chromatin, indicating that the antisense RNA, or its transcription, suppresses *p21* promoter activity by directing silent-state epigenetic marks. Interestingly, epigenetic modification of the *p21* promoter by the antisense RNA is dependent on Argonaute-1 protein, a key player in the small RNA induced gene silencing pathway, suggesting a functional role for the RNAi pathway in the antisense RNA-mediated epigenetic silencing of the *p21* promoter. Like in the majority of promoters, the *p21* promoter is also associated with promoter-associated RNAs (pRNAs) in both the sense and antisense directions. The appearance of the antisense pRNA is dependent on Argonaute-1, indicating a strong link between the *p21* antisense RNA and the Argonaute-1 protein in the generation of the antisense pRNA, which could play an important role in the epigenetic modification of the *p21* promoter. However, it remains to be seen whether the Argonaute-1-dependent silencing of the *p21* promoter has any resemblance to RNAi-mediated heterochromatinization in eukaryotes such as fission yeast *S. pombe*. The above studies clearly demonstrate that the *p21* and *p15* antisense RNAs exploit different strategies in executing silencing of their sense promoters.

A functional role for RNA *per se* is evident in the case of *p15* silencing through an RNAi-independent pathway. However in the case of *p21*, the functional role played by the act of ncRNA transcription or the ncRNA itself, as well as the possible existence of a functional link between ncRNA and RNAi components, have not yet been determined.

LONG ncRNAs AND HOMEOBOX GENES

Homeobox genes (*Hox*) genes (which control embryonic body development in all bilaterally symmetrical animals), are organized into gene clusters and are coordinately regulated by common long-range *cis* acting regulatory elements. Epigenetic mechanisms, involving histone methylation, demethylation and ncRNAs, have recently been shown to be critical for the fidelity of *Hox* gene expression in a spatio-temporal manner along the developmental axis. Epigenetic regulation of *Hox* gene expression occurs primarily due to the interplay between two protein complexes: trithorax- and polycomb-group proteins. Trithorax-group proteins, which possess enzymatic activity to mediate the methylation of the lysine 4 residue of histone H3 (H3K4), are required to maintain *Hox* gene activation, whereas polycomb-group proteins, comprising PRC2 complex members, maintain the repressive state of the *Hox* genes by mediating the methylation of histone H3 at lysine 27 (H3K27). Although *cis* acting DNA sequences (polycomb responsive elements, PRE) that recruit polycomb/trithorax proteins have been well-characterized in *Drosophila melanogaster*, in other organisms the recruitment of these proteins to the *Hox* gene cluster in a spatio-temporal manner is poorly understood.[35,36] The interplay between polycomb and trithorax proteins generates active or inactive chromatin domains in the *Hox* gene clusters along the developmental axis. Evidence from the last few years indicates that noncoding transcripts are prevalent among the *Hox* gene clusters, interweaving with genic and intergenic regions and that they have a key role in configuring this epigenetic landscape.[26,37]

Although ncRNAs were detected over two decades ago in *Drosophila Hox* gene clusters, their mode of action in regulating the spatio-temporal expression of *Hox* genes remained obscure. However, recent studies addressing the role of ncRNA in *Hox* gene regulation have revealed that ncRNA itself or its transcription, affect *Hox* gene activation and silencing through multiple pathways. The functional role of ncRNAs has been thoroughly characterized in the *Drosophila* bithorax complex (*BX-C*) region, which is implicated in two-thirds of the body plan. Two recent reports investigated the functional role of short processed transcripts from the *bxd* region, which flanks the Ultrabithorax (*Ubx*) *Hox* gene.[38,39] Both studies implicate *bxd* ncRNA as playing a functional role in the regulation of *Ubx* expression, but they contradict each other regarding how the regulation is brought about. One report suggests that *bxd* ncRNAs encoded from polycomb/trithorax elements upstream of the *Ubx* gene are retained at the site of transcription through sequence homology and this sequence-based interaction is required for the RNA mediated recruitment of trithorax protein ASH1 to the downstream *Ubx* gene promoter.[38] The RNA-mediated recruitment of ASH1 results in the formation of active chromatin around the *Ubx* promoter, thus making it more accessible to transcription initiation machinery. Interestingly, this study further shows that ectopic expression of *bxd* transcripts in larval imaginal discs increases the activation of the *Ubx* promoter, suggesting that the *bxd* transcript itself and not just its transcription, is crucial for *Ubx* promoter activation (Fig. 1A). In contrast, the other report suggests that ncRNA transcription from the *bxd* region extends into the *Ubx* promoter region, which negatively regulates *Ubx* promoter activity.[39] Importantly, ectopic expression of the regions that span the *bxd* transcripts do not activate *Ubx* transcription, suggesting that the act of transcription interferes with the initiation of transcription at the *Ubx* promoter (Fig. 1B). From these investigations, the mode of action of *bxd* transcripts in *Ubx* gene regulation is difficult to infer, but nevertheless, both studies clearly implicate a functional role for ncRNAs.

Similarly, several ncRNAs have been mapped to the genic and intergenic regions of the mouse and human *Hox* gene clusters.[26,37] In many cases their expression correlated with the active status of the neighboring *Hox* gene, indicating that *Hox* ncRNAs in mammals probably mediate the activation of flanking genes. However, a detailed examination of one of the human *Hox* ncRNAs, *HOTAIR* encoded from the *HOXC* locus, revealed that it regulates gene expression of the *HOXD* locus *in trans* by targeting one of the PRC2 complex members, Ezh2.[26] Although *trans* function of small ncRNAs such as microRNAs is mechanistically well defined, the *trans* function of long ncRNAs is largely unknown. Thus, this observation is of considerable importance given the fact that *HOTAIR* mediates gene silencing of the entire *HOXD* locus *in trans* despite a lack of sequence homology with

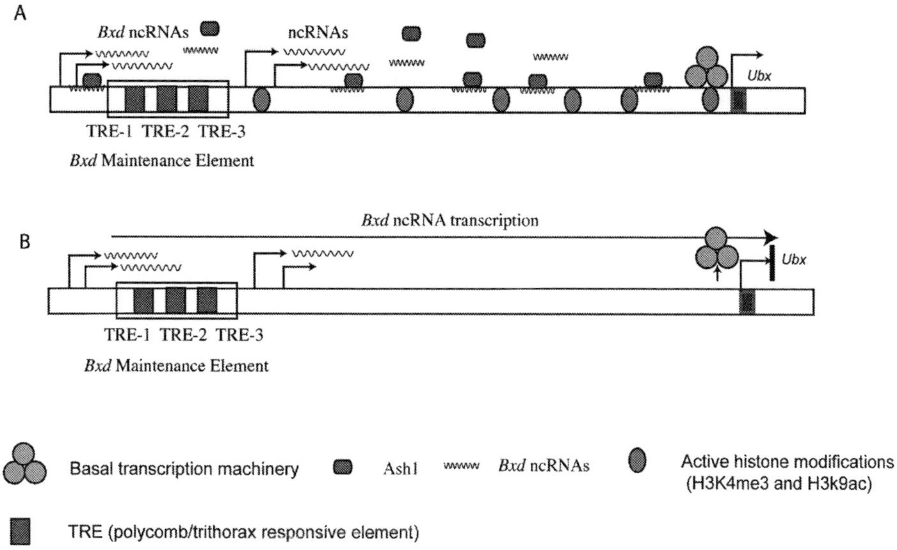

Figure 1. Models explaining the functional role of long ncRNAs from the *Bxd* region on the flanking *Ubx* gene expression. A) *Bxd* ncRNAs are retained at the *Ubx* locus via RNA-DNA interactions and recruit trithorax protein ASH1. The RNA dependent recruitment of ASH1 facilitates the formation of active chromatin structure and *Ubx* transcription. B) This model explores a functional role for noncoding transcription, rather than ncRNA per se, in the control of *Ubx* gene expression. According to this model, *Bxd* ncRNA transcription through the 5′ regulatory elements of the *Ubx* promoter results in the transcriptional repression of the *Ubx* gene via transcriptional interference or promoter occlusion.

the target sequences. The *trans* function of *HOTAIR* ncRNA mirrors the trans-activation of the *Ubx* promoter by *bxd* transcripts through the targeting of the trithorax protein ASH1.

Likewise, a recent study in mice has functionally linked two long antisense ncRNAs, *Evx1as* and *Hoxb5/6as*, to *Hox* gene regulation.[40] The *Evx1as* and *Hoxb5/6as* ncRNAs showed concordant expression with their sense partners *Evx1* and *Hox5/6*, respectively, during the primitive streak phase of EB differentiation. These two ncRNAs were enriched in H3K4me3-precipitated chromatin. Moreover, both of these ncRNAs have been shown to interact with MLL1 histone methyltransferase, which is responsible for the establishment of H3K4me3, indicating that these antisense ncRNAs activate their associated sense protein-coding genes by establishing active chromatin structures.

Taken together, the above observations from various organisms imply that homeobox gene regulation by long ncRNAs is a conserved transcriptional regulatory mechanism across the evolutionary ladder, involving ncRNA and/or the act of its transcription.

LONG ncRNAs AND DOSAGE COMPENSATION

Dosage compensation is an epigenetic process that ensures equal dosage of X-linked gene products between male and female organisms and it is achieved by either hyper-activation or inactivation of the X chromosome. In *Drosophila*, in which, like mammals, males have one X chromosome and females have two X chromosomes, equal levels of X-linked gene products are achieved between males and females through hyper-activation of the X chromosome in males.[41,42] In mammals, the equal dosage of X-linked gene products between males and females is achieved through inactivation of one of the two X chromosomes in females during early embryonic development.[6,43,44] As *Drosophila* and mammals are distantly related organisms, it is interesting to note that the dosage compensation process is brought about by ncRNAs in both species, while the underlying mechanisms seem to differ significantly. In mammals, X-chromosome inactivation has been investigated in depth using the mouse as a model system. In XX female mammals, genes on one of the X chromosomes undergo transcriptional silencing in a chromosome-wide manner during early embryonic development, which occurs in nonrandom or random fashion and in a stage- and tissue-specific manner. Nonrandom or imprinted X-chromosome inactivation (the paternal chromosome is always selected for inactivation) occurs in very early preimplantation embryos and extra-embryonic tissues, whereas random X-chromosome inactivation (where both parental chromosomes have an equal probability of being inactivated) is restricted to the epiblast lineage i.e., the embryo.[45]

A 500-kb complex locus known as the X-inactivation center (*Xic*) is implicated in both random and nonrandom X-chromosome inactivation pathways (Fig. 2A). This region is replete with genes that transcribe ncRNAs and a few protein-coding RNAs. The ncRNA genes, which span a region of approximately 100-200 kb, control most of the X-chromosome inactivation pathways.[14] Five ncRNA genes have been identified in the *Xic*: *Xist*, *Tsix*, *Xite*, *RepA* and *DXPas34*. *Xist* is 15-17 kb in length and transcribed exclusively from the future inactive X chromosome.[46] *Tsix*, the antisense partner of *Xist*, is transcribed from the 3′ end of the *Xist* gene and covers the entire *Xist* coding region as well as its promoter region. It is about 40 kb in length and exclusively expressed by the future active X chromosome (Fig. 2C).[47] *Xite*, an upstream activator sequence, is located 10 to 15 kb upstream of the *Tsix* promoter and encodes several ncRNAs.[48] *RepA* is a 1.6-kb transcript encoded from the Repeat A region of *Xist*. *RepA* selectively transactivates the *Xist* promoter on the future inactive X chromosome (Fig. 2B).[49] A microsatellite repeat element, *DXPas34*, at the 5′ end of the *Tsix* transcript encodes short bidirectional transcripts which have been shown to regulate *Tsix* transcription.[50,51] In addition, bidirectional transcription over the *Xist* locus leads to the generation of several 25-45-bp small RNAs (xiRNAs).[52] Below, the functional interplay between these ncRNAs at various stages of the X-inactivation process, resulting in chromosomal asymmetry leading to active and inactive X chromosomes, will be discussed in greater detail (see Chapter 1, Figure 5, page 12 for a summary of XCI mechanism).

In the mouse, initiation of X-chromosome inactivation coincides with the onset of *Xist* expression. *Xist* expression is first detected on the paternal X chromosome at the two-cell stage of embryogenesis and in subsequent stages of embryogenesis *Xist* RNA spreads from its site of synthesis and blankets the entire X chromosome in cis (Fig. 2B).[53] The paternal-specific *Xist* expression coincides with transcriptional gene silencing and accumulation of heterochromatic marks along the paternal X chromosome, indicating a link between *Xist* expression and imprinted X inactivation. However, the functional role of *Xist* in the establishment of imprinted of X chromosome inactivation is not very clear, as a recent investigation documented that transcriptional silencing of the paternal chromosome occurs independent of *Xist*, because the genes on paternal chromosomes lacking the *Xist* gene were still repressed.[54] However, the maintenance of the imprinted form of X-chromosome inactivation in extra-embryonic tissues requires the presence of

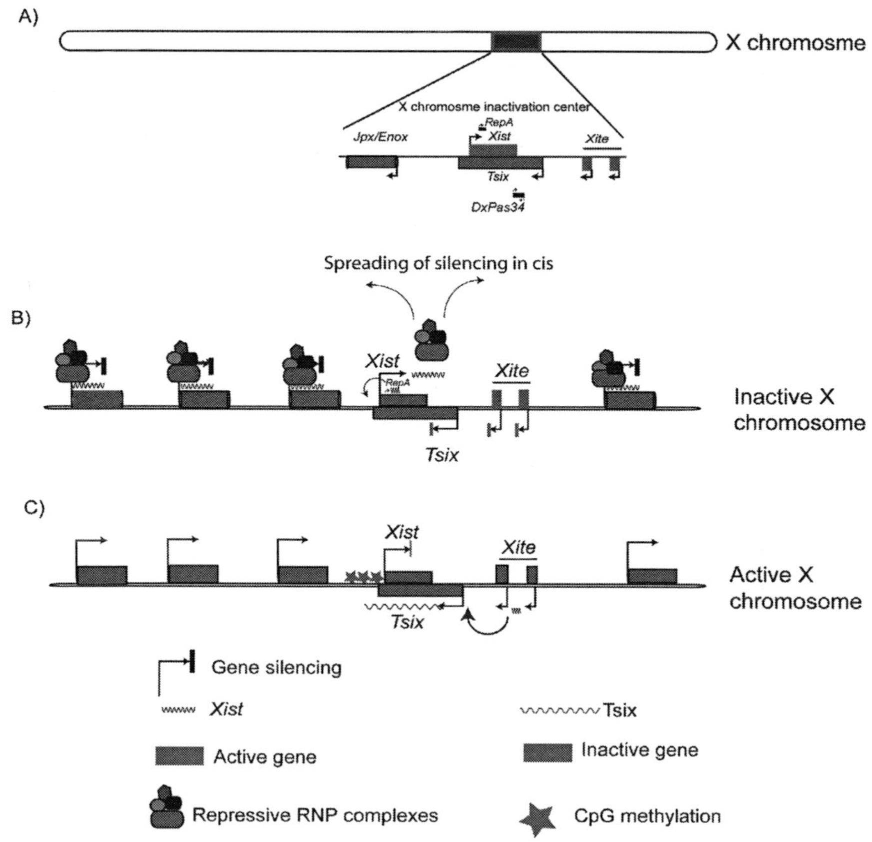

Figure 2. A) Doage compensation in mammals. X chromosome showing the X inactivation center (*Xic*). B) Inactive X chromosome showing *Xist* expression and its accumulation along the X chromsome. C) Active X chromosome showing *Tsix* activation, *Xist* inactivation and gene activation along the future active X chromosome. Maps are not drawn to scale.

Xist. In the embryonic lineage, the paternally inactivated X chromosome is reactivated and one of the parental chromosomes is randomly chosen for inactivation.

The random X inactivation process has been investigated using differentiating mouse ES cells as a model system, which faithfully recapitulates most of the X chromosome inactivation steps that occur during postimplantation development. In female ES cells prior to X-chromosome inactivation, *Tsix* is expressed at high levels and *Xist* is expressed at low levels on both chromosomes. Pluripotent transcription factors such as *Oct4*, *Nanog* and *Sox2* have been implicated in the repression of *Xist*, but it is not clear how *Tsix* transcript levels are maintained at higher levels in cells prior to X-chromosome inactivation.[55] Abundant *Tsix* expression across the *Xist* locus is correlated with establishment of active chromatin over the *Xist* gene body, which is thought to be involved in resetting the epigenetic marks prior to the onset of random X inactivation.[56] In addition, a recent investigation has demonstrated that *Tsix* transcription in cells prior to X-chromosome activation helps to limit the spread of heterochromatin into the *Xist* promoter region from the so-called heterochromatic hot spot, which is located further upstream of the *Xist* promoter (Fig. 2), indicating that *Tsix* transcription performs multiple tasks which set the stage for the onset of random X chromosome inactivation.[57] *Xite* ncRNA positively regulates *Tsix* transcription. The truncation of *Xite* ncRNA using transcription termination signals results in downregulation of *Tsix* transcription,[48] indicating that *Xite* transcription probably creates transcriptionally competent chromatin at the *Tsix* promoter.

The onset of random inactivation involves two important steps: counting the number of X chromosomes and selection of the future active and inactive X chromosomes. These steps ensure inactivation of only one X chromosome while keeping the other X chromosome in an active form. The counting step is initiated only when the ratio of X chromosomes to autosomes is one or more (X:A ≥ 1) and hence X chromosome inactivation occurs only in females but not males, in which the X chromosome to autosome ratio is less than one (X:A = 0.5). A recent investigation implicated the X chromosome-encoded transcription factor E3 ubiquitin ligase RNF12 in the counting step of X chromosome inactivation. It has been shown that over expression of RNF12 in male and female ES cells resulted in inactivation of one and two X chromosomes, respectively, indicating that the RNF12 is an activator of X chromosome inactivation process.[58] Based on these observations, RNF12 has been speculated to directly activate *Xist* transcription in female ES cells during X-chromosome inactivation. However, no evidence to date has documented a direct interaction between RNF12 and the *Xist* promoter, let alone its mode of action.

Transient X chromosome pairing and eventual molecular cross-talk between the X chromosomes during the initiation of X chromosome inactivation have been suggested to play an important role in the selection of the active and inactive X chromosomes. This process has been shown to be mediated by autosome-encoded transcription factors such as OCT4, CTCF and YY1 by trans-activating *Tsix* transcription.[59] Importantly, *Tsix* RNA synthesis is critical for chromosome pairing,[60,61] indicating that RNA-protein interactions determine the transient pairing of two X chromosomes. However, the mechanisms by which this physical association leads to the selection of active and inactive X chromosomes in a mutually exclusive manner in the same nuclear milieu are unclear. In addition to *Xite*, *Xist* and *Tsix* ncRNAs, *RepA* also been shown to play an important role in X chromosome inactivation. Prior to X chromosome inactivation, *RepA* is expressed by both X chromosomes. At the onset of X chromosome inactivation, *Tsix* RNA is downregulated on the future inactive X chromosome by as yet unidentified mechanisms, while *RepA* remains active, interacts with the PRC2 complex and tethers it to the *Xist* promoter on the future inactive X chromosome to establish transient local heterochromatin structure with H3K27me3 marks. Paradoxically, this tethering aids in the *trans*-activation of the *Xist* promoter, leading to accumulation of *Xist* along the future inactive X chromosome and the subsequent recruitment of heterochromatin modifiers, resulting in an RNA polymerase II-deficient heterochromatin compartment.[49] However, on the future active X-chromosome, *Tsix* is exclusively expressed and represses *in cis* the *Xist* and *RepA* promoters by targeting CpG methylation through interactions with heterochromatin machinery containing Dnmt3a.[62]

The functional link between ncRNA and X-chromosome inactivation has been investigated in more depth than other biological pathways. However, several outstanding questions remain to be answered. For example, the factors involved in guiding *Xist* along the future inactive X-chromosome *in cis* have yet to be determined. Furthermore, how does the expression of *Xist* and *Tsix* become exclusively restricted to inactive and active X chromosomes, respectively? Addressing this important issue will have implications in our understanding of the chromosomal counting and selection processes of X-chromosome inactivation.

Dosage compensation in flies has the similar objective of equalizing X-linked gene products between male and female flies with unequal numbers of X chromosomes. However, in contrast to silencing one of the two X chromosomes in female mammals, dosage compensation in the flies is achieved through hyper-transcription of the single X chromosome in males (Fig. 3A-B). Nevertheless, like in mammals, the dosage compensation process in flies is carried out by long ncRNAs, here *rox1* and *rox2*, along with a defined set of male-specific lethal proteins: MSL1, MSL2, MSL3, MLE (maleless) and H4K16 acetyltransferase MOF. This ribonucleoprotein complex is referred to as the Dosage Compensation Complex or the MSL complex.[63] Like *Xist*, *rox* ncRNA blankets the male X chromosome *in cis* (Fig. 3A). Surprisingly, *rox* can coat the X chromosome *in trans* even when the *rox* gene has been translocated onto an autosome, indicating marked mechanistic differences in the mechanisms by which *Xist* and *rox* effect dosage compensation. *Rox* ncRNAs are encoded by RNA polymerase II and localized to the nuclear compartment. Although both *rox* ncRNAs are located on the X chromosome and perform similar functions, they differ structurally in size and sequence. *rox1*, the longer transcript and *rox2* are 3.7 kb and 0.6 kb in length, respectively.

The *rox* ncRNAs are exemplary of the ability of diverse ncRNAs to carry out a similar function despite a lack of sequence similarity. Single mutations involving one of the *rox* ncRNAs have no apparent phenotypic effect on either the RNA chromosomal coating or hyper-transcription of the male X chromosome, indicating that they are functionally redundant. However, flies with a deletion or mutation of *both rox* ncRNAs demonstrated mislocalization of the SL complex and reduction in the transcription of the X chromosome, indicating that the *rox* ncRNAs are integral parts of the MSL complex.[64] Both *rox1* and *rox2* harbor evolutionarily conserved short sequence stretches at the 3′end which can form stem-loop structures. Deletion or mutation of the putative stem-loop structures from *rox1* or *rox2* resulted in the reduction of male viability.[44] These stem-loop structures are thought to play an important role in the recruitment of the MSL complex, but evidence is still lacking as to the actual role of *rox* ncRNAs in the functional organization of the MSL complex.

The question remains as to how *rox* ncRNAs target the MSL complex along the male X chromosome. Using ChIP-on-chip technology with antibodies directed against components of the MSL complex, several high-affinity MSL binding sites were mapped along the X chromosome and found to be twice as prevalent there in comparison to autosomes.[65] If the MSL complex is recruited through high-affinity DNA sequences, then *rox* ncRNA simply acts as a RNA scaffold for integrating the MSL complex. Interestingly, most MSL high-affinity binding sites map to the 3′ end of protein-coding genes. In light of these observations, tethering of the MSL complex to the 3′ end of genes has been proposed to facilitate transcriptional elongation, resulting in hyper-transcription of the targeted genes.[66] The MSL complex could also be targeted by *rox* ncRNA through RNA-RNA or RNA-DNA interactions with the nascent transcripts or the promoter DNA sequences of the X-linked genes *in cis* to achieve hyper-transcription. However, experimental support for the latter hypothesis is lacking. Understanding the role of *rox* ncRNA in the functional integration of the MSL complex and its targeting along

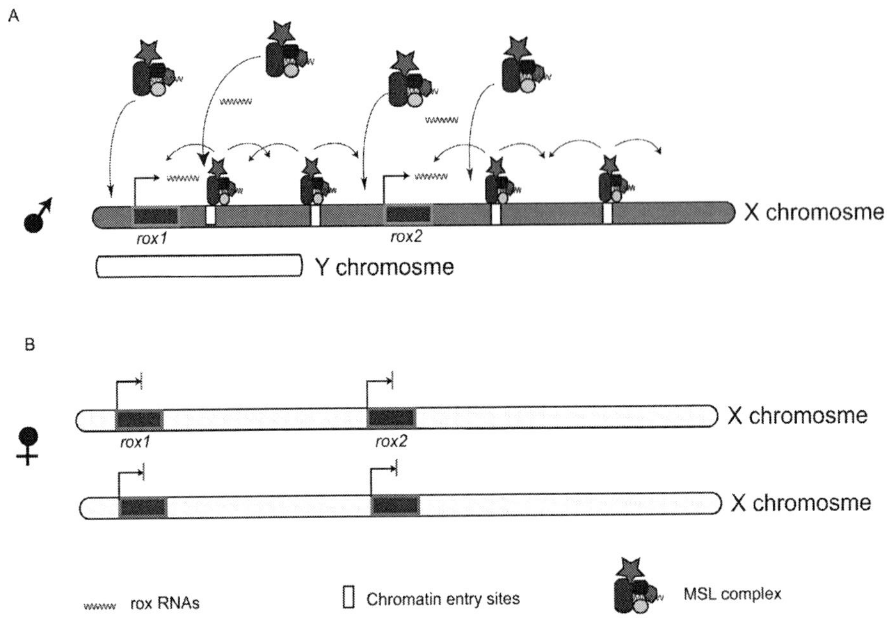

Figure 3. Dosage compensation in Drosophila. A) Single X chromosome in males showing transcriptionally active *rox* genes (*rox1* and *rox2*) and chromatin entry sites. The MSL complex specifically activates *rox* genes on the male X chromosome. *rox* RNAs are incorporated into the MSL complexes and the *rox*-MSL complex spread along the male X chromosomes via chromatin entry sites. The hyper-transcriptional activity on the male X chromosome is shown in dark color. B) Drosophila females have two X chromosomes. The *rox* genes are inactive on the female X chromosome and the low level of transcriptional activity on two female X chromosomes is shown in light color.

the male X chromosome will provide greater insight into how long ncRNAs achieve transcriptional activation.

LONG ncRNAs AND PLURIPOTENCY

The transcriptional network centered on a cocktail of transcription factors including Oct4, Nanog, c-Myc and Sox2, has been well-established as playing an essential role in the establishment and maintenance of the pluripotent state of embryonic stem cells. Interestingly, some of these transcription factors have been implicated in the regulation of small ncRNAs whose level is critical for maintaining pluripotency. For example, c-Myc regulates microRNAs such as miR-141, miR-200 and miR-429 in ES cells, but not in differentiated cells. Downregulation of these miRNAs accelerates the process of ES cell differentiation,[67] indicating that they are critical for maintaining ES cell pluripotency and that c-Myc regulates the pluripotent state of ES cells through ncRNA levels. Likewise, miR-145 represses pluripotency in human embryonic stem cells by regulating OCT4, SOX2 and KLF4 transcription factors.[68] Loss of miR-145 impairs differentiation and elevates OCT4, KLF4 and SOX2 levels. Intriguingly, miR-145 is, in turn, repressed in ES cells by OCT4 and downregulation of OCT4 in ES cells increases the level of miR-145 and promotes differentiation. These findings indicate that small ncRNAs and the pluripotent transcriptional program maintain pluripotency through a double-negative feedback regulatory loop that switches embryonic stem cells between self-renewal and differentiation (Fig. 4).

Emerging evidence suggests a functional link between long ncRNAs and pluripotency. In a recent analysis, four conserved long ncRNAs were identified based on their proximity to Oct4 and Nanog binding sites; two of these were found to be direct targets of Oct4 and Nanog.[69] Similar to miRNAs, long ncRNAs and pluripotent transcription factors form a regulatory feedback loop. For example, Oct4 positively regulates one of the long ncRNAs, which in turn activates Oct4 transcription. These data establish another hidden regulatory loop in the maintenance of pluripotency. The handful of long ncRNAs mentioned above only represents the tip of the iceberg and a comprehensive analysis of long ncRNA transcription across the ES cell genome is needed in order to determine the extent of ncRNA involvement in the maintenance of pluripotency. Along these lines, a recent investigation using genome-wide analyses identified 945 long ncRNAs in differentiating ES cells of which 174 were differentially expressed. In-depth analysis of two novel differentially expressed ncRNAs revealed that they play an important

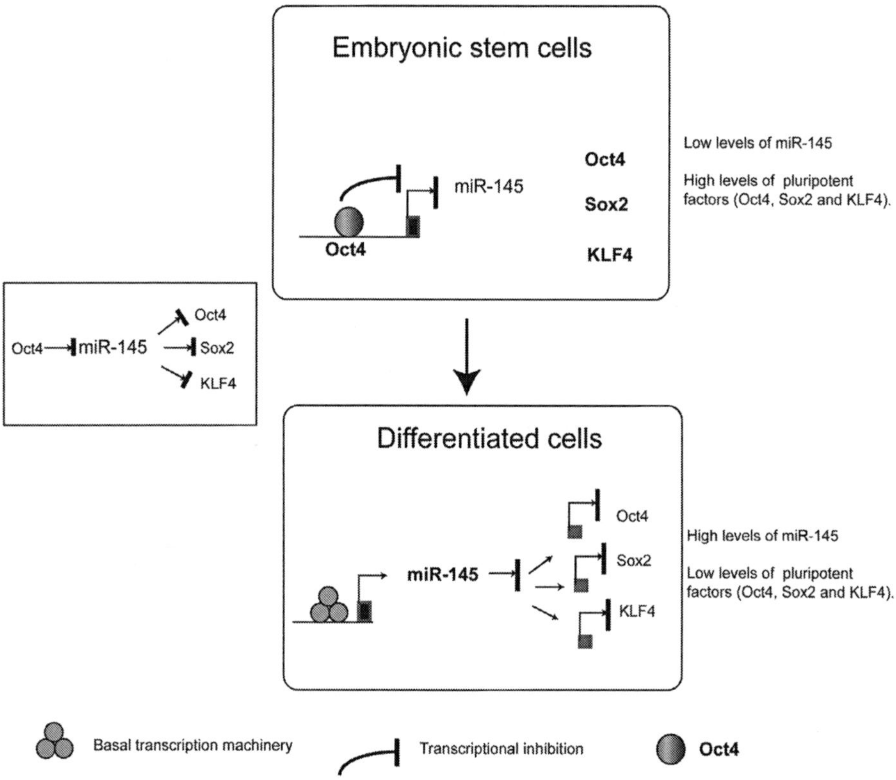

Figure 4. Model depicting the double-negative feedback loop by miR-145 and three pluripotent transcription factors Oct4, Sox2 and KLF4. miR-145 maintains the differentiated state of a cell via repressing the pluripotency transcription factors. Like-wise, Oct4 controls the pluripotency of embryonic stem cells by repressing miR-145. Bold font represents higher levels, whereas regular font represents low levels.

role in pluripotency and differentiation by regulating the chromatin structure through interactions with chromatin-modifying complexes.[40] This study is indeed encouraging and future research along these lines will likely uncover more pluripotent transcriptional programs involving long ncRNAs.

LONG ncRNAs AND GENOMIC IMPRINTING

Genomic imprinting is an epigenetically controlled gene regulatory mechanism by which expression of a sub-group of autosomal genes is restricted to either the maternal or paternal allele. So far, more than 100 imprinted genes have been identified in mouse and the majority of imprinted genes are organized in clusters containing both protein-coding and noncoding mRNA genes. The noncoding mRNA genes contain both small and long ncRNAs. The small RNAs include miRNAs, C/D small RNAs and piRNAs which range in size from 19 to 71 nucleotides and the long ncRNAs measure from 100 bp to several hundred kb.[70,71] Genes in imprinted clusters are coregulated by long-range *cis*-acting elements, which are short stretches of DNA sequence (1 to 3 kb) that often carry epigenetic marks such as CpG methylation and/or histone modifications on only one parental chromosome. These differentially marked regions are called DMRs or ICRs (imprinting control regions), as their deletion in mice was shown to affect imprinting in clusters.[72] The parent-specific epigenetic profile at the ICR or DMR is established during the germline (in the majority of cases it occurs in female germline but in some cases in male germline) or postimplantation stages of mouse development. Although imprinted clusters carry common epigenetic features, the mechanisms by which they attain these features differ across clusters. Among the 13-14 imprinted clusters identified so far, only the mechanisms of seven that contain differentially methylated imprinting control regions have been investigated. The mechanisms underlying the ICR-mediated long-range gene regulatory mechanism have been investigated intensely over the last several years. These studies have shown that ICRs regulate genomic imprinting in clusters either by acting as chromatin insulators[73,74] or by harboring the promoters for ncRNA.[24] Since ncRNA and chromatin architecture is the theme of this chapter, only the imprinted clusters regulated by ncRNAs will be discussed further (Fig. 5).

Kcnq1, *Igf2r* and *Gnas* imprinted clusters contain very long ncRNAs encoded from the promoters that map to ICRs (Fig. 5).[70] For example, *Kcnq1ot1* ncRNA is a 91-kb, poorly spliced, moderately stable transcript localized to the nucleus. Its promoter maps to an imprinting control region (*Kcnq1* ICR), which lies in intron 10 of the *Kcnq1* gene on mouse chromosome 7.[75] *Kcnq1ot1* transcribes in the antisense direction to the *Kcnq1* gene, starting in intron 10 and ending in intron 11 and thus it does not span across the promoter region of the *Kcnq1* gene (Fig. 5A). Targeted deletion of the *Kcnq1ot1* promoter or truncation of the *Kcnq1ot1* transcript from 91 kb to a size range of 0.7-2.6 kb resulted in the loss of silencing of 8 to 9 maternally expressed protein-coding mRNA genes spread over one mega-base region on either side of the *Kcnq1ot1* promoter in several different studies, indicating that *Kcnq1ot1* RNA is a bidirectional silencer.[76-80] However, these experiments did not distinguish between the functional role of the ncRNA itself and the act of ncRNA transcription. By interfering with the stability of episome-encoded *Kcnq1ot1* transcript in human cells by flanking *Kcnq1ot1* with the destabilizing domain of highly unstable *c-fos* transcript, *Kcnq1ot1* was also found to play an important role in bidirectional silencing.[75] Recent data from several independent investigations have

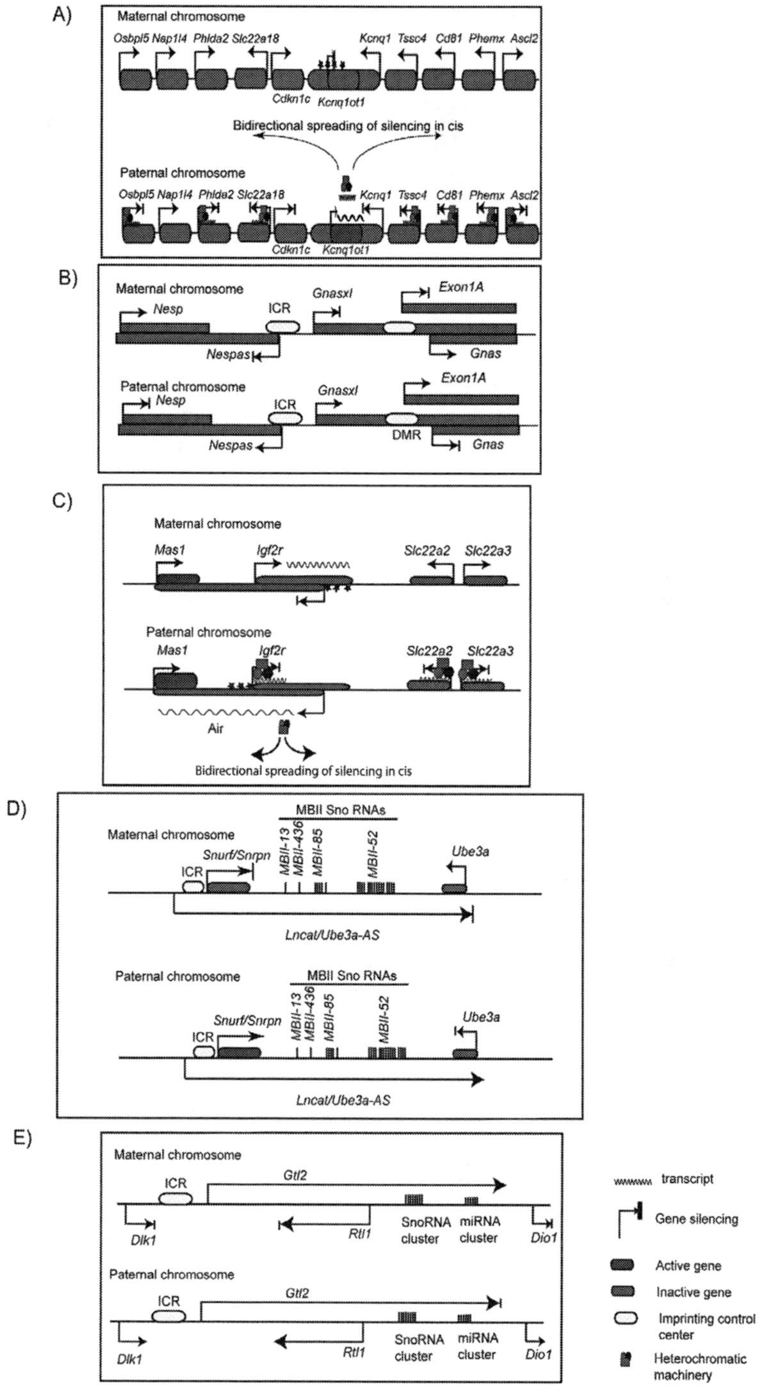

Figure 5. A-E) The mouse imprinted gene clusters *Kcnq1* (A), *Gnas* (B), *Igf2r* (C), *Snurf/Snrpn* (D) and *Dlk1/Gtl2* (E) containing small and/or long noncoding RNAs. Arrows show the direction of transcription. Maps are not drawn to scale.

shown that *Kcnq1ot1* interacts with chromatin and recruits heterochromatin complexes containing PRC2 and G9a in a gene-specific manner.[75] This mode of action of *Kcnq1ot1* has parallels with the mode of action of *Xist* ncRNA, which mediates transcriptional gene silencing by coating the future inactive X chromosome.

Airn, an antisense transcript to *Igf2r*, is 108 kb in length and localized to the nuclear compartment. *Airn* is spliced, but the splice variants constitute an insignificant portion of the total *Airn* transcription and moreover are localized to the cytoplasmic compartment.[81] The *Airn* promoter maps to a differentially methylated imprinting control region located in intron 2 of the *Igf2r* gene. Similar to *Kcnq1ot1*, selective deletion of the *Airn* promoter or truncation of its transcript from 108 kb to 3 kb resulted in loss of silencing of 3 protein-coding genes, *Slc22a3*, *Slc22a2* and *Igf2r*, on either side of the *Airn* promoter, indicating that *Airn* is also a bidirectional silencer (Fig. 5C).[82] It is not clear whether the RNA itself or its transcription is crucial for bidirectional silencing. However, a recent investigation has documented that *Airn* interacts with G9a histone methyltransferase and specifically targets the nonoverlapping *Slc22a3* promoter but not the overlapping *Igf2r* promoter, indicating that *Airn* probably uses different mechanisms to silence the overlapping and nonoverlapping promoters.[83] *Airn* is transcribed through the overlapping *Igf2r* promoter in all tissues and, unlike nonoverlapping genes silenced only in extra-embryonic tissues in a developmentally regulated manner, the overlapping *Igf2r* gene is constitutively silenced on the paternal chromosome. Based on these observations, the act of transcription has been proposed to silence the overlapping gene *in cis* through transcriptional interference, whereas *Airn* itself mediates nonoverlapping gene silencing by acting as a heterochromatin scaffold.[16]

The *Gnas* imprinted cluster is approximately 70 kb in length and accommodates three protein-coding (*Gnasxl*, *Nesp* and *Gnas*) and two noncoding mRNA (*Exon1A* and *Nespas*) genes (Fig. 5B).[70] Unlike the *Kcnq1* and *Igf2r* clusters, in which protein-coding mRNA is expressed only by the maternal allele, the *Gnas* cluster contains both paternally and maternally expressed protein-coding mRNAs. Its two ncRNAs are expressed exclusively by the paternal allele. The promoters of these two ncRNAs map to germline-derived differentially methylated regions. Targeted deletion of exon 1A DMR affected the imprinting of the flanking *Gnasxl* gene and deletion of the DMR containing the *Nespas* promoter resulted in loss of imprinting of all genes, indicating that *Nespas* could play an important role in the imprinting of the *Gnas* cluster.[84] However, no experiments have been published to date that implicate a functional role of *Nespas* and/or its transcriptional process in imprint acquisition and maintenance.

The *PWS/AS* and *Gtl2/Dlk1* imprinted clusters encode long ncRNAs containing large repeated arrays of C/D small nucleolar RNA (snoRNA) and miRNA (Fig. 5D and E). In mice, the PWS/AS locus encodes a huge paternally expressed 1-Mb poly-cistronic long ncRNA, Lncat.[85] The locus has approximately 148 exons that include ICR/DMR, paternally expressed protein-coding Snrpn/Snurf mRNAs, two repeated large arrays of MBII-52 and MBII-85 C/D SnoRNAs and an antisense RNA (Ube3a-as) to maternally expressed Ube3a transcript (Fig. 5D). The existence of a 1-Mb long ncRNA in the PWS/AS locus is controversial. Using high-resolution fluorescence *in situ* hybridization with oligonucleotide probes, a recent investigation has demonstrated that MBII-52 and MBII-85 C/D snoRNAs are not part of the large Lncat RNA but are instead derived from two independent noncoding transcripts.[86] This evidence was ascertained based on the fact that spliced MBII-52 and MBII-85 accumulate as "RNA clouds" in a non-overlapping fashion near the site of transcription. However, this observation cannot rule out the possibility that Lncat is pervasively transcribed through protein-coding and noncoding transcription units at a very low level. The formation of RNA clouds by MBII-85 and MBII-52 near their transcription units is indicative of a functional role in the chromatin-based transcriptional regulation of neighboring genes. The functional significance of paternally expressed Ube3a-as in the paternal repression of its sense counterpart Ube3a has not yet been investigated.

Numerous imprinted microRNA genes are organized in two clusters at the *Dlk1-Gtl2* domain (Fig. 5E). A small group of 5-6 miRNAs (miR-431, miR-433, miR-127, miR-434 and miR-136) are processed from the maternally expressed antisense transcript that overlaps the paternally expressed *Rtl1* (retrotransposon-like 1) gene.[71,87] The miRNAs show perfect complementary to paternally expressed *Rtl1* mRNA and, importantly, several cleavage sites have been experimentally mapped within the *Rtl1* mRNA using 5′ RACE technology, suggesting that these miRNAs may mimic the mode of action of siRNA.[88] It remains to be seen whether these miRNAs have a role in the maternal-specific repression of *Rtl1 in cis*. In addition, the mechanism by which paternally expressed *Rtl1* transcript escapes the actions of miRNA *in trans* must still be established. Another larger cluster of nearly 40 miRNAs maps to the region downstream of the C/D snoRNA cluster. The functional role of this microRNA cluster in genomic imprinting of the *Dlk1/Gtl2* locus has not yet been investigated. The C/D snoRNA cluster is located between the two miRNA clusters upstream of the *Rtl1* gene. Its functional significance in the imprinting of *Dlk1/Gtl2* is also currently unknown.

LONG ncRNAs AND CENTROMERIC HETEROCHROMATIN

Spatial organization of heterochromatin and euchromatin in the mammalian genome is crucial for normal development and differentiation. While the euchromatic environment is conducive to active transcription, heterochromatin plays an important role in genome stability, chromosomal segregation and gene regulation. Defects in the molecular pathways that functionally demarcate the genome could lead to abnormal chromosome segregation and genome instability.[89] For example, the centromeric heterochromatin formation is crucial for faithful chromosome segregation during cell division and defective centromeres lead to abnormal chromosome segregation and aneuploidy.[90] The centromere is an epigenetic regulatory framework that plays a supportive role in mitotic spindle formation and sister chromatid cohesion. The centromere is made up of an array of repeat elements thought to organize heterochromatic structures by recruiting repressive chromatin remodeling complexes. Recent investigations have demonstrated that small and long ncRNAs play a functional role in the establishment as well as maintenance of centromeric heterochromatin.[20,91,92] In *S. pombe*, RNAi-based silencing mechanisms play a pivotal role in the establishment and maintenance of centromeric heterochromatin.[93-95] The centromere in *S. pombe* is compartmentalized into a central core (cnt) flanked by inner (*imr*) and outer (*otr*) repeats (Fig. 6). CENP-A (the centromere-specific histone H3 variant) nucleosomes in the central core region act as an anchor point for the kinetochore, which connects the mitotic spindle to the centromere. The outer repeat region contains dg-dh repeats which upon bidirectional transcription produce long double-stranded RNAs that are tethered to RNAi substrates to form heterochromatin at the *Otr* region (Fig. 6).[94] The heterochromatin at the *Otr*

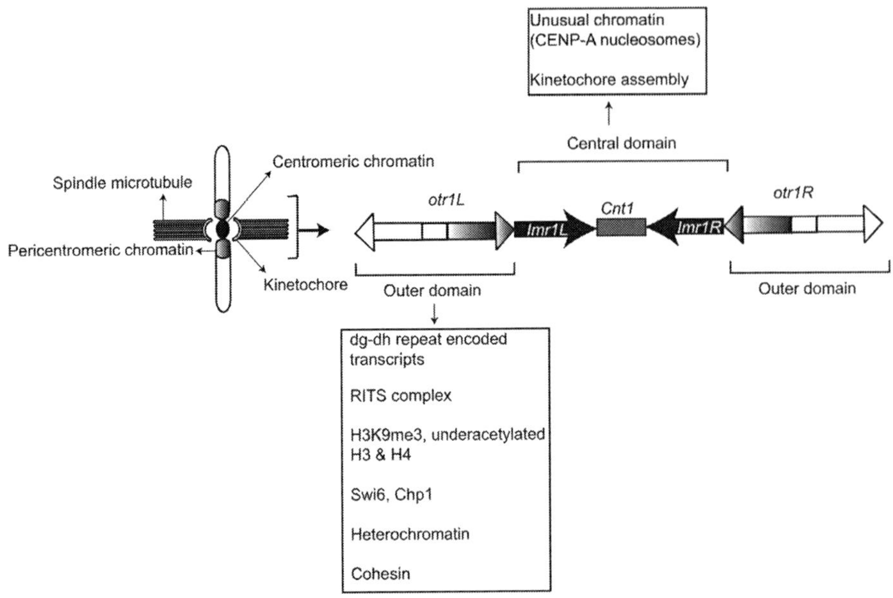

Figure 6. Centromere structure and organization in *S.pombe*. A-B) Schematic depiction of *S.pombe* centromere and pericentromere regions. Centromere1 (*cnt1*) contains a central core of nonrepetitive sequences flanked by innermost repeats (*imr1L and imr1R*) and outer repeats (*otr1L and otr1R*). *Cnt1* acts as an anchor point for kinetochore assembly. The *otr* region composed of dg-dh repeats and are transcribed in forward and reverse directions, thus giving rise to double stranded RNA (dsRNA). dsRNAs, derived from the dg-dh repeats, have been shown to take part in heterochromatin organization of the *otr* region by triggering RNAi via recruiting RITS-RDRP complexes. The heterochromatic modifications such as H3K9me2 and H3K9me3 in the *otr* region recruit Swi6 (HP1) and Chp1 heterochromatic proteins and they, in turn, are responsible for the recruitment of Cohesin. Cohesin is required for the proper orientation of centromeres during mitosis.

region is not only critical for sister chromatid cohesion but also helps in CENP-A chromatin formation in the *Cnt* region through provision of RNAi substrates.[96]

In eukaryotes, however, the molecular mechanisms by which the centromeric heterochromatin is established and maintained are not clear. In mammals, the centromere contains two domains: a central domain required for the kinetochore assembly and flanking pericentric heterochromatin required for sister chromatid cohesion. Thus, both structural and functional features of centromeres are highly conserved across the evolutionary spectrum. Previously, RNA was shown to define the higher order chromatin structure at pericentric heterochromatin by organizing heterochromatic components such as H3K9me3 and HP1 proteins. Recent analyses have further linked both RNAi components and long ncRNAs in the maintenance of pericentric heterochromatin. Chicken cell lines lacking Dicer, a dsRNA RNA-specific endonuclease and a crucial component of the RNA-induced silencing complex (RISC), showed reduced CpG methylation and H3K9me3 in the pericentric heterochromatin.[97] Whether Dicer plays a direct or indirect role in the maintenance of pericentric heterochromatin has not yet been determined. In Dicer mutants, a decrease in Dnmt1, a maintenance methyltransferase, levels occurs due to overexpression of Rbl2, which is caused by downregulation of its negative regulator miR-290.[98] Since DNA methylation is associated with H3K9me3 levels and Dicer depletion does not have any effect on CENPA and CENPC1 localization to the centromeres,[20,97] the functional role of the RNAi components in centromeric heterochromatin formation requires a thorough investigation in eukaryotes. However, recent analyses have identified a functional role for long ncRNAs in centromeric heterochromatin formation. In mammals, the centromere contains two classes of tandemly organized repeat arrays, termed minor and major satellite sequences. Major satellite sequences map to the pericentric portion of the centromere; minor satellite sequences are located at the primary constriction of condensed mitotic chromosomes. Minor (alpha) satellite sequences are transcribed, with transcripts ranging between 100-500 nt in size and localized specifically to the centromere.[92] Overexpression of 120-nt RNA molecules, blocking the 120-bp unit minor satellite sequence in mammalian cells, or treatment of mitotic cells with single-stranded RNA-specific nucleases leads to mislocalization of centromere-associated heterochromatin proteins, indicating that the minor-satellite repeat RNA acts as a scaffold in recruiting heterochromatin proteins.[92] In addition, minor satellite repeat RNA has been shown to directly interact with centromere-associated proteins such as CENPC1 and the centromere associated chromosomal passenger complex, which contains Aurora B kinase, Survivin and inner centromere protein (INCENP), indicating that this ribonucleoprotein complex is critical for proper centromere formation.[20,21] Taken together, the above data implicate ncRNA in mammalian centromere formation and thereby, genome stability.

CONCLUSION

Functional demarcation of the eukaryotic genome into active and inactive chromatin compartments enables spatio-temporal gene expression, which is crucial for the normal development of an organism. Accumulating evidence in recent years suggests that long ncRNAs and/or the process of their transcription act as scaffolds in the spatial organization of active and inactive chromatin compartments. Although the latter two processes have been implicated in the functional organization of the genome, a number of gaps remain in our understanding of the mechanisms they govern. In most cases the effects of ncRNA transcription are restricted to overlapping genes, whereas the effects of ncRNAs themselves can range from a single gene to an entire chromosome. Functional dissection of these mechanisms is imperative to gain further understanding of the contribution of ncRNAs to organismal complexity. Although, in the majority of instances, ncRNA sequences and their target genes differ across the evolutionary spectrum, their physical location, promoter sequences and, in part, their modes of action are conserved. For example, well-investigated ncRNAs such as *Xist* and *Kcnq1ot1* significantly differ in their primary sequence across different species, but their modes of action remain unchanged between mice and humans. This consistency indicates that the secondary structures of ncRNAs likely play a role in their functions.

Misexpression of several long ncRNAs deregulates clinically significant genes, as evidenced by the formation of focal heterochromatin structures in several diseases, For example, aberrant expression of antisense ncRNAs through the *p15* and *p21* promoters in certain diseases results in the inactivation of these promoters via CpG methylation. What triggers aberrant antisense ncRNA expression in certain diseases is not clear, nor, importantly, is how their transcription across an overlapping or neighboring gene causes specific methylation of their promoters. Similarly, recent investigations have also exposed a link between genome stability and some ncRNAs, which play an important role in the maintenance of higher order chromatin structure at the centromere. Defects in the pathways that lead to functional centromeres result in loss of chromosome segregation and aneuploidy. The loss of genome stability triggers chromosome deletions and translocations, which, in turn, result in aberrant ncRNA transcription by cryptic promoters, through several genes with clinical significance.

Though the objectives of certain ncRNA-mediated epigenetic pathways are similar, the mechanisms used to achieve these objectives differ considerably depending on the organism. For example, dosage compensation pathways in flies and mammals employ the common theme of ncRNA but use different mechanisms, i.e., hyperactivation of one X chromosome in male flies and transcriptional silencing of one X chromosome in female mammals, to ensure an equal dosage of X-linked gene products in males and females. Interestingly, both actions involve coating the entire X chromosome with ncRNA.

The majority of long ncRNAs described here constitute a network that regulates various biological functions via modifying the chromatin structure in a spatio-temporal manner. Hence this network of long ncRNAs is a part of the RNA infrastructure dedicated towards maintaining chromatin structure in time and space.

Though the evidence collected over the last few years has enabled us to understand ncRNA as a hidden gene regulatory mechanism participating in diverse biological functions, much more remains to be learned about their actions before their real impact on organismal complexity can be fully appreciated.

ACKNOWLEDGEMENTS

We thank Radha Raman Pandey and Natalia Guseva for comments on the chapter. This work was supported by the grants from the Swedish Cancer Research foundation (Cancerfonden), Swedish Medical Research Council (VR-M) and Swedish Childhood Cancer Society (Barncancerfonden) to CK. CK is a Senior Research Fellow supported by VR-M.

REFERENCES

1. Mattick JS. RNA regulation: a new genetics? Nat Rev Genet 2004; 5(4):316-323.
2. Kapranov P, Cheng J, Dike S et al. RNA maps reveal new RNA classes and a possible function for pervasive transcription. Science 2007; 316(5830):1484-1488.
3. Cheng J, Kapranov P, Drenkow J et al. Transcriptional maps of 10 human chromosomes at 5-nucleotide resolution. Science 2005; 308(5725):1149-1154.
4. Prasanth KV, Spector DL. Eukaryotic regulatory RNAs: an answer to the 'genome complexity' conundrum. Genes Dev 2007; 21(1):11-42.
5. Mercer TR, Dinger ME, Mattick JS. Long noncoding RNAs: insights into functions. Nat Rev Genet 2009; 10(3):155-159.
6. Whitehead J, Pandey GK, Kanduri C. Regulation of the mammalian epigenome by long noncoding RNAs. Biochim Biophys Acta 2008.
7. Faghihi MA, Wahlestedt C. Regulatory roles of natural antisense transcripts. Nat Rev Mol Cell Biol 2009; 10(9):637-643.

8. Collins LJ, Penny D. The RNA infrastructure: dark matter of the eukaryotic cell? Trends Genet 2009; 25(3):120-128.
9. Collins LJ, Chen XS. Ancestral RNA: the RNA biology of the eukaryotic ancestor. RNA Biol 2009; 6(5):495-502.
10. Guttman M, Amit I, Garber M et al. Chromatin signature reveals over a thousand highly conserved large noncoding RNAs in mammals. Nature 2009; 458(7235):223-227.
11. Brosnan CA, Voinnet O. The long and the short of noncoding RNAs. Curr Opin Cell Biol 2009; 21(3):416-425.
12. Seto AG, Kingston RE, Lau NC. The coming of age for Piwi proteins. Mol Cell 2007; 26(5):603-609.
13. Aravin AA, Hannon GJ, Brennecke J. The Piwi-piRNA pathway provides an adaptive defense in the transposon arms race. Science 2007; 318(5851):761-764.
14. Kanduri C, Whitehead J, Mohammad F. The long and the short of it: RNA-directed chromatin asymmetry in mammalian X-chromosome inactivation. FEBS Lett 2009; 583(5):857-864.
15. Lee JT. Lessons from X-chromosome inactivation: long ncRNA as guides and tethers to the epigenome. Genes Dev 2009; 23(16):1831-1842.
16. Mohammad F, Mondal T, Kanduri C. Epigenetics of imprinted long noncoding RNAs. Epigenetics 2009; 4(5):277-286.
17. Clemson CM, Hutchinson JN, Sara SA et al. An architectural role for a nuclear noncoding RNA: NEAT1 RNA is essential for the structure of paraspeckles. Mol Cell 2009; 33(6):717-726.
18. Sasaki YT, Ideue T, Sano M et al. MENepsilon/beta noncoding RNAs are essential for structural integrity of nuclear paraspeckles. Proc Natl Acad Sci USA 2009; 106(8):2525-2530.
19. Schoeftner S, Blasco MA. A 'higher order' of telomere regulation: telomere heterochromatin and telomeric RNAs. EMBO J 2009; 28(16):2323-2336.
20. Wong LH, Brettingham-Moore KH, Chan L et al. Centromere RNA is a key component for the assembly of nucleoproteins at the nucleolus and centromere. Genome Res 2007; 17(8):1146-1160.
21. Ferri F, Bouzinba-Segard H, Velasco G et al. Non-coding murine centromeric transcripts associate with and potentiate Aurora B kinase. Nucleic Acids Res 2009; 37(15):5071-5080.
22. Willingham AT, Orth AP, Batalov S et al. A strategy for probing the function of noncoding RNAs finds a repressor of NFAT. Science 2005; 309(5740):1570-1573.
23. Mondal T, Rasmussen M, Pandey GK et al. Characterization of the RNA content of chromatin. Genome Res 2010; 20:899-907.
24. Kanduri C. Functional insights into long antisense noncoding RNA Kcnq1ot1 mediated bidirectional silencing. RNA Biol 2008; 5(4):208-211.
25. Wilusz JE, Sunwoo H, Spector DL. Long noncoding RNAs: functional surprises from the RNA world. Genes Dev 2009; 23(13):1494-1504.
26. Rinn JL, Kertesz M, Wang JK et al. Functional demarcation of active and silent chromatin domains in human HOX loci by noncoding RNAs. Cell 2007; 129(7):1311-1323.
27. Katayama S, Tomaru Y, Kasukawa T et al. Antisense transcription in the mammalian transcriptome. Science 2005; 309(5740):1564-1566.
28. Yu W, Gius D, Onyango P et al. Epigenetic silencing of tumour suppressor gene p15 by its antisense RNA. Nature 2008; 451(7175):202-206.
29. Houseley J, Rubbi L, Grunstein M et al. A ncRNA modulates histone modification and mRNA induction in the yeast GAL gene cluster. Mol Cell 2008; 32(5):685-695.
30. Ligtenberg MJ, Kuiper RP, Chan TL et al. Heritable somatic methylation and inactivation of MSH2 in families with Lynch syndrome due to deletion of the 3′ exons of TACSTD1. Nat Genet 2009; 41(1):112-117.
31. Tufarelli C, Stanley JA, Garrick D et al. Transcription of antisense RNA leading to gene silencing and methylation as a novel cause of human genetic disease. Nat Genet 2003; 34(2):157-165.
32. Hekimoglu B, Ringrose L. Non-coding RNAs in polycomb/trithorax regulation. RNA Biol 2009; 6(2):129-137.
33. Hirota K, Miyoshi T, Kugou K et al. Stepwise chromatin remodelling by a cascade of transcription initiation of noncoding RNAs. Nature 2008; 456(7218):130-134.
34. Morris KV, Santoso S, Turner AM et al. Bidirectional transcription directs both transcriptional gene activation and suppression in human cells. PLoS Genet 2008; 4(11):e1000258.
35. Ringrose L, Paro R. Polycomb/Trithorax response elements and epigenetic memory of cell identity. Development 2007; 134(2):223-232.
36. Schuettengruber B, Cavalli G. Recruitment of polycomb group complexes and their role in the dynamic regulation of cell fate choice. Development 2009; 136(21):3531-3542.
37. Sessa L, Breiling A, Lavorgna G et al. Noncoding RNA synthesis and loss of Polycomb group repression accompanies the colinear activation of the human HOXA cluster. RNA 2007; 13(2):223-239.
38. Sanchez-Elsner T, Gou D, Kremmer E et al. Noncoding RNAs of trithorax response elements recruit Drosophila Ash1 to Ultrabithorax. Science 2006; 311(5764):1118-1123.

39. Petruk S, Sedkov Y, Riley KM et al. Transcription of bxd noncoding RNAs promoted by trithorax represses Ubx in cis by transcriptional interference. Cell 2006; 127(6):1209-1221.
40. Dinger ME, Amaral PP, Mercer TR et al. Long noncoding RNAs in mouse embryonic stem cell pluripotency and differentiation. Genome Res 2008; 18(9):1433-1445.
41. Kelley RL, Kuroda MI. Noncoding RNA genes in dosage compensation and imprinting. Cell 2000; 103(1):9-12.
42. Hallacli E, Akhtar A. X chromosomal regulation in flies: when less is more. Chromosome Res 2009; 17(5):603-619.
43. Heard E, Clerc P, Avner P. X-chromosome inactivation in mammals. Annu Rev Genet 1997; 31:571-610.
44. Ilik I, Akhtar A. roX RNAs: noncoding regulators of the male X chromosome in flies. RNA Biol 2009; 6(2):113-121.
45. Wutz A, Gribnau J. X inactivation Xplained. Curr Opin Genet Dev 2007; 17(5):387-393.
46. Brockdorff N, Ashworth A, Kay GF et al. The product of the mouse Xist gene is a 15 kb inactive X-specific transcript containing no conserved ORF and located in the nucleus. Cell 1992; 71:515-526.
47. Lee JT, Davidow LS, Warshawsky D. Tsix, a gene antisense to Xist at the X-inactivation centre. Nat Genet 1999; 21(4):400-404.
48. Lee JT. Regulation of X-chromosome counting by Tsix and Xite sequences. Science 2005; 309(5735):768-771.
49. Zhao J, Sun BK, Erwin JA et al. Polycomb proteins targeted by a short repeat RNA to the mouse X chromosome. Science 2008; 322(5902):750-756.
50. Cohen DE, Davidow LS, Erwin JA et al. The DXPas34 repeat regulates random and imprinted X inactivation. Dev Cell 2007; 12(1):57-71.
51. Vigneau S, Augui S, Navarro P et al. An essential role for the DXPas34 tandem repeat and Tsix transcription in the counting process of X chromosome inactivation. Proc Natl Acad Sci USA 2006; 103(19):7390-7395.
52. Ogawa Y, Sun BK, Lee JT. Intersection of the RNA interference and X-inactivation pathways. Science 2008; 320(5881):1336-1341.
53. Rougeulle C, Avner P. Controlling X-inactivation in mammals: what does the centre hold? Semin Cell Dev Biol 2003; 14(6):331-340.
54. Kalantry S, Purushothaman S, Bowen RB et al. Evidence of Xist RNA-independent initiation of mouse imprinted X-chromosome inactivation. Nature 2009; 460(7255):647-651.
55. Navarro P, Chambers I, Karwacki-Neisius V et al. Molecular coupling of Xist regulation and pluripotency. Science 2008; 321(5896):1693-1695.
56. Navarro P, Pichard S, Ciaudo C et al. Tsix transcription across the Xist gene alters chromatin conformation without affecting Xist transcription: implications for X-chromosome inactivation. Genes Dev 2005; 19(12):1474-1484.
57. Navarro P, Chantalat S, Foglio M et al. A role for noncoding Tsix transcription in partitioning chromatin domains within the mouse X-inactivation centre. Epigenetics Chromatin 2009; 2(1):8.
58. Monkhorst K, Jonkers I, Rentmeester E et al. X inactivation counting and choice is a stochastic process: evidence for involvement of an X-linked activator. Cell 2008; 132(3):410-421.
59. Donohoe ME, Silva SS, Pinter SF et al. The pluripotency factor Oct4 interacts with Ctcf and also controls X-chromosome pairing and counting. Nature 2009; 460(7251):128-132.
60. Xu N, Donohoe ME, Silva SS et al. Evidence that homologous X-chromosome pairing requires transcription and Ctcf protein. Nat Genet 2007; 39(11):1390-1396.
61. Augui S, Filion GJ, Huart S et al. Sensing X chromosome pairs before X inactivation via a novel X-pairing region of the Xic. Science 2007; 318(5856):1632-1636.
62. Sun BK, Deaton AM, Lee JT. A transient heterochromatic state in Xist preempts X inactivation choice without RNA stabilization. Mol Cell 2006; 21(5):617-628.
63. Gilfillan GD, Dahlsveen IK, Becker PB. Lifting a chromosome: dosage compensation in Drosophila melanogaster. FEBS Lett 2004; 567(1):8-14.
64. Franke A, Baker BS. The rox1 and rox2 RNAs are essential components of the compensasome, which mediates dosage compensation in Drosophila. Mol Cell 1999; 4(1):117-122.
65. Alekseyenko AA, Peng S, Larschan E et al. A sequence motif within chromatin entry sites directs MSL establishment on the Drosophila X chromosome. Cell 2008; 134(4):599-609.
66. Kind J, Vaquerizas JM, Gebhardt P et al. Genome-wide analysis reveals MOF as a key regulator of dosage compensation and gene expression in Drosophila. Cell 2008; 133(5):813-828.
67. Lin CH, Jackson AL, Guo J et al. Myc-regulated microRNAs attenuate embryonic stem cell differentiation. EMBO J 2009; 28(20):3157-3170.
68. Xu N, Papagiannakopoulos T, Pan G et al. MicroRNA-145 regulates OCT4, SOX2 and KLF4 and represses pluripotency in human embryonic stem cells. Cell 2009; 137(4):647-658.

69. Sheik Mohamed J, Gaughwin PM, Lim B et al. Conserved long noncoding RNAs transcriptionally regulated by Oct4 and Nanog modulate pluripotency in mouse embryonic stem cells. RNA 2009.
70. Peters J, Robson JE. Imprinted noncoding RNAs. Mamm Genome 2008; 19(7-8):493-502.
71. Royo H, Cavaille J. Non-coding RNAs in imprinted gene clusters. Biol Cell 2008; 100(3):149-166.
72. Lewis A, Reik W. How imprinting centres work. Cytogenet Genome Res 2006; 113(1-4):81-89.
73. Kanduri C, Pant V, Loukinov D et al. Functional interaction of CTCF with the insulator upstream of the H19 gene is parent of origin-specific and methylation-sensitive. Curr Biol 2000; 10:853-856.
74. Kanduri C, Holmgren C, Franklin G et al. The 5′-flank of the murine H19 gene in an unusual chromatin conformation unidirectionally blocks enhancer-promoter communication. Curr Biol 2000; 10:449-457.
75. Pandey RR, Mondal T, Mohammad F et al. Kcnq1ot1 antisense noncoding RNA mediates lineage-specific transcriptional silencing through chromatin-level regulation. Mol Cell 2008; 32(2):232-246.
76. Pandey RR, Ceribelli M, Singh PB et al. NF-Y regulates the antisense promoter, bidirectional silencing and differential epigenetic marks of the Kcnq1 imprinting control region. J Biol Chem 2004; 279(50):52685-52693.
77. Thakur N, Tiwari VK, Thomassin H et al. An antisense RNA regulates the bidirectional silencing property of the kcnq1 imprinting control region. Mol Cell Biol 2004; 24(18):7855-7862.
78. Kanduri C, Thakur N, Pandey RR. The length of the transcript encoded from the Kcnq1ot1 antisense promoter determines the degree of silencing. EMBO J 2006; 25(10):2096-2106.
79. Mancini-Dinardo D, Steele SJ, Levorse JM et al. Elongation of the Kcnq1ot1 transcript is required for genomic imprinting of neighboring genes. Genes Dev 2006; 20(10):1268-1282.
80. Shin JY, Fitzpatrick GV, Higgins MJ. Two distinct mechanisms of silencing by the KvDMR1 imprinting control region. EMBO J 2008; 27(1):168-178.
81. Koerner MV, Pauler FM, Huang R et al. The function of noncoding RNAs in genomic imprinting. Development 2009; 136(11):1771-1783.
82. Sleutels F, Zwart R, Barlow DP. The noncoding Air RNA is required for silencing autosomal imprinted genes. Nature 2002; 415(6873):810-813.
83. Nagano T, Mitchell JA, Sanz LA et al. The Air Noncoding RNA Epigenetically Silences Transcription by Targeting G9a to Chromatin. Science. 2008.
84. Williamson CM, Ball ST, Nottingham WT et al. A cis-acting control region is required exclusively for the tissue-specific imprinting of Gnas. Nat Genet 2004; 36(8):894-899.
85. Landers M, Bancescu DL, Le Meur E et al. Regulation of the large (approximately 1000 kb) imprinted murine Ube3a antisense transcript by alternative exons upstream of Snurf/Snrpn. Nucleic Acids Res 2004; 32(11):3480-3492. Print 2004.
86. Vitali P, Royo H, Marty V et al. Long nuclear-retained noncoding RNAs and allele-specific higher-order chromatin organization at imprinted snoRNA gene arrays. J Cell Sci 123(Pt 1):70-83.
87. Lin SP, Youngson N, Takada S et al. Asymmetric regulation of imprinting on the maternal and paternal chromosomes at the Dlk1-Gtl2 imprinted cluster on mouse chromosome 12. Nat Genet 2003; 35(1):97-102.
88. Davis E, Caiment F, Tordoir X et al. RNAi-mediated allelic trans-interaction at the imprinted Rtl1/Peg11 locus. Curr Biol 2005; 15(8):743-749.
89. Bernstein E, Allis CD. RNA meets chromatin. Genes Dev 2005; 19(14):1635-1655.
90. Pezer Z, Ugarkovic D. Role of noncoding RNA and heterochromatin in aneuploidy and cancer. Semin Cancer Biol 2008; 18(2):123-130.
91. Chen ES, Zhang K, Nicolas E et al. Cell cycle control of centromeric repeat transcription and heterochromatin assembly. Nature 2008; 451(7179):734-737.
92. Bouzinba-Segard H, Guais A, Francastel C. Accumulation of small murine minor satellite transcripts leads to impaired centromeric architecture and function. Proc Natl Acad Sci USA 2006; 103(23):8709-8714.
93. Djupedal I, Ekwall K. Epigenetics: heterochromatin meets RNAi. Cell Res 2009; 19(3):282-295.
94. Motamedi MR, Verdel A, Colmenares SU et al. Two RNAi complexes, RITS and RDRC, physically interact and localize to noncoding centromeric RNAs. Cell 2004; 119(6):789-802.
95. Verdel A, Jia S, Gerber S et al. RNAi-mediated targeting of heterochromatin by the RITS complex. Science 2004; 303(5658):672-676.
96. Kagansky A, Folco HD, Almeida R et al. Synthetic heterochromatin bypasses RNAi and centromeric repeats to establish functional centromeres. Science 2009; 324(5935):1716-1719.
97. Fukagawa T, Nogami M, Yoshikawa M et al. Dicer is essential for formation of the heterochromatin structure in vertebrate cells. Nat Cell Biol 2004; 6(8):784-791.
98. Sinkkonen L, Hugenschmidt T, Berninger P et al. MicroRNAs control de novo DNA methylation through regulation of transcriptional repressors in mouse embryonic stem cells. Nat Struct Mol Biol 2008; 15(3):259-267.

CHAPTER 12

PROMOTER-ASSOCIATED LONG NONCODING RNAs REPRESS TRANSCRIPTION THROUGH A RNA BINDING PROTEIN TLS

Riki Kurokawa

Division of Gene Structure and Function, Research Center for Genomic Medicine, Saitama Medical University, Yamane, Hidaka-shi, Saitama-Ken, Japan
Email: rkurokaw@saitama-med.ac.jp

Abstract: The majority of the human genome is found to be transcribed and generates mostly noncoding (nc) RNAs that do not possess protein information. MicroRNAs are one of the well-identified small ncRNAs, but occupy merely a fraction of ncRNAs. Long (large) ncRNAs are emerging as a novel class of ncRNAs, but knowledge of these ncRNAs is far less accumulated. Long ncRNAs are tentatively classified as an ncRNA species containing more than 200 nucleotides. Recently, a long promoter-associated ncRNA (pncRNA) has been identified to be transcribed from the cyclin D1 promoter upon induction by genotoxic factors like ionizing-irradiation. The cyclin D1 pncRNA is specifically bound with an RNA-binding protein TLS (Translocated in liposarcoma) and exerts transcriptional repression through histone acetyltransferase (HAT) inhibitory activity. Analysis of TLS and the pncRNAs could provide a model for elucidating their roles in regulation of mammalian transcriptional programs. The pncRNA binding to TLS turns out to be an essential event for the HAT inhibitory activity. A key consensus sequence of the pncRNA is composed of GGUG, while not every RNA sequence bearing GGUG is targeted by TLS, suggesting that a secondary structure of the GGUG-bearing RNAs is also involved in recognition by TLS. Taken together, TLS is a unique mediator between signals of the long ncRNAs and transcription, suggesting that RNA networking functions in living cells.[1-3]

RNA Infrastructure and Networks, edited by Lesley J. Collins .

INTRODUCTION

Ninety percent of the human genome appears to be expressed to transcripts mainly consisting of noncoding (nc) RNAs.[4-7] Lists of ncRNAs have been rapidly growing especially regarding micro (mi) RNAs and their related small ncRNAs including siRNAs of length 20 to 25 nucleotides.[8,9] These small ncRNAs share core machinery composed of the RNA induced silencing complex (RISC) for exerting their function.[5,6] miRNAs repress translation by blocking activity of target mRNAs with the RISC. Translational repression by miRNAs has been well analyzed, while long ncRNAs (also with their specific RNA binding proteins) have been emerging to play a role in the regulation of transcription. Long ncRNAs are tentatively defined as the ncRNAs that have a length of more than two hundred nucleotides. There have been reported divergent species of long ncRNAs. Some long ncRNAs function as coactivator and corepressor of transcription. Indeed, the ncRNAs, SRA and Evf-2 work as a coactivator,[10,11] while HOTAIR and P15AS work as a corepressor,[12,13] Furthermore, we found that long ncRNAs transcribed from the cyclin D1 promoter with the 200- and 330-nucleotides (nt) length represses transcription of the cyclin D1 gene.[14] The cyclin D1 promoter-associated ncRNAs (pncRNAs) bind an RNA binding protein TLS (Translocated in Liposarcoma) and exert inhibition of histone acetyltransferase (HAT) activity to execute the transcriptional repression.[14] We have performed extensive biochemical analysis on the pncRNA-TLS transcriptional repression and obtained a suitable model for studying long ncRNA functions.

TLS was initially identified as TLS-CHOP, a fusion protein arising from a chromosomal translocation presumed to be causative of myxoid liposarcoma.[15] It has been shown to play roles in RNA processing, homologous DNA pairing and transcription.[8,9] Deletion of the TLS gene in mice resulted in male infertility, chromosomal instability and increased sensitivity to DNA damaging agents,[9] with the data from analysis of TLS null mice showing that TLS plays a pivotal role in maintaining genomic integrity.[16,17]

TLS has been reported to possess homologs referred to as the TET (TLS, EWS and TAF15) protein family.[18,19] Intriguingly, EWS and TAF15 also have been identified as partners in fusion genes that are linked to sarcomas. EWS was initially identified as EWS-FLI1 in Ewing's sarcoma.[18] TAF15 was initially described as a component of core transcriptional machinery[19] but was also found to be a partner in a fusion gene (TAFII68-CHN) in a chondrosarcoma patient[20,21] in which the fusion partner is an orphan nuclear hormone receptor. In each case, the relevant fusion proteins generated through the translocation event combine the N-terminus of the TET family member with the DNA binding domain of its corresponding partner. These fusion proteins cause mostly sarcomas.[21-23]

Here, I review recent publications regarding the RNA binding of TLS and focus on transcriptional regulation by newly identified long ncRNAs transcribed from the cyclin D1 promoter.[14] This chapter reviews long ncRNA-dependent transcriptional repression and presents an overview of biological consequences of the long ncRNAs, suggesting that the transcriptional repression is one of examples of RNA network in mammalian cells.[1-3]

THE TARGET RNA SEQUENCES RECOGNIZED BY TLS

This section details our experiments to investigate the molecular mechanism of TLS and its target RNA sequences. We use this as a model of how long ncRNA-based mechanisms can be examined. Our quest for RNA molecules that specifically bind TLS began just after the identification of TLS as a HAT inhibitory molecule, because of the possibility of the regulatory function of RNA specifically binding to TLS.[14] These following experiments indicated the specific RNA sequences recognized by TLS.[24] The SELEX (systematic evolution of ligands by exponential enrichment) experiments used twenty-five nucleotides with randomized sequences.[24] Four rounds of the SELEX through binding of bacterially expressed GST-TLS protein captured on GST-agarose gel beads with a ^{32}P-labeled randomized RNA pool were performed and a subsequent three rounds of a selected RNA pool with a gel-shift assay were processed. After selection, the three clones (9, 60 and 75) turned out to possess the best bindings to TLS and the GGUG consensus TLS binding site (Table 1).[24] Clone 9 is referred to as the GGUG-RNA and was selected for further analysis. The SELEX system was assembled artificially with synthetic RNA oligos and bacterially expressed GST-TLS. However, the selected consensus sequence GGUG possesses a robust binding to TLS and also to other TET family proteins in biochemical assay systems.[24] The GGUG RNA is bound with TLS to enhance the inhibitory activity of TLS,[14] and is a useful tool to assess the RNA dependency of the TLS HAT inhibitory activity.

After identification of the GGUG RNA as a regulatory RNA sequence, we explored naturally occurring RNA sequences with the ability to enhance the HAT inhibition by TLS,[14] especially since the expression of cyclin D could be repressed by ionizing irradiation (IR).[14] The cyclin D1 promoter DNA sequence region was found to contain the GGUG sequence at six regions (A, B, C, D, E and F). The RT-PCR experiments showed enhancement of the transcripts from regions A, B, D and E, while they showed no increment of transcript from the regions C and F, upon IR. Sequences of region A and B have the GGUG sequence (Table 1) and bind TLS at gel-shift assay. The expression of the region A and B is induced by IR. Region C is not induced by IR, but contains the GGUG sequence (Table 1). The oligos containing GGUG sequence designed from the region C, however, have no binding to TLS and no HAT inhibitory effect. These results showed that not all GGUG sequences are a determinant for their binding specificity to TLS and that sequences outside the GGUG and also a secondary structure of the GGUG RNAs might be involved in their binding specificity to TLS. Actually, the sequence outside of the GGUG consensus has been proved to be important. I discuss this issue in detail later.

Extensive biochemical experiments demonstrated that the cyclin D1 pncRNAs are transcribed through RNA polymerase II and have a poly A tail, but no CAP structure.[14] The transcription is induced by IR and other DNA-damaging reagent treatments, although the mechanism of the induction of the cyclin D1 pncRNAs remains unclear.[25] The transcripts appeared at distinct bands of 200 and 330 nucleotides, but always exhibited diffused electrophoresis patterns around these two major bands, suggesting that the transcription might be started from multiple sites or terminated as premature products. Elucidation of the transcriptional mechanism of the long ncRNAs would pose a crucial clue to understanding of the biology of these molecules. More than 1000 long ncRNAs have been identified, but transcriptional mechanisms of most of the long ncRNAs have not been well investigated.[26-28] Understanding the transcriptional mechanisms of the ncRNAs will provide us what kind of cellular signals induces the ncRNA transcription, suggesting biological functions of the ncRNAs. So far our knowledge of the cyclin D1 pncRNAs make this of the best models to execute the analysis of the ncRNA transcription.[14]

Table 1. RNA sequences related to the binding of TLS

	SEQUENCE	TLS Binding	HAT Inhibition	References
Clone 9 (GGUG-RNA)	UUUUGUAUUUUGAGCUAGUUUGGUGAUGU	+	+	24
Clone 60	UUGCGUAUCAUAUGGUGUGGCGUGAUAGU	+	NT	24
Clone 75	UUUUGAAGGAGGGUGAAGGGGCCUUAGGU	+	NT	24
CUC-GGUG	UUGUAUUUUGAGCUAGCUCGGUGAU	+/-	NT	24
ΔUAG-GGUG	UUGUAUUUUGAGC---UUUGGUGAU	+/-	NT	24
The wild type IVSB7 RNA	GGCGAAUUCGCUGGGGCUGGGCAGAGCGCGCAGGGUUGAGGGGAGCAGGGUCCUUCA-CUGGGGUGAA	+	NT	24
-341a Region A	AGAGCCCAAAAGCCAUCCCUGAGGCGCCGCGCCUGCCCUGCGGCGGAGUUGCCCCU-GUAGUCCGGUUUUCAUAGAAAUGCAAAUCGCCCCGCUGCAGCCUUUCUCCCCCGC-CAGGGAAGAGGGGUGCAGGGGGCCCCGUCCGCCUCGAGUGGGUCCCCCGGGAU-UUAGGGGGUGAGGUGGAGGGAUGGCUUUUGGGCUCUGAUUUAGGGGGUGAGGUGGAG	+	+	14
-454s Region B	CUGAGAUUCUUUGGCCGUCUGCUCUGCAGUAGGGGACAACCUGAGAUUCUUUGGCCGUCU-GUCCGCCCGUGGGUGCCCUCGUGGCGUUCUUGGAAAUGCGCCCAUUCUGCCGGCUUGGAU-AUGGGGUGUCGCCGCGCCCCAGUCACCCCUUCUCGUGGUCUCCCCAGGCUGCGUGUGGC-CUGCCGGCCUUCCUAGUUGUCCCCUACUGCAGAGC	+	+	14
-764a Region C	ACAACCCCUGUGCAAGUUUCAUUCCGGCGCACAGGGGCGUCGUUG-CAAAUGCCCAAGGGGGUAACCCUAAAAGUUAAAGGGAUUUCAGCUUAGCAUGCGCUCG-CUCAAAAAAUAAAAUAAAAUAAAUGCCCGAAAAUUCCAGCAGCAGCCCAAGAUG-GUG-GCCAGCAUUUCCUUCAUCUUGUCCUUCUAGCCUGGAGACUCUUCGGGCUGCCUUC-CUACCUUGACCAGUCGGGAAACUUGCACAGGGGUUGUCUUCCUACCUUGACCAGUCGG	-	-	14

RNA sequences that contain the GGUG consensus are shown with or without their binding ability to TLS and inhibitory activity against the CBP/p300 HAT. The SELEX result indicated that 39 among 79 selected clones contained the GGUG sequence as a consensus binding site of TLS.[24] The sequences of clone 9, 60, 75, CUC-GGUG, ΔUAG-GGUG and the wild type IVSB7 RNA are from reference 24. The sequences of regions A, B and C are from the cyclin D1 pncRNAs.[14] The sequences of regions A and B bind TLS, while that of region C does not bind TLS.[14] The binding to TLS was examined with gel-shift assays described in each reference.[14,24] Underlined: GGUG; Dotted lines: G rich regions.

It has been reported based upon analyses of the RNA binding regions, that TLS binds single stranded RNA molecules and their possession of a RNA recognition motif (RRM) and the zinc finger domain, indicates that they can bind single stranded RNAs.[14,24,29] Indeed, the single stranded CCND1-pncRNA bound TLS and had inhibitory effect on HAT activity.[14] Treatment of HeLa cells with IR induced bi-directional transcription of the pncRNAs from the cyclin D1 promoter.[14] Preliminary experiments showed that both sense and antisense strands of pncRNAs are active on the HAT inhibition (Kurokawa, unpublished data). In case that the double stranded pncRNAs are not bound with TLS, transcription of both strands of the pncRNAs might be regulatory mechanism of the inhibitory activity of the pncRNAs, although the formation of double stranded pncRNA with sense-antisense strand has not been examined.[14] The sense-antisense pairing ncRNAs have been found also in Alu repeats and the untranslated regions of protein coding genes,[30,31] although biological meaning of formation of these double stranded RNAs has not been well understood. Another example of antisense ncRNA is a tumor suppressor gene p15.[13] The antisense RNA of the p15 gene was reported to exert epigenetic silencing of this gene itself, but not by formation of the double stranded RNAs.[13] Formation of the double stranded long ncRNAs is possible mechanism for regulation of the function of these ncRNAs, although specific examples are not widely characterized. Furthermore, subcellular fractionation studies revealed that the long pncRNAs and TLS are mainly localized to nuclear fraction.[14] The nuclear localization of the long pncRNAs suggests that the pncRNAs and their binding partner TLS play a role in transcriptional regulation.

DOMAIN STRUCTURE OF TLS

TLS is a key molecule to execute the signal from the cyclin D1 ncRNAs. Dissection of the TLS domain structure presents the molecular mechanism of its binding to RNA sequences. TLS forms a modular structure consisting of several functional domains between an acidic N-terminus and a basic C-terminal region. Analysis of the amino acid composition of TLS indicates that the N-terminus of TLS (1-211) is an acidic peptide (pI: 2.6) and that the C-terminus (211-526) is a basic one (pI: 10.45) (Fig. 1A,B). The protein binding assay showed that the N-terminus of TLS interacts with the C-terminus and that this interaction is blocked by binding of the GGUG-RNA to the RNA binding regions of the C-terminus.[14]

The N-terminus region of TLS has high content of three amino acids, serine (S), glutamine (Q) and tyrosine (Y) and has been reported as a putative transcriptional activation domain (Fig. 1A).[19,32] Pasting the N-terminus to a heterogenic DNA binding domain of the yeast transcription factor GAL4 results in transcriptional activation measured with a GAL4 reporter assay, indicating that the N-terminus is a transcriptional activation domain.[32] The N-terminal region of the TET proteins also contains a glutamine-rich domain that is a common structure of transcriptional activation domains, as described above (Fig. 2).[14] These data suggest that the fusion of the transcriptional activation domain of a TET protein to the DNA-binding domain of CHOP, Fli or CHN results in aberrant transcriptional activation that promotes the development of their related sarcomas. The N-terminus is bound by CREB-binding protein CBP and adenovirus p300 that are potent transcriptional coactivators with the potent HAT activity.[14] The binding of CBP/p300 to the N-terminus at least partly contributes to its transcription activity.

Figure 1. Domain structure of TLS. A. Amino acid sequence of human TLS. The amino acid sequences of the QSY domain, RGG1, RRM, RGG2, zinc finger domain and RGG3. This tells us that TLS contains highly repetitive amino acid sequences. Arrows show the domains of TLS. Boxes: the cysteine residues forming a zinc finger. B. Domain structure of TLS. Approximate boundaries of the CBP binding region and RNA binding region are shown. Abbreviation: "Zn" represents the zinc finger domain.

The C-terminus of TLS and also the TET protein is markedly a basic region that fits a criterion for nucleic acid binding domain (Fig. 1B and Fig. 2). Actually, the C-terminus has several putative RNA interacting regions: RRM; the RGG-repeats that have multiple tandem repeats of arginine, glycine and glycine: RGG1, RGG2, RGG3; a zinc finger domain (Fig. 1A,B).[32,33] Lerga et al showed that the GGUG-RNA bound to the RRM (271-392; amino acid numbers shown), activation domain (AD)-RGG1 (1-271) and RGG2-3 containing zinc finger (392-526). Extensive experiments with gel-shift assays using mutants of the GGUG-RNA (Clone 9, Table 1) indicated that RRM and RGG2-3 with the zinc finger (RGG2-3/zinc finger) are able to bind the mutants at the outside of core GGUG sequence (CUC-GGUG RNA, ΔUAG-RNA), while AD-RGG1 is not able to bind these mutants (Table 1). These data imply that AD-RGG1, RRM and RGG2-3/zinc finger play distinctive roles in recognition of the GGUG-RNA and

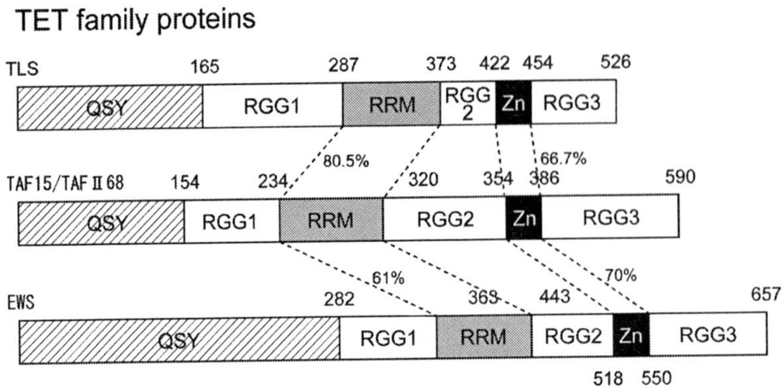

Figure 2. Alignment of TLS homologues. Alignment of the TET family protein with homology. The homology percentages are calculated compared to the TLS sequence. The sequences are from human genes. Abbreviation: "Zn" represents the zinc finger domain.

that sequences outside the GGUG motif support the binding of full-length TLS to the GGUG-RNA. Contrast to the data, NMR analysis showed that the RRM fragment (266-375) failed to bind the RNA oligos (5′-UAGUUUGGUGAU-3′) that contain the core GGUG motif trimmed down to 12 nucleotides, while the zinc finger fragment (398-468) bound to the RNA oligos.[29] There is a remarkable discrepancy regarding the binding of the RRM to the GGUG RNA between these two data. The discrepancy could be explained by the difference in the RRM fragments used at these two experiments.[24,29] Iko et al employed the RRM fragment (266-375) that does not contain the RGG repeat at the C-terminus,[29] while Lerga et al utilized the RRM (271-392) that contains the C-terminal RGG repeat.[24] Addition to this explanation, shorter RNA oligos used by Iko et al might lose their ability to bind to the RRM. The binding of the zinc finger domain of TLS to the GGUG RNA has been confirmed by the data from both groups.[24,29] Moreover, the NMR analysis showed that the RRM (266-375) and the zinc finger have no interaction with or without the GGUG RNA. These data have indicated that TLS has multiple regions of interaction to the GGUG-RNA and each region could have a distinctive role to support the RNA binding.

The zinc finger domain of TLS plays a crucial role in its binding to the GGUG-RNA and belongs to the RanBP2-type zinc finger family.[34] The RanBP2-type zinc finger has been analyzed regarding the RNA-binding protein ZRANB2 showing that ZRANB2 possesses two zinc finger domains. These zinc fingers, F1 (9-41) and F2 (65-95) bind an AGGUAA sequence with dissociation constants (Kds) of 3 μM and 2 μM, respectively.[34] The AGGUAA sequence has been identified by a SELEX experiment using gel-shift assays and randomized 25 mer RNA oligonucleotides.[34] Both F1 and F2 can bind the AGGUAA RNA sequence. These two zinc fingers in a fragment F12 (1-95) synergistically bind the tandem repeat of AGGUAA with higher Kd of 0.19μM, indicating that bipartite fragments of the zinc fingers coordinately recognize the target RNA.[34] NMR analysis of the structure of the F2 of ZRANB2 dictated that the RanBP2-type zinc finger domain recognizes the AGGUAA sequence and also revealed a core sequence of AGGUAA as the GGU of the target mRNA for splicing.[34] A protein sequence alignment of the zinc fingers of ZRANB2 and TET family proteins like TLS,

EWS and TAF15 indicates significant homology.[34] The sequence alignment between ZRANB2-F2 and TET proteins also indicates exact matching at the amino acids (D425, N435, F438, R440, N445: amino acid numbers of TLS are shown) contacting the core sequence, GGU and confirms that the zinc finger of the TET family proteins binds the GGU tract of the GGUG-RNA. Therefore, TET family proteins need to recognize the additional "G" residue at the end of GGU"G". This indicates that more residues in the TET proteins than the homologous amino acids to ZRANB2-F2 are involved in recognition of the GGUG-RNA.

Gel-shift binding assay using ^{32}P-GGUG-RNA indicated that Kd of full-length TLS is 250 nM,[24] while the NMR analysis of the zinc finger fragment (398-468) of TLS with the truncated GGUG-RNA (5′-UAGUUUGGUGAU-3′) showed that Kd is 10000 nM.[29] The zinc finger domain alone displays 40 times lower affinity to the GGUG consensus although affinity to the zinc finger domain might be underestimated because of possible loss of the affinity to shorter RNA oligos used. This suggests that the zinc finger of TLS cooperates with other domains to achieve the higher affinity to grasp the GGUG-RNA. These data also imply that TLS is able to bind multiple RNA molecules simultaneously because of its having multiple binding regions, to have synergistic and allosteric effect on its protein binding ability. Therefore, the multiple binding regions to RNA sequences give TLS versatility in its regulatory functions.

TRANSCRIPTION-REGULATORY LONG ncRNA, CYCLIN D1 pncRNA

After characterising the HAT-inhibitory effect of the GGUG RNA, we can then explore the naturally occurring RNA sequence bound by TLS and the cyclin D1 pncRNA can be identified. The cyclin D1 pncRNA is a regulatory long ncRNA for transcription and requires TLS as a compulsory binding partner to execute its inhibitory effect on the HAT activity to repress the transcription (Fig. 3). There have been a few other ncRNAs transcribed from 5′-untranslated regulatory regions, promoters and enhancers.[35,36] One example of a long ncRNA transcribed from a promoter is that of dehydrofolate reductase (DHFR).[35] In quiescent mammalian cells, expression of DHFR is repressed. It has been reported that a transcript of a minor promoter located upstream of a major promoter is involved in the repression of DHFR.[35] In the quiescent cells, the transcript of this minor promoter inhibits transcriptional initiation from the major promoter through direct binding to TFIIB of the pre-initiation complex. There are various genes that possess major and minor promoters like DHFR gene.[37,38] Therefore, this could be considered a general mechanism in that the transcripts from a minor promoter have a regulatory role in transcription of the major promoter. The ncRNAs transcribed from this kind of alternative promoters could have a significant impact on the regulation of the genes in these promoter contexts. The DHFR pncRNA does not need to bind any RNA binding protein, but is directly interacted with the target molecule of the basic core transcription factor TFIIB.

Recently, novel ncRNAs (eRNA) transcribed from enhancers have been reported in primary culture of mouse cortical neurons.[36] Genome-wide sequencing analysis with a KCl-dependent enhancer function in mouse cortical neurons, indicated 12,000 neuronal activity-regulated enhancers that are bound by CBP in a neuronal activity-dependent manner. A function of CBP is to recruit RNA polymerase II. Actually, activity-regulated binding of RNA polymerase II with thousands of enhancers was observed. Intriguingly,

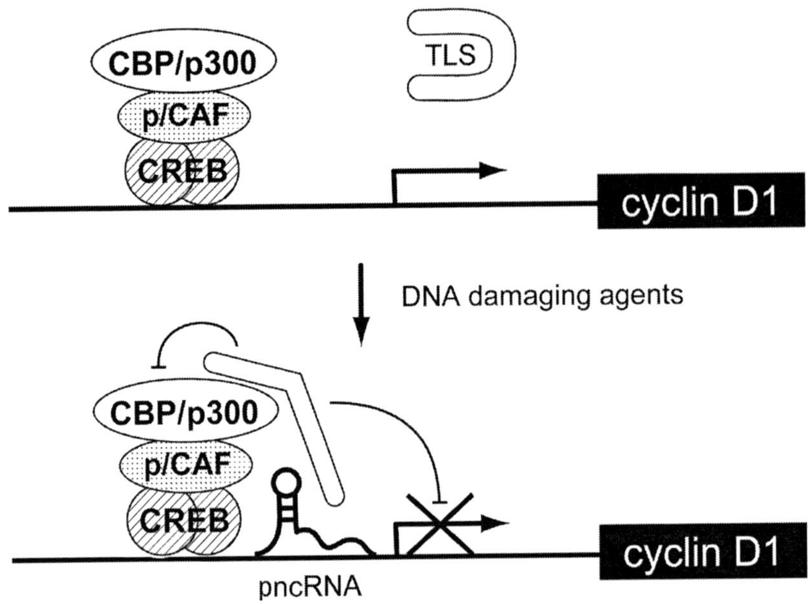

Figure 3. Promoter-associated ncRNA-dependent transcriptional repression. At the cyclin D1 promoter, CBP/p300 is bound with cAMP response element binding protein (CREB) on the CRE site of the promoter and activates transcription through its HAT activity. Upon DNA damage from agents like ionizing radiation, the transcription of the pncRNAs from the promoter is induced. TLS is bound with the pncRNAs and recruited to the CBP/p300-CREB complex where pCAF is a structural component, to exert transcriptional repression by inhibition of the HAT of CBP/p300.[14]

RNA polymerase II at enhancers transcribes bi-directionally eRNA within enhancer regions defined by the presence of histone H3 monomethylated at lysine 4 (H3K4me1). The level of eRNA expression at neuronal enhancers positively correlates with the level of mRNA synthesis at nearby genes. This suggests that eRNA expression is engaged in inducing mRNA synthesis. Experiments with the *ARC* gene, which regulates synaptic functions, indicate that truncation of the *ARC* promoter resulted in no eRNA synthesis although the *ARC* promoter was bound with transcription factor SRF and RNA polymerase II. Thus, the recruitment of RNA polymerase II to the *ARC* enhancer is not sufficient for eRNA synthesis and suggests that eRNA synthesis might require an interaction of the enhancer with a promoter. Given that the enhancers are interacted with the promoter to tether them to each other, eRNA might be bound with an RNA-binding protein like TLS to exert transcriptional regulations. Contrast to these two ncRNAs, the cyclin D1 pncRNA has been proved to require TLS as a binding partner to exert transcriptional regulation, although the DHFR ncRNAs and the eRNAs may have unidentified specific RNA-binding protein. Thus far, TLS and the cyclin D1 pncRNA are characteristic of a unique machinery to exert transcriptional repression in eukaryotic transcriptional programs (Fig. 3).

TLS is the distinctive enzyme inhibitor with specific target genes. Molecular mechanisms that enable TLS to target specific sets of genes have been explored. Our biochemical experiment showed that TLS inhibits the HAT activity of CBP/p300. This data prompted us to examine if TLS could repress the CREB-target genes that have been reported for more than 240 genes, because CBP/p300 is a major component of the CREB coactivator complex.[39] Unexpectedly, TLS only repressed two genes, cyclin D1 and E1. Further analysis has been performed just on the cyclin D1 promoter indicating that TLS could specifically recognize the CBP just on the cyclin D1 gene promoter. The mechanism of the specific recognition of TLS to the CBP located on the cyclin D1 promoter remains unsolved. One possible explanation is that the cyclin D1-pncRNA specifically localizes TLS to the promoter sites at which the cyclin D1-pncRNAs are transcribed and also CBP/p300 is bound. In this case, transcription of the cyclin D1-pncRNA itself should function as a signal for recruiting TLS to the cyclin D1 promoter site. Another possible mechanism is that TLS could recognize specific histone-modifications near by the binding sites of the cyclin D1 promoter. Targeting only cyclin D1 and E2 by TLS indicates that every CREB binding site is not equivalent regarding specific binding of TLS. Upon binding of the ncRNAs, TLS specifically recognizes the cyclin D1 promoter to repress its expression. This might be another regulatory mechanism for selective repression of a gene during embryonic development or processes requiring differential expression of genes.

A significant percentage of the human genome is transcribed. Many of them are ncRNAs which localize to both nuclei and cytoplasm, although copy numbers of each ncRNA are not abundant. The cyclin D1-pncRNAs are also low copy number ncRNAs. For instance, the cyclin D1-pncRNA-D that is transcribed from the region D of the cyclin D1 promoter was found as low as two copies in a single cell without the treatment of ionizing radiation and induced up to four copies after the treatment. Amounts of other pncRNAs from the cyclin D1 promoter are supposed to be at a similar level based upon observations with quantitative PCR.[14] Low abundance of the cyclin D1-pncRNA-D suggests that it functions as a regulatory agent instead of a structural component in living cells. Indeed, the cyclin D1-pncRNAs has a regulatory role in transcription of the cyclin D1 gene.[14] The low abundance of the cyclin D1-pncRNAs limits their effective area or forces their function just around a milieu where their transcription occurs. This enforces the cyclin D1-pncRNAs to work as a cis-acting agent, but not as a trans-acting one. The benefit of the cis-acting mode of TLS is that the regulatory effect is limited to the promoter region in which the cyclin D1 pncRNAs are generated. Hence, the localization of the effect enhances the specificity of the cyclin D1 pncRNAs just towards the cyclin D1 expression.

The originally selected GGUG-RNA oligos (25 mers) have a relatively low affinity to TLS. TLS binds the GGUG-RNA oligos with a Kd of 250 nM.[24] Gel-shift assays with the GGUG-RNA and the cyclin D1-pncRNAs A and B showed that TLS has similar level of the Kd to the pncRNAs.[14] However, the estimated concentration of the cyclin D1 pncRNA is much lower than the level of the Kd. Indeed, a single HeLa cell has only four molecules of the cyclin D1 pncRNA-D even induced by IR. Given that the cell volume is 1pL, concentration of the cyclin D1 pncRNA-D is merely 0.0066 nM. Therefore, TLS needs to enhance its affinity to the pncRNAs in the cells to capture the cyclin D1 pncRNAs. A possible mechanism to enhance the TLS affinity to the cyclin D1-pncRNAs is allosteric effect by complex formation of TLS with other RNA-binding proteins. Protein-protein interactions during formation of the TLS complex may enhance the affinity of TLS to the pncRNAs and it is noteworthy that other TET family proteins, EWS and TAF15 also form distinctive complexes. Considering that EWS and TAF15 have a similar HAT inhibitory effect upon binding with RNAs,[14] this complex formation could contribute to enhancement of affinities to each target ncRNA.

Protein modifying enzymes in the complex could modify TLS to induce higher affinity to the pncRNAs. Indeed, our mass spectrometric analysis showed that the protein arginine methyltransferase1 (pRMT1) that has been reported to have a role in transcription[40] is a component of the TLS complex and pRMT1 methylates TLS and modifies its function (Kun Du and Kurokawa, unpublished data). Taken together, the formation of the TLS complex could be an essential process to enable TLS to play a pivotal role in mammalian transcription regulatory programs.

CONCLUSION

Long ncRNAs are extremely heterogeneous molecules transcribed from numerous sites of the eukaryotic genome. It seems unlikely that all of these long ncRNAs utilize a common mechanism to exert their biological functions. A potential clue to unravel the functions of long ncRNAs is to explore their binding partners, RNA-binding proteins like TLS.

I have presented here a detailed description of the structures and functions of TLS as a model system for understanding biological functions of long ncRNAs and corresponding RNA binding proteins in transcription regulatory programs. This discussion might assist analysis of the biological consequence of other classes of long ncRNAs. Characterisation of the function of RNA-binding proteins specific to their target long ncRNAs should include efficient strategies to explore their biological consequence in the human genome.

However, biological function of long ncRNAs still remains largely unanalyzed. The cyclin D1 pncRNAs and TLS is a model system for establishing a biological significance of long ncRNAs.[41] Indeed, TLS has been discovered as a new causative gene of amyotrophic lateral sclerosis (ALS).[42,43] TLS has been reported to have multiple mutants to cause ALS. ALS and frontotemporal degeneration (FTLD) are neurodegenerative diseases with clinical and pathological similarity.[44] Remarkable discoveries of more than 20 mutations including H517Q and R521G in TLS as causatives of ALS, combined with the abnormal aggregation of the protein, have initiated a shifting paradigm for the underlying pathogenesis of multiple neurodegenerative diseases.[42,43,45] This TLS association with ALS and other neurodegenerative diseases is redirecting research efforts toward understanding the role of RNA metabolism in neurodegenerative diseases and provide a biological significance of TLS and related ncRNAs.

ACKNOWLEDGEMENTS

The author thanks Ms. R. Tanji for preparation of the manuscript and Dr. C.K.Glass for critical discussion. This work was supported by Takeda Science Foundation, the Naito foundations, Astellas Foundation for Research on Metabolic Disorders Foundation and also by Grant-in-Aid for Scientific Research (B: nos22390057) and Grant-in-aid for "Support Project of Strategic Research Center in Private Universities" from the Ministry of Education, Culture, Sports, Science and Technology to Saitama Medical University Research Center for Genomic Medicine.

REFERENCES

1. Pawlicki JM, Steitz JA. Nuclear networking fashions premessenger RNA and primary microRNA transcripts for function. Trends Cell Biol 2009; 20:52-61.
2. Collins LJ, Penny D. The RNA infrastructure: dark matter of the eukaryotic cell? Trends Genet 2009; 25:120-128.
3. Siomi H, Siomi MC. On the road to reading the RNA-interference code. Nature 2009; 457:396-404.
4. ENCODE-consortium. The ENCODE (ENCyclopedia Of DNA Elements) Project. Science 2004; 306:636-640.
5. Carninci P, Kasukawa T, Katayama S et al. The transcriptional landscape of the mammalian genome. Science 2005; 309:1559-1563.
6. Bertone P, Stolc V, Royce TE et al. Global identification of human transcribed sequences with genome tiling arrays. Science 2004; 306:2242-2246.
7. Willingham AT, Orth AP, Batalov S et al. A strategy for probing the function of noncoding RNAs finds a repressor of NFAT. Science 2005; 309:1570-1573.
8. Fire A, Xu S, Montgomery MK et al. Potent and specific genetic interference by double-stranded RNA in Caenorhabditis elegans. Nature 1998; 391:806-811.
9. Filipowicz W, Bhattacharyya SN, Sonenberg N. Mechanisms of posttranscriptional regulation by microRNAs: are the answers in sight? Nat Rev Genet 2008; 9:102-114.
10. Lanz RB, McKenna NJ, Onate SA et al. A steroid receptor coactivator, SRA, functions as an RNA and is present in an SRC-1 complex. Cell 1999; 97:17-27.
11. Feng J, Bi C, Clark BS et al. The Evf-2 noncoding RNA is transcribed from the Dlx-5/6 ultraconserved region and functions as a Dlx-2 transcriptional coactivator. Genes Dev 2006; 20:1470-1484.
12. Rinn JL, Kertesz M, Wang JK et al. Functional demarcation of active and silent chromatin domains in human HOX loci by noncoding RNAs. Cell 2007; 129:1311-1323.
13. Yu W, Gius D, Onyango P et al. Epigenetic silencing of tumour suppressor gene p15 by its antisense RNA. Nature 2008; 451:202-206.
14. Wang X, Arai S, Song X et al. Induced ncRNAs allosterically modify RNA-binding proteins in cis to inhibit transcription. Nature 2008; 454:126-130.
15. Crozat A, Aman P, Mandahl N et al. Fusion of CHOP to a novel RNA-binding protein in human myxoid liposarcoma. Nature 1993; 363:640-644.
16. Kuroda M, Sok J, Webb L et al. Male sterility and enhanced radiation sensitivity in TLS(–/–) mice. EMBO J 2000; 19:453-462.
17. Hicks GG, Singh N, Nashabi A et al. Fus deficiency in mice results in defective B-lymphocyte development and activation, high levels of chromosomal instability and perinatal death. Nat Genet 2000; 24:175-179.
18. Delattre O, Zucman J, Plougastel B et al. Gene fusion with an ETS DNA-binding domain caused by chromosome translocation in human tumours. Nature 1992; 359:162-165.
19. Bertolotti A, Lutz Y, Heard DJ et al. hTAF(II)68, a novel RNA/ssDNA-binding protein with homology to the pro-oncoproteins TLS/FUS and EWS is associated with both TFIID and RNA polymerase II. EMBO J 1996; 15:5022-5031.
20. Bertolotti A, Bell B, Tora L. The N-terminal domain of human TAFII68 displays transactivation and oncogenic properties. Oncogene 1999; 18:8000-8010.
21. Attwooll C, Tariq M, Harris M et al. Identification of a novel fusion gene involving hTAFII68 and CHN from a t(9;17)(q22;q11.2) translocation in an extraskeletal myxoid chondrosarcoma. Oncogene 1999; 18:7599-7601.
22. Ladanyi M. The emerging molecular genetics of sarcoma translocations. Diagn Mol Pathol 1995; 4:162-173.
23. Riggi N, Cironi L, Provero P et al. Expression of the FUS-CHOP fusion protein in primary mesenchymal progenitor cells gives rise to a model of myxoid liposarcoma. Cancer Res 2006; 66:7016-7023.
24. Lerga A, Hallier M, Delva L et al. Identification of an RNA binding specificity for the potential splicing factor TLS. J Biol Chem 2001; 276:6807-6816.
25. Miyakawa Y, Matsushime H. Rapid downregulation of cyclin D1 mRNA and protein levels by ultraviolet irradiation in murine macrophage cells. Biochem Biophys Res Commun 2001; 284:71-76.
26. Guttman M, Amit I, Garber M et al. Chromatin signature reveals over a thousand highly conserved large noncoding RNAs in mammals. Nature 2009; 458:223-227.
27. Khalil AM, Guttman M, Huarte M et al. Many human large intergenic noncoding RNAs associate with chromatin-modifying complexes and affect gene expression. Proc Natl Acad Sci USA 2009; 106:11667-11672.
28. Huarte M, Guttman M, Feldser D et al. A large intergenic noncoding RNA induced by p53 mediates global gene repression in the p53 response. Cell 2010; 142:409-419.
29. Iko Y, Kodama TS, Kasai N et al. Domain architectures and characterization of an RNA-binding protein, TLS. J Biol Chem 2004; 279:44834-44840.

30. Kapranov P, Cheng J, Dike S et al. RNA maps reveal new RNA classes and a possible function for pervasive transcription. Science 2007; 316:1484-1488.
31. Wang P, Yin S, Zhang Z et al. Evidence for common short natural trans sense-antisense pairing between transcripts from protein coding genes. Genome Biol 2008; 9:R169.
32. Zinszner H, Albalat R, Ron D. A novel effector domain from the RNA-binding protein TLS or EWS is required for oncogenic transformation by CHOP. Genes Dev 1994; 8:2513-2526.
33. Prasad DD, Ouchida M, Lee L et al. TLS/FUS fusion domain of TLS/FUS-erg chimeric protein resulting from the t(16;21) chromosomal translocation in human myeloid leukemia functions as a transcriptional activation domain. Oncogene 1994; 9:3717-3729.
34. Loughlin FE, Mansfield RE, Vaz PM et al. The zinc fingers of the SR-like protein ZRANB2 are single-stranded RNA-binding domains that recognize 5′ splice site-like sequences. Proc Natl Acad Sci USA 2009; 106:5581-5586.
35. Martianov I, Ramadass A, Serra Barros A et al. Repression of the human dihydrofolate reductase gene by a noncoding interfering transcript. Nature 2007; 445:666-670.
36. Kim TK, Hemberg M, Gray JM et al. Widespread transcription at neuronal activity-regulated enhancers. Nature 2010; 465:182-187.
37. Ayoubi TA, Van De Ven WJ. Regulation of gene expression by alternative promoters. FASEB J 1996; 10:453-460.
38. Martens JA, Laprade L, Winston F. Intergenic transcription is required to repress the Saccharomyces cerevisiae SER3 gene. Nature 2004; 429:571-574.
39. Impey S, McCorkle SR, Cha-Molstad H et al. Defining the CREB regulon: a genome-wide analysis of transcription factor regulatory regions. Cell 2004; 119:1041-1054.
40. An W, Kim J, Roeder RG. Ordered cooperative functions of PRMT1, p300 and CARM1 in transcriptional activation by p53. Cell 2004; 117:735-748.
41. Kurokawa R, Rosenfeld MG, Glass CK. Transcriptional regulation through noncoding RNAs and epigenetic modifications. RNA Biol 2009; 6:233-236.
42. Kwiatkowski TJ Jr, Bosco DA, Leclerc AL et al. Mutations in the FUS/TLS gene on chromosome 16 cause familial amyotrophic lateral sclerosis. Science 2009; 323:1205-1208.
43. Vance C, Rogelj B, Hortobágyi T et al. Mutations in FUS, an RNA processing protein, cause familial amyotrophic lateral sclerosis type 6. Science 2009; 323:1208-1211.
44. Van Langenhove T, van der Zee J, Sleegers K et al. Genetic contribution of FUS to frontotemporal lobar degeneration. Neurology 2010; 74:366-371.
45. Lagier-Tourenne C, Cleveland DW. Rethinking ALS: the FUS about TDP-43. Cell 2009; 136:1001-1004.

CHAPTER 13

RNA NETWORKS IN PROKARYOTES I: CRISPRs and Riboswitches

Patrick J. Biggs[1]* and Lesley J. Collins[2]

[1]Institute of Veterinary, Animal and Biomedical Sciences, Massey University, Palmerston North, New Zealand; [2]Institute of Fundamental Sciences, Massey University, Palmerston North, New Zealand

**Corresponding Author: Patrick J. Biggs—Email: p.biggs@massey.ac.nz*

Abstract: As with eukaryotes, prokaryotes employ a variety of mechanisms to allow the various types of RNA to interact and perform complex functions as a network. This chapter will detail prokaryotic molecular systems, such as riboswitches and CRISPRs, to show how they perform unique functions within the cell. These systems can interact with each other to gain a higher level of control and here we highlight some examples of such interactions including the cleavage of certain riboswitches by RNaseP, and endoribonuclease cleavage of pre-crRNAs in the CRISPR system. Thanks to such insights, we are beginning to get a glimpse of the prokaryotic RNA infrastructure, just as we have done with eukaryotes.

INTRODUCTION

This chapter and the next will look at certain groups and types of RNAs within the bacterial and archaeal cell, with some examples illustrating points and principles for their modes and mechanisms of action. This chapter begins with riboswitches, sequences with untranslated regions that can control the expression of genes and operons through the binding of metabolites and other ligands. Clustered regularly interspersed short palindromic repeats (CRISPRs) are considered next. These are a class of RNA molecules that are key elements of what can be termed 'bacterial immunity'. Complex sequence recognition mechanisms allow the produced and processed RNAs to act in concert with proteins to elicit the response to invading phage. Through the analyses of these systems, and more yet to be discovered, we are beginning to get a glimpse of the prokaryotic RNA infrastructure, just as we have done with the eukaryotic RNA infrastructure described in other chapters in this book.

RNA Infrastructure and Networks, edited by Lesley J. Collins.

RIBOSWITCHES CONTROL GENE AND OPERON EXPRESSION VIA METABOLITE BINDING

Riboswitches are RNA elements found in the untranslated regions (UTRs) of bacterial genes, and function to control the expression of those genes by sensing metabolite and other small molecule concentrations within the cell. The mechanism can be regarded as an adaptation of protein biosynthesis to detectable environmental conditions.[1] This regulation works by a mutually exclusive conformational change in the secondary structure of the RNA molecule on the binding of a small ligand, of which there is a broad range, for example amino acids, vitamin cofactors, metal ions, purine nucleobases and second messenger molecules (for a review see ref. 2 and references therein). Notable examples of such ligands will be discussed later in this section. There are at least 20 classes of riboswitches that have been identified where the ligand is known,[2] but there are also classes that can be regarded as "orphan", as their ligands are as yet undetected. Riboswitches can be regarded as a regulatory mechanism for gene expression, and are involved in bacterial RNA networks.

Riboswitches are now seen as a major prokaryotic gene regulation mechanism, for example they control ~4% of the genes in *Bacillus subtilis*.[3] They are found in the 5′-UTRs of bacteria, and recently have also been found in the 3′-UTRs of some bacterial genes.[4] Riboswitches have also been seen in the 3′-UTRs of eukaryotic genes, for example the thiamine pyrophosphate (TPP) riboswitch in the *THIC* gene in plants.[5] A new riboswitch called crcB has been reported in archaea, making it only the second example of a riboswitch found across two domains of life,[4] the other example being the TPP riboswitch mentioned above. There is also evidence of riboswitches being found in fungi and algae, for example the three TPP riboswitches in the fungus *Neurospora crassa*.[6] Studies on the *NMT1* gene which is a gene involved in TPP metabolism, and known to be repressed by excess thiamine, showed that thiamine caused alternative splicing of this gene. The decrease in expression of the main open reading frame (ORF) and concomitant increase in alternative transcripts are the direct result of thiamine addition to the cells.[6]

Structurally, riboswitches have two regions, an aptamer domain containing the ligand receptor that can have very complex 3D structures, and an expression platform whose secondary structure regulates the response.[7] The aptamer domain is transcribed first, allowing immediate sensing of the cellular environment.[2] Most riboswitches exist in two different conformations. In a ligand-bound state adjacent UTR sequence is sequestered into a tightly folded domain. Without the ligand this same portion is involved in transcriptional or translational control, by the use of an intrinsic terminator stem, and changing the availability of the Shine-Dalgarno (SD) sequence respectively. Therefore these regions can be regarded as different structural states of an RNA segment,[7] rather than regions. An example of the TPP riboswitch under the conditions of ligand presence and absence is shown in Figure 1. There is a much greater flexibility in the structure of the aptamer than the expression platform, due to the wide variety of ligands for the former, and the structural constraints that nucleic acids provide to the latter.

There are some very well characterised riboswitches that will be mentioned briefly. Cobalamin (co-enzyme B_{12}) has been known to repress expression of the *cob* operon in a variety of bacteria at the level of translation for nearly two decades.[8] This operon encodes nearly all 25 genes required for cobalamin de novo synthesis. The *cob* 5′ UTR is 462 nucleotides (nt) long, and highly structured, containing elements including a ribosomal binding site (RBS).[8] It has been shown that in the absence of cobalamin, a long range interaction of RNA element ~200 bp upstream of the initiating AUG suppresses the

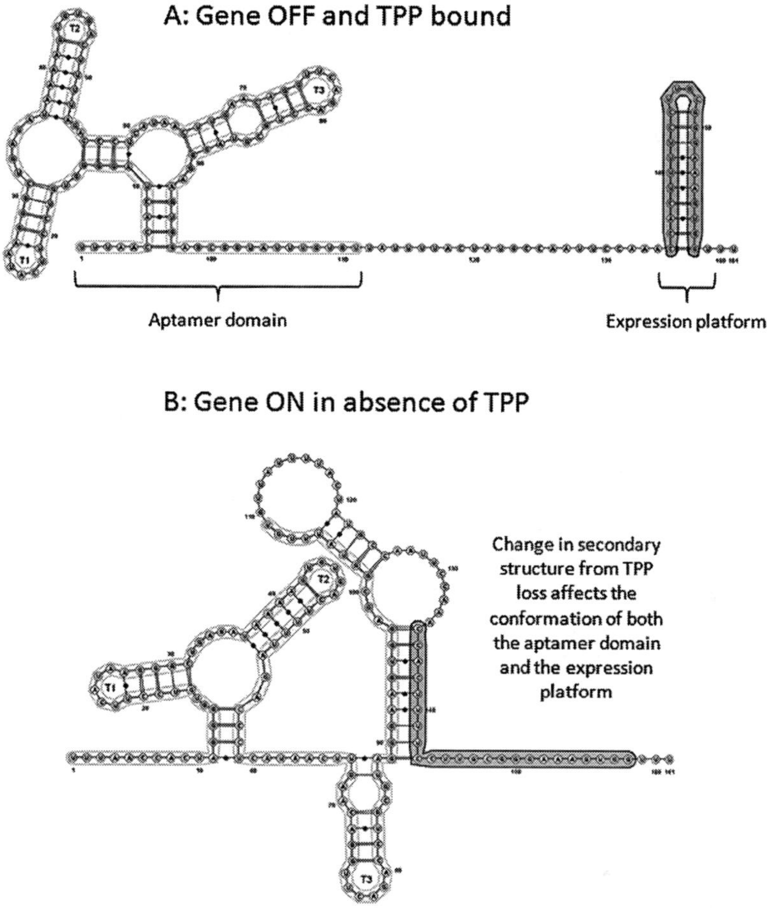

Figure 1. An example of a riboswitch in *Bacillus subtilis subsp. subtilis* str. 168 (GenBank accession AL009126).[59] The TPP riboswitch structure was retrieved from the RFAM[60] family RF0059 (TPP riboswitch) and drawn with the Java® applet VARNA.[61] A) Riboswitch structure with TPP bound, resulting in the inhibition of translation of the thiamine biosynthesis genes due to the high TPP concentration. The aptamer domain binding TPP is shown on the left, and the hairpin expression domain is on the right. B) With lower TPP concentrations, translation of the thiamine biosynthesis genes is initiated as the conformational change induced from the loss of TPP binding results in a new secondary structure. A color version of this figure is available at www.landesbioscience.com/curie.

formation of a short local RNA hairpin that normally sequesters the cob RBS to inhibit the translation of cob mRNA.[8] Interestingly, *btu*B mRNA, needed for extracellular cobalamin import, is controlled by a similar riboswitch mechanism,[9] indicating that the control of cellular function goes beyond the level of the operon, and into the RNA network proper.

The co-enzyme S-adenosylmethionine (SAM) is an important cellular metabolite and is also a riboswitch ligand, being found initially in Gram-positive bacteria, and consequently controlling the expression of 26 genes involved in sulphur metabolism.[10] A high level of discrimination for SAM and its analogues has been shown for the SAM-I riboswitch, with even the change of methyl group accounting for an affinity change of 100-fold. A second structure for a riboswitch binding SAM (SAM-II) has also been found, for example in the *met*A gene 5′UTR in *Agrobacterium tumefaciens*.[11] Despite a far simpler structure, SAM-I and SAM-II show similar levels of discrimination for SAM and related compounds, albeit with the latter having a lower affiity.[11] The SAM-I structure has been solved, and it shows that nearly all functional groups in SAM are recognized by the riboswitch.[12]

Whilst nearly all riboswitches function through the binding of metabolites or ligands, some do not. Examples include noncoding RNA elements, Mg^{2+} concentration and temperature. Other riboswitches are also known to act as ribozymes, RNA molecules that are catalysts in RNA cleavage reactions. An example of this, the *glm*S riboswitch, will be discussed in the next chapter (see Chapter 14, pags 226-227 for more detail on the glmS system) highlighting the way various systems within bacteria work together as a network. Considering tRNAs, T-box riboswitches modulate the expression of amino acid metabolic genes in Gram-positive bacteria, and use uncharged tRNAs as their signature molecule.[13] They show the classic two conformations depending on ligand binding, with the correct pairing of an uncharged tRNA with the leader sequences promoting the antiterminator structure, and allowing subsequent downstream gene expression.[13]

The pathogen *Salmonella enterica* has a variety of genes to control intracellular cation concentrations. The Mg^{2+} transporter *mtg*A has been shown to have a riboswitch in its 5′-UTR providing a mechanism for control.[14] At high Mg^{2+} concentrations, the conformation acts as a transcriptional terminator.[14] Mg^{2+} is required for RNA folding, as well as other cellular functions, so its discovery as a determinant in an RNA switch mechanism was important in expanding the variety of ligands riboswitches could use. It has also been shown that a single base mutation in the *mtg*A riboswitch, in a region not thought to be important in regulation, causes a high level constitutive expression of *mgt*A.[15] The surprising consequence of this expression was an enhanced thermotolerance.[15]

The *ags*A gene in *Salmonella* has been shown to be an 'RNA thermometer'.[16] In this case, the conformational changes in this very simple riboswitch are induced by temperature. A small heat shock protein, induced at high temperatures, is encoded by *ags*A, and a stable hairpin within the 5′-UTR blocks the SD sequence at normal temperatures (e.g., 30°C), but at higher temperatures (e.g., 45°C), the hairpin is open, and a translation initiation complex forms.[16]

Most riboswitches exist on their own, but occasionally tandem arrays are found, allowing a higher degree of regulation. The simplest arrangement is a tandem array of the same riboswitch, as can be seen in the glycine riboswitch that controls the *gcv*T operon that in turn codes for proteins in the glycine cleavage system. Analysis of this riboswitch in *B. subtilis* showed that the two riboswitch copies (two aptamers and one expression platform) are functionally linked to each other,[17] each being able to bind a single molecule of glycine. This arrangement allows for a greater response to ligand concentration changes, as there was co-operative ligand binding observed i.e., the binding of a glycine to one aptamer influenced the binding of another glycine to the other aptamer.[17]

However, more complex arrangements in tandem riboswitches have also been detected, where expressions can be affected by different metabolites, as is the case for the 5′-UTR of the *Bacillus clausii met*E gene that contains riboswitches for both SAM and vitamin B_{12} (also known as adenosylcobalamin, AdoCbl).[18] *met*E and *met*H both catalyse the formation of methionine from homocysteine, whilst *met*H requires an AdoCbl derivative for its function. Hence this architecture allows the two ligands to independently repress the *met*E gene, and for two metabolites to control protein production.[18] So in effect, this tandem arrangement allows complex genetic decisions to be made by the RNA without the need for proteins.

It should also be noted that as well as acting in tandem, riboswitches can also interact with other parts of the bacterial RNA network, for example their cleavage by RNase P is described in more detail in the next chapter. It is thus likely that more combinations of riboswitches will be found in the future as more bacterial genomes are sequenced, and computational methodologies to find them improve. As network and systems biology approaches also mature, the interactions between what are now considered to be discrete RNA systems will increase to show that the control of bacterial cellular function is subject to control by a variety of synergistic mechanisms.

THE CRISPR/Cas SYSTEM IS ANALOGOUS, BUT NOT HOMOLOGOUS, TO EUKARYOTIC RNAi

Clustered regularly interspersed short palindromic repeats (CRISPRs) are a feature of the bacterial and archaeal genome that were first noticed as genomic repeats nearly a decade ago.[19] As a genetic entity, CRISPRs have a longer history, having first been reported in the *Escherichia coli* genome two decades earlier as part of the 3′ flanking region of the *iap* gene.[20] The locus was shown to have a number of repeats, with nonrepeating sequences (spacers) between them. CRISPRs have had a wide variety of pseudonyms during their short life, but now CRISPR is the acronym by which these features are most commonly known. With the advent of computational searching of bacterial and archaeal genomes, it has become possible to perform large scale analyses to see how many bacteria and archaea have CRISPRs present.[21] Current estimates, depending on the software used, put the level at ~45% for bacteria and ~90% for archaea. This frequency discrepancy has been suggested to be due to the sequencing of cultured bacterial strains that may have lost CRISPR arrays from a lack of exposure to phage.[22] Software for CRISPR detection is briefly discussed later. It should also be noted however that CRISPRs are absent from both viruses and eukaryotes.

CRISPRs have been thought to be involved in replicon partitioning,[23] or to be mobile elements,[19] or to be involved in DNA repair.[24] However, it was shown independently by three groups in 2005 that the spacer sequences contained plasmid or phage derived DNA.[25-27] It was thus proposed that CRISPRs were involved in bacterial immunity against infection, primarily from phage.[27] Two years later, the immunity hypothesis was shown to be correct with the addition of new spacers to phage-challenged bacteria, and the resistance to phage upon exposure.[28-30] Our understanding of the molecular mechanism behind CRISPR action in still in its infancy. Broadly, there are two stages to the process, as shown in Figure 2. First, there is adaptation or immunization, which is the incorporation of sequences (proto-spacers) from the invading organism into the CRISPR locus through the action of the Cas protein complex. The second stage is interference or immunity, which is the activation of the Cas complex to destroy incoming DNA, and protect the bacteria in the process.

In structure, a typical CRISPR locus displays four key characteristics; CRISPR-associated genes (Cas), an AT-rich leader sequence, and then the array of direct repeats (DRs), and intervening spacer sequences. The CRISPR locus which contains only the spacers and DRs, and thus no ORF, is transcribed from inside the leader sequence.[31,32] Approximately half of genomes with CRISPRs have more than one locus,[33] and they can thus comprise up to 1% of the genome. Observations of sequences within CRISPRs indicate that there is a fair degree of horizontal gene transfer involving the CRISPR loci.[33-35] It has been observed that in a given CRISPR array, the DRs are nearly identical in size and sequence.[19] DRs can vary between 24

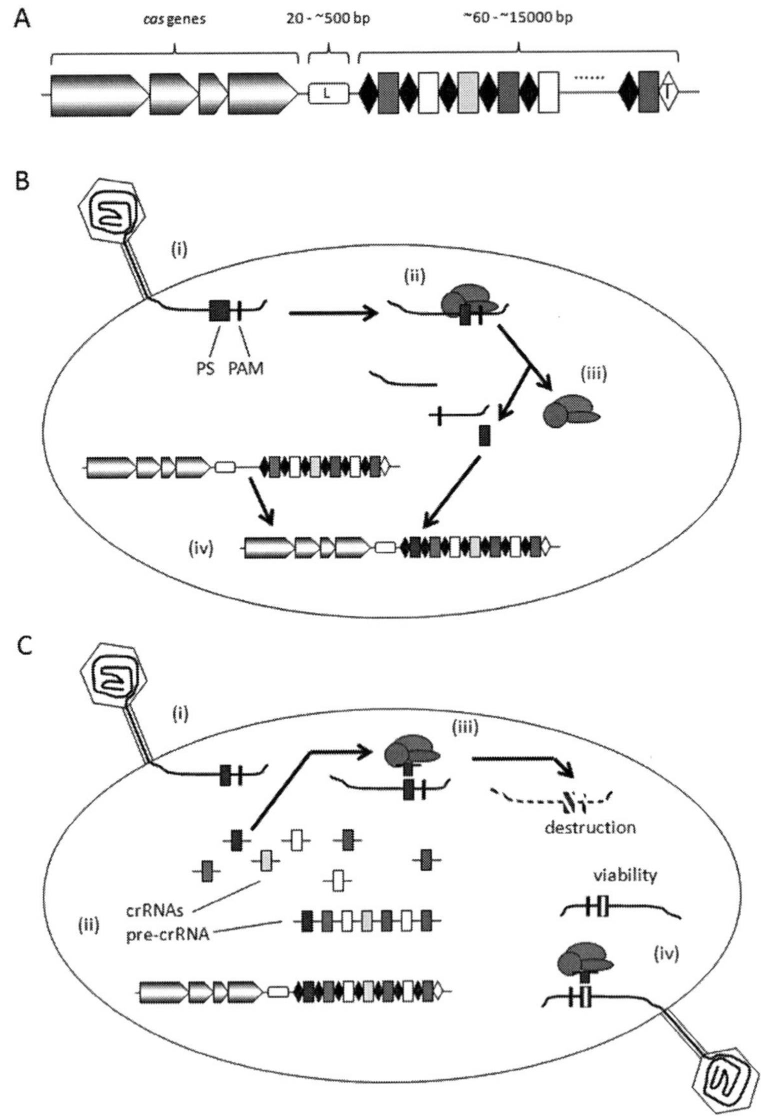

Figure 2. Description of the CRISPR system. A) A schematic CRISPR system is shown with *cas* genes indicated. The direct repeats (DRs) are in black diamonds, and the spacers are in boxes. The Leader sequence (L) is also shown, as is the often mutated terminal DR (T). B) CRISPR adaptation: An example showing how a protospacer (PS) from a phage is incorporated into the CRISPR locus. The cas complex is shown as a group of circles and the protospacer-associated-motif (PAM) is indicated. (i) Phage DNA enters the cell. (ii) The PS is recognised by the cas complex. (iii) Cleavage of the phage DNA results in the generation of a new spacer. (iv) The new spacer is inserted into the CRISPR locus near the leader sequence. C) CRISPR interference: (i) The same phage infect the cell at a later point in time. (ii) crRNAs, along with the cas genes are produced from the CRISPR locus. (iii) The appropriate spacer targets the homologous region in the phage, leading to interference and the destruction of the foreign DNA. (iv) A phage with a mutant sequence in the protospacer (shown by the box on the strand in iv) now evades the CRISPR/cas system, as perfect base pairing is required for interference. This new material can replicate inside the cell. Figure is based on reference 62. A color version of this figure is available at www.landesbioscience.com/curie.

and 47 bp, and they can be grouped into 12 major groups.[36] A large study analysed the secondary structure of the DRs and found them to be heterogeneous, though certain groups contained short palindromes.[36] The finding of G:U base pairs also gave a hint at the mechanism of the CRISPR system using an RNA intermediate. Clustering showed that groups with well defined secondary structures clustered together better than those with less well defined structures.[36]

Whilst being highly divergent, CRISPRs have been defined in 12 groups based on sequence similarity.[36] It is also known that spacers will have a homologue in extrachromosomal material. To look at why such spacer sequences are integrated, Mojica et al[37] have analysed the sequences around proto-spacers (so-called proto-spacer adjacent motifs; PAMs) with the aim of finding some sequence homology around them.[37] This has been found, with two or three nucleotides conserved on the 3′ side of the proto-spacer, as well as a degree of conservation with whether the CRISPR-type is folded or not. For example, CRISPR types 1, 7 and 10, which are unfolded show an NGG motif in their PAM, whereas palindromic CRISPRs such as types 2, 3 and 4 have unique PAM motifs (CWT, GAA and GG respectively).[37] Hence the choosing of the proto-spacer sequence for integration is not seen as being random.

The nomenclature surrounding the *cas* genes is complex. Initially four 'cas' genes were identified, but an extensive genomic study increased the number to six, naming the genes *cas1* to *cas6*.[34] There are eight CRISPR/Cas system subtypes defined by Haft et al[34] named after the genome in which they were first discovered, a nomenclature which has now been widely adopted. There was also a group of modular genes found within the genome of bacteria/archaea having a CRISPR array, but not always close to the CRISPR locus. Each of these eight subtypes have multiple genes in their arrays,[34] however nearly all have *cas1* and *cas2* present, making these genes universal markers for CRISPR defence mechanisms.[35] The Haft et al[34] analysis showed that 45 'guilds' of proteins were identifiable, and furthermore, this large gene family encode for proteins carrying different functional domains involved in DNA interaction.[34] The reader is directed towards this paper (ref. 34) for a thorough discussion of CRISPR/Cas subtypes.

It has been shown that the CRISPR-Cas system prevents phage infection in *Streptococcus thermophilus*,[28] but nothing similar could be seen in *E. coli,* which was considered to be a slight oddity for that, the most studied of bacteria. No well-known phage sequences were seen in *E. coli* CRISPRs, meaning either that the *E. coli* CRISPR system is nonfunctional, or *E. coli* have a very large number of phage, most of which are unknown. Recently, it has been shown that the histone-like nucleoid structuring protein (H-NS), a dual regulator of gene expression in Gram-negative bacteria, binds and represses CRISPR arrays, indicating resistance is conferred when H-NS silencing is circumvented.[38] In *E. coli hns* disrupted strains, a protection to phage λ is conferred by phage matching spacers.[39] It was also shown that those spacers farthest from the leader sequence, i.e., the oldest, showed no decrease in abundance compared to newer spacers, and therefore were likely to be as functional as the newer spacers.[39] This is the current situation in *E. coli*, but whether it is universal in bacteria with very long spacers remains to be seen. The spacers tend to be unique in a given bacterial or archaeal genome, but show high homology to phage sequence or other extrachromosomal elements,[25-27] and can come from either the sense or antisense strand.[29] Analyses have shown that only a small fraction of the spacers match to known sequences, highlighting the vast lack of knowledge we have about the potential size of phage sequence space.[40] The leader sequence can be over 500 bp long, is upstream of most CRISPR loci, and tends to be AT-rich.[19] Leader sequences are conserved within species, but not necessarily between species, and play a key role in the CRISPR system as new repeat-spacer units are introduced between the leader and the previous unit.[28]

Because of their occurrence within a large proportion of bacterial genomes, and the increasing relative use of sequencing bacterial genomes with high-throughput sequencing methodologies, software for detecting CRISPRs is an area of active research. Software dedicated to CRISPRs has only been relatively recently developed, and are available for users either to download, e.g., PILER-CR,[41] CRISPR Recognition Tool,[42] to use on websites e.g., CRISPRFinder as part of CRISPRdb,[43] including precomputed databases to search against,[43-45] or as part of larger sequence motif detection tools.[46] Different researchers have employed slightly different methodologies in their software, resulting for example in a different frequency of occurrence amongst the bacteria. Recently, it has been suggested that a combinatorial approach of using CRISPRFinder and CRISPI has its advantages to produce a thorough analysis of a given genome,[47] and in so doing, partially overcomes some of these differences.

It is becoming clear that CRISPR spacers can also be from the host's genome. One immediate consequence of this was that the CRISPR system had some role in gene regulation,[48] thereby adding complexity to the network. However, a review of self-targeting CRISPR spacers in bacterial genomes has shown that the explanation is more likely to be leaky incorporation of self nucleic acids leading to autoimmunity,[49] i.e., the inability of an organism to distinguish between what is self and foreign, thereby resulting in a response against self. Approximately 0.4% of spacers are self-targeting, but 18% of organisms with CRISPRs have at least one self-targeting spacer, none of which were the same.[49] About half of these spacers came from sequences of possible exogenous origin, but the other half did not. Moreover, these spacers were significantly likely to be the most recently incorporated spacers into the array. Incorporation would require shutting down of the host's CRISPR system, potentially by a variety of mechanisms. An alternative hypothesis for the low self-targeting fraction being due to autoimmunity is that phage could occasionally capture their host's DNA, and use it to hijack that same host's CRISPR system. There have been examples seen of CRISPR loci being found on prophages within bacterial genomes, such as in *Clostridium difficile*,[50] thereby suggesting the acquisition of CRISPR loci by phage.

For any organism autoimmunity is something to be avoided. In the case of CRISPRs, the CRISPR genomic locus and proto-spacer in the invading phage will be the same sequence, and detectable through the crRNA (small RNAs from the CRISPR locus). Using *Staphylococcus epidermis*, it has been shown that the detection of mismatches 5′ to the spacer marks the foreign DNA for degradation, and is thus the discriminatory factor, sparing the bacterial chromosome from any interference in the process.[51]

A more specific example of host interaction concerns the interference of his-tRNA synthetase (*his*S) in *Pelobacter carbinolicus*,[52] and the role this had has in the evolution of the species. *P. carbinolicus* is a member of the *Geobacter* genus and it cannot reduce Fe(III) directly, unlike other members of the genus. It has been hypothesised that this observation is due to the fact that the evolution of *P. carbinolicus* has been influenced by a spacer within the bacteria that matches a proto-spacer within *hisS*, that has in turn resulted in the loss of ancestral genes that contain multiple histidines.[52]

An early prediction linked CRISPRs to RNAi[35] meant that the RNA from the system was of prime interest to study. Analysis of archaeal noncoding RNAs showed homology to the CRISPR locus,[32] providing the first evidence that crRNAs came from a larger CRISPR transcript. Tang et al[31] went on to show that crRNAs had approximately half a DR, the spacer and another half DR. The mechanisms around this process have started to be investigated in *E. coli*, showing that a protein complex called Cascade (CRISPR-associated complex for antiviral defence) cleaves the long CRISPR transcript into crRNAs.[53] It has been commented by many authors that while it is useful to consider the CRISPR/Cas system as being analogous to eukaryotic RNAi, it is becoming clear that there are differences in the two systems, i.e., they are not homologous.[22] For example, the two systems have distinct protein machineries,[35] and the CRISPR-Cas complex binds DNA, suggesting little phylogenetic relationship between the two systems.[37] Also, no spacers from RNA viruses have been found, again adding weight to the CRISPR system recognising DNA.[37]

It is thus of great interest that recent discoveries in archaea add another dimension to this apparent duality. Archaea show morphological similarity to bacteria, but also show genetic similarity to eukaryotes. A study of a large number of archaeal genomes has shown that some species have genes involved in eukaryotic RNAi pathways, which is extremely interesting given that nearly all archaea have CRISPR systems present.[54]

It has already been noted that the production of crRNAs from the CRISPR locus is central to the CRISPR immunity system. Recently, an endoribonuclease called Csy4 has been found that is responsible for pre-crRNA processing in *Pseudomonas aeruginosa*.[55] Csy4 is a member of the Ypest subtype.[34] The structure shows interactions within the major group of the crRNA DR stem loop, allowing cleavage of the pre-crRNA at the downstream ssRNA-dsRNA junction.[55] Mutational analysis also showed that there is a requirement for there to be a C-G pair at the base of the stem loop structure. Csy4 has two other homologues in other subtypes, again indicating that co-evolution has occurred in shaping CRISPR recognition mechanisms.[55]

It has been realised that CRISPRs can be used to study epidemiological and co-evolutionary dynamics in microbial communities. Indeed, CRISPRs may provide the molecular record of co-evolution of host and phage pathogens in their natural environments.[56] Along with restriction modification and abortive infective systems, CRISPRs could therefore provide a record of what has happened in the recently termed "phage-host arms race", and how these have shaped microbial evolution.[57] As phage-derived spacers are integrated at the CRISPR 5′ leader sequence, the order of spacers gives a temporal record of the infection history, and as such is unparalleled in host-pathogen systems.[56] It might be that phage from certain environments are likely to co-exist in that environment, and so bacterial CRISPRs to those phage would be far more efficient if they were arranged in a way that kept the same phage cohort exposures together. Again, considering the microbial community of an environment, one outcome of metagenomics is that CRISPR-mediated co-evolution can be studied in noncultivable bacteria, important given how few bacteria (~1%) are currently cultivable. However, CRISPR incorporation might have a price for bacteria. For example, spacer incorporation would take longer to replicate, and result in a reduced growth rate. Finally, there might also be mutation accumulation on the spacers themselves, resulting in lower fitness for the organism that has the CRISPRs.[56] However, as it is at present unknown how many CRISPRs are expressed during infection, the actual organismal cost could be hard to determine.

A recent analysis of the two CRISPR loci in 51 genomes from *Escherichia* and *Salmonella* showed one locus was specialised in plasmid-based genes.[58] In addition, relatively recently diverged species had almost identical CRISPRs, suggesting contrary to recent studies that the use of the CRISPR locus as an epidemiological tool may be of limited value, as its rate of evolution for an immune system is quite slow.[58] It was also proposed that one of the CRISPR loci could be an anti-CRISPR, i.e., activation of this locus by native *cas* genes could lead to protection from any mobile elements or plasmids that might themselves harbour other *cas* genes. In eukaryotes we see a host-parasite war of mechanism and anti-mechanism so the discovery of such antagonism in bacteria is perhaps not so surprising. However, this recent discovery indicates that there is still much to learn about the fascinating CRISPR system, and its mode of action.

CONCLUSION

In this chapter, we have given some examples of how bacteria and archaea use RNA in various ways to provide a wealth of functions to the cell, often acting in conjunction with other RNAs, and therefore working in a true network of RNAs. These functions are diverse, and show just how diverse a molecule of RNA can be. We have described in some detail riboswitches and CRISPRs. The expression of gene control amenable through riboswitch action is immense; complete multi-gene operons can be controlled by the sensing of certain metabolites. CRISPRs have been shown to be a very powerful method for bacteria to combat the continual onslaught of bacteriophages that are present in the environment. The mode of action is not completely understood, but what we know shows a remarkable system relying on reasonably short sequence to elicit a major response.

As both the sequencing of bacterial genomes becomes a more routine activity, out of the purview of large sequencing centres, and computer algorithms continue to improve, the ability to find other regulatory or important RNA motifs outside of bacterial genes continues apace. It is thus a very exciting time for bacterial genomics, and no doubt, more players in the bacterial RNA infrastructure will be revealed in the not-too-distant future, providing even greater insight into how sophisticated bacteria really are.

ACKNOWLEGEMENTS

LJC would like to thank Prof. David Penny for continued support, financial and intellectual and stimulating discussions around this topic. PJB would like to thank Prof. Peter Lockhart for his support, and is part funded by the Royal Society of New Zealand Marsden Fund (08-MAU-099).

REFERENCES

1. Serganov A. The long and the short of riboswitches. Curr Opin Struct Biol 2009; 19(3):251-259.
2. Garst AD, Batey RT. A switch in time: detailing the life of a riboswitch. Biochim Biophys Acta 2009; 1789(9-10):584-591.
3. Mandal M, Boese B, Barrick JE et al. Riboswitches control fundamental biochemical pathways in Bacillus subtilis and other bacteria. Cell 2003; 113(5):577-586.
4. Weinberg Z, Wang JX, Bogue J et al. Comparative genomics reveals 104 candidate structured RNAs from bacteria, archaea, and their metagenomes. Genome Biol 2010; 11(3):R31.
5. Wachter A, Tunc-Ozdemir M, Grove BC et al. Riboswitch control of gene expression in plants by splicing and alternative 3′ end processing of mRNAs. Plant Cell 2007; 19(11):3437-3450.
6. Cheah MT, Wachter A, Sudarsan N et al. Control of alternative RNA splicing and gene expression by eukaryotic riboswitches. Nature 2007; 447(7143):497-500.

7. Edwards TE, Klein DJ, Ferre-D'Amare AR. Riboswitches: small-molecule recognition by gene regulatory RNAs. Curr Opin Struct Biol 2007; 17(3):273-279.
8. Ravnum S, Andersson DI. An adenosyl-cobalamin (coenzyme-B12)-repressed translational enhancer in the cob mRNA of Salmonella typhimurium. Mol Microbiol 2001; 39(6):1585-1594.
9. Nahvi A, Barrick JE, Breaker RR. Coenzyme B12 riboswitches are widespread genetic control elements in prokaryotes. Nucleic Acids Res 2004; 32(1):143-150.
10. Winkler WC, Nahvi A, Sudarsan N et al. An mRNA structure that controls gene expression by binding S-adenosylmethionine. Nat Struct Biol 2003; 10(9):701-707.
11. Corbino KA, Barrick JE, Lim J et al. Evidence for a second class of S-adenosylmethionine riboswitches and other regulatory RNA motifs in alpha-proteobacteria. Genome Biol 2005; 6(8):R70.
12. Montange RK, Batey RT. Structure of the S-adenosylmethionine riboswitch regulatory mRNA element. Nature 2006; 441(7097):1172-1175.
13. Gutierrez-Preciado A, Henkin TM, Grundy FJ et al. Biochemical features and functional implications of the RNA-based T-box regulatory mechanism. Microbiol Mol Biol Rev 2009; 73(1):36-61.
14. Cromie MJ, Shi Y, Latifi T et al. An RNA sensor for intracellular Mg(2+). Cell 2006; 125(1):71-84.
15. O'Connor K, Fletcher SA, Csonka LN. Increased expression of Mg(2+) transport proteins enhances the survival of Salmonella enterica at high temperature. Proc Natl Acad Sci U S A 2009; 106(41):17522-17527.
16. Waldminghaus T, Heidrich N, Brantl S et al. FourU: a novel type of RNA thermometer in Salmonella. Mol Microbiol 2007; 65(2):413-424.
17. Mandal M, Lee M, Barrick JE et al. A glycine-dependent riboswitch that uses cooperative binding to control gene expression. Science 2004; 306(5694):275-279.
18. Sudarsan N, Hammond MC, Block KF et al. Tandem riboswitch architectures exhibit complex gene control functions. Science 2006; 314(5797):300-304.
19. Jansen R, Embden JD, Gaastra W et al. Identification of genes that are associated with DNA repeats in prokaryotes. Mol Microbiol 2002; 43(6):1565-1575.
20. Ishino Y, Shinagawa H, Makino K et al. Nucleotide sequence of the iap gene, responsible for alkaline phosphatase isozyme conversion in Escherichia coli, and identification of the gene product. J Bacteriol 1987; 169(12):5429-5433.
21. Mojica FJ, Diez-Villasenor C, Soria E et al. Biological significance of a family of regularly spaced repeats in the genomes of Archaea, Bacteria and mitochondria. Mol Microbiol 2000; 36(1):244-246.
22. Marraffini LA, Sontheimer EJ. CRISPR interference: RNA-directed adaptive immunity in bacteria and archaea. Nat Rev Genet 2010; 11(3):181-190.
23. Mojica FJ, Ferrer C, Juez G et al. Long stretches of short tandem repeats are present in the largest replicons of the Archaea Haloferax mediterranei and Haloferax volcanii and could be involved in replicon partitioning. Mol Microbiol 1995; 17(1):85-93.
24. Makarova KS, Aravind L, Grishin NV et al. A DNA repair system specific for thermophilic Archaea and bacteria predicted by genomic context analysis. Nucleic Acids Res 2002; 30(2):482-496.
25. Bolotin A, Quinquis B, Sorokin A et al. Clustered regularly interspaced short palindrome repeats (CRISPRs) have spacers of extrachromosomal origin. Microbiology 2005; 151(Pt 8):2551-2561.
26. Mojica FJ, Diez-Villasenor C, Garcia-Martinez J et al. Intervening sequences of regularly spaced prokaryotic repeats derive from foreign genetic elements. J Mol Evol 2005; 60(2):174-182.
27. Pourcel C, Salvignol G, Vergnaud G. CRISPR elements in Yersinia pestis acquire new repeats by preferential uptake of bacteriophage DNA, and provide additional tools for evolutionary studies. Microbiology 2005; 151(Pt 3):653-663.
28. Barrangou R, Fremaux C, Deveau H et al. CRISPR provides acquired resistance against viruses in prokaryotes. Science 2007; 315(5819):1709-1712.
29. Deveau H, Barrangou R, Garneau JE et al. Phage response to CRISPR-encoded resistance in Streptococcus thermophilus. J Bacteriol 2008; 190(4):1390-1400.
30. Horvath P, Romero DA, Coute-Monvoisin AC et al. Diversity, activity, and evolution of CRISPR loci in Streptococcus thermophilus. J Bacteriol 2008; 190(4):1401-1412.
31. Tang TH, Polacek N, Zywicki M et al. Identification of novel non-coding RNAs as potential antisense regulators in the archaeon Sulfolobus solfataricus. Mol Microbiol 2005; 55(2):469-481.
32. Tang TH, Bachellerie JP, Rozhdestvensky T et al. Identification of 86 candidates for small nonmessenger RNAs from the archaeon Archaeoglobus fulgidus. Proc Natl Acad Sci U S A 2002; 99(11):7536-7541.
33. Godde JS, Bickerton A. The repetitive DNA elements called CRISPRs and their associated genes: evidence of horizontal transfer among prokaryotes. J Mol Evol 2006; 62(6):718-729.
34. Haft DH, Selengut J, Mongodin EF et al. A guild of 45 CRISPR-associated (Cas) protein families and multiple CRISPR/Cas subtypes exist in prokaryotic genomes. PLoS Comput Biol 2005; 1(6):e60.
35. Makarova KS, Grishin NV, Shabalina SA et al. A putative RNA-interference-based immune system in prokaryotes: computational analysis of the predicted enzymatic machinery, functional analogies with eukaryotic RNAi and hypothetical mechanisms of action. Biol Direct 2006; 1:7.

36. Kunin V, Sorek R, Hugenholtz P. Evolutionary conservation of sequence and secondary structures in CRISPR repeats. Genome Biol 2007; 8(4):R61.
37. Mojica FJ, Diez-Villasenor C, Garcia-Martinez J et al. Short motif sequences determine the targets of the prokaryotic CRISPR defence system. Microbiology 2009; 155(Pt 3):733-740.
38. Mojica FJ, Diez-Villasenor C. The on-off switch of CRISPR immunity against phages in Escherichia coli. Mol Microbiol 2010; 77(6):1341-1345.
39. Pougach K, Semenova E, Bogdanova E et al. Transcription, processing and function of CRISPR cassettes in Escherichia coli. Mol Microbiol 2010; 77(6):1367-1379.
40. Edwards RA, Rohwer F. Viral metagenomics. Nat Rev Microbiol 2005; 3(6):504-510.
41. Edgar RC. PILER-CR: fast and accurate identification of CRISPR repeats. BMC Bioinformatics 2007; 8:18.
42. Bland C, Ramsey TL, Sabree F et al. CRISPR recognition tool (CRT): a tool for automatic detection of clustered regularly interspaced palindromic repeats. BMC Bioinformatics 2007; 8:209.
43. Grissa I, Vergnaud G, Pourcel C. The CRISPRdb database and tools to display CRISPRs and to generate dictionaries of spacers and repeats. BMC Bioinformatics 2007; 8:172.
44. Grissa I, Vergnaud G, Pourcel C. CRISPRcompar: a website to compare clustered regularly interspaced short palindromic repeats. Nucleic Acids Res 2008; 36(Web Server issue):W145-W148.
45. Rousseau C, Gonnet M, Le Romancer M et al. CRISPI: a CRISPR interactive database. Bioinformatics 2009; 25(24):3317-3318.
46. Mrazek J, Xie S, Guo X et al. AIMIE: a web-based environment for detection and interpretation of significant sequence motifs in prokaryotic genomes. Bioinformatics 2008; 24(8):1041-1048.
47. Deveau H, Garneau JE, Moineau S. CRISPR/Cas system and its role in phage-bacteria interactions. Annu Rev Microbiol 2010; 64:475-493.
48. Sorek R, Kunin V, Hugenholtz P. CRISPR—a widespread system that provides acquired resistance against phages in bacteria and archaea. Nat Rev Microbiol 2008; 6(3):181-186.
49. Stern A, Keren L, Wurtzel O et al. Self-targeting by CRISPR: gene regulation or autoimmunity? Trends Genet 2010; 26(8):335-340.
50. Sebaihia M, Wren BW, Mullany P et al. The multidrug-resistant human pathogen Clostridium difficile has a highly mobile, mosaic genome. Nat Genet 2006; 38(7):779-786.
51. Marraffini LA, Sontheimer EJ. Self versus nonself discrimination during CRISPR RNA-directed immunity. Nature 2010; 463(7280):568-571.
52. Aklujkar M, Lovley DR. Interference with histidyl-tRNA synthetase by a CRISPR spacer sequence as a factor in the evolution of Pelobacter carbinolicus. BMC Evol Biol 2010; 10:230.
53. Brouns SJ, Jore MM, Lundgren M et al. Small CRISPR RNAs guide antiviral defense in prokaryotes. Science 2008; 321(5891):960-964.
54. Li Y, Liu X, Huang L et al. Potential coexistence of both bacterial and eukaryotic small RNA biogenesis and functional related protein homologs in Archaea. J Genet Genomics 2010; 37(8):493-503.
55. Haurwitz RE, Jinek M, Wiedenheft B et al. Sequence- and structure-specific RNA processing by a CRISPR endonuclease. Science 2010; 329(5997):1355-1358.
56. Touchon M, Rocha EP. The small, slow and specialized CRISPR and anti-CRISPR of Escherichia and Salmonella. PLoS One 2010; 5(6):e11126.
57. Stern A, Sorek R. The phage-host arms race: shaping the evolution of microbes. Bioessays 2011; 33(1):43-51.
58. Lai LB, Vioque A, Kirsebom LA et al. Unexpected diversity of RNase P, an ancient tRNA processing enzyme: challenges and prospects. FEBS Lett 2010; 584(2):287-296.
59. Kunst F, Ogasawara N, Moszer I et al. The complete genome sequence of the gram-positive bacterium Bacillus subtilis. Nature 1997; 390(6657):249-256.
60. Gardner PP, Daub J, Tate JG et al. Rfam: updates to the RNA families database. Nucleic Acids Res 2009; 37(Database issue):D136-D140.
61. Darty K, Denise A, Ponty Y. VARNA: Interactive drawing and editing of the RNA secondary structure. Bioinformatics 2009; 25(15):1974-1975.
62. Horvath P, Barrangou R. CRISPR/Cas, the immune system of bacteria and archaea. Science 2010; 327(5962):167-170.

CHAPTER 14

RNA NETWORKS IN PROKARYOTES II: tRNA Processing and Small RNAs

Lesley J. Collins*,[1] and Patrick J. Biggs[2]

[1]Institute of Fundamental Sciences, Massey University, Palmerston North, New Zealand; [2]Institute of Veterinary, Animal and Biomedical Sciences, Massey University, Palmerston North, New Zealand
**Corresponding Author: Lesley J Collins—Email: lesleycollins.nz@gmail.com*

Abstract: It is becoming clear that in prokaryotes RNAs interact and perform complex functions as a network similar to what we have uncovered in eukaryotes. This chapter will continue the discussion of prokaryotic molecular systems, showing how these systems can interact with each other to gain a higher level of control within the cell. Our examples include RNase P, the tRNA cleaving molecule that, as well as performing other functions, also cleaves certain riboswitches; and the glmS gene under the control of both a ribozyme in its 5′ untranslated region and two small RNAs. With further investigation of nonprotein coding RNA interactions (i.e., the RNA infrastructure), in bacteria and archaea, we gain greater understanding of the influence that small strands of RNA sequence can have over the entire cell.

INTRODUCTION

The previous chapter began our discussion of the RNA networks within the prokaryotic cell with an overview of the CRISPR system of viral defence, and riboswitch regulation. Although both of these systems have equivalents within eukaryotes, it is clear that prokaryotes operate these mechanisms in their own way. In a similar way, this chapter will look at some further RNA-based networks in the prokaryotic cell, some of which are unique to prokaryotes and some with similar networks in eukaryotes.

A key endoribonuclease RNase P is examined first in relation to its role in RNA cleavage and the removal of leader sequences, and since it potentially affects many hundreds of genes, it can be considered central to a large bacterial RNA network. Small RNAs and sigma factors are discussed next, with a focus on a key bacterial regulator,

RNA Infrastructure and Networks, edited by Lesley J. Collins.

the sigma factor RpoS and its role in sensing the cellular environment and detecting stress. Finally, a relatively newcomer to bacterial RNA network studies, transfer messenger RNA, a cross between tRNA and mRNA responsible for releasing stalled ribosomes, is discussed in light of its role in translation and bacterial sporulation. With new studies on prokaryotic RNAs being released almost daily, it is impossible to cover all the angles surrounding RNA networks. These examples and those from the previous chapter therefore, give a taste of what is likely to become in the near future, a complex RNA-infrastructure similar to that seen in eukaryotes, but with its own unique flavour.

RNase P AND tRNA PROCESSING IN BACTERIA AND ARCHAEA

RNase P is a site-specific endoribonuclease which plays a central role in the prokaryotic RNA-processing network (Fig. 1), (for a review see refs. 1-4). In addition to its primary action of cleaving pre-tRNAs to progress their maturation, RNase P has been shown to cleave other substrates in bacteria such as pre-4.5S RNA, pre-tmRNA, polycistronic operons, viral transcripts and riboswitches. Although present in all three kingdoms of life (i.e., eukaryotes, bacteria and archaea) the RNase P holoenzyme has a very different

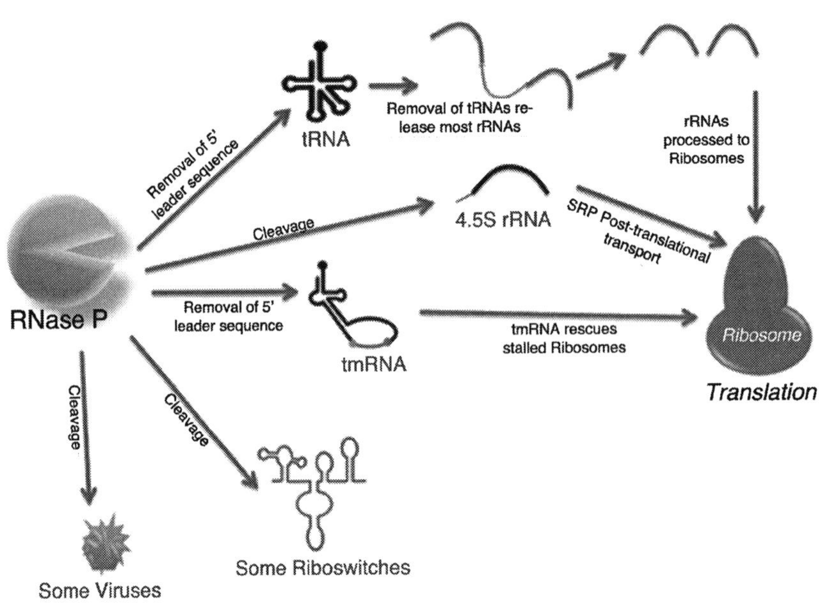

Figure 1. The role of RNase P in bacterial RNA processing. RNase P not only removes the 5′ leader sequence from pre-tRNAs but affects other key transcripts within bacteria. In a similar manner it cleaves the 5′ leader sequence from tmRNAs which rescues stalled ribosomes, as well having been shown to cleave some viruses and some riboswitches. A downstream effect of tRNA cleavage is the release of pre-rRNAs which are then processed to ribosomes. Cleavage of the 4.5S rRNA permits the mature RNA to be part of the SRP post-translational transport complex. A color version of this figure is available at www.landesbioscience.com/curie.

macromolecular structure in each kingdom, and there are differences in how RNase P interacts with other RNAs in each kingdom's RNA infrastructure.

In bacteria, RNase P consists of a single RNA (mass ~120 kDa) and a single protein (mass ~15 kDa).[2] Although this basic protein is small, it plays diverse roles such as enhancing substrate binding, altering substrate recognition, stabilising RNA conformation, and aiding catalysis by discriminating between the substrate and product by binding to the 5′ leader sequence of the pre-tRNA.[1,2] In archaea multiple proteins (five including the ribosomal protein L7Ae[5]),bind to the single RNA.[6] These archaeal proteins show some evolutionary relationship to eukaryotic RNase P proteins and most were identified by sequence similarity to yeast proteins (summarized in ref. 7). However, the Mth687 (pop5) protein from *Methanothermobacter thermoautotrophicus* adopts a fold similar to that of the bacterial protein and this protein may carry out some of the functions found in the bacterial protein.[3,7] The RNase P RNA from some representatives from each kingdom can be induced to perform weak catalysis without its accompanying proteins, but only with high salt and high cation conditions in vitro (summarized in ref. 7).

Within the bacterial RNase P holoenzyme, the protein and RNA subunits function synergistically in both substrate recognition and catalysis, to process the products of more than 80 pre-tRNA genes (*Escherichia coli* and *Bacillus subtilis)*.[8] In all cases the action of RNase P is straightforward, it cleaves sequences at a designated site and most interactions are RNA-RNA between the RNase P RNA and the substrate RNA.

Outside of cleaving 5′ leader sequences of tRNAs and tmRNAs (discussed later) there are a number of other targets for RNase P cleavage. Some polycistronic mRNA transcripts can be processed with RNase P in *E. coli* independent of RNase E (which is typically required).[9] For example, with the valV-valW and LeuQ-leuP-leuV transcripts only the RNase P activity is necessary to generate pre-tRNAs with mature 5′ ends.[9] The 4.5S RNA substrate which is part of the Signal Recognition Particle (SRP) responsible for post-translational transport, is also cleaved by RNase P. Interestingly the structure of the 4.5S RNA is very different from the tRNA clover-leaf shape, forming instead a long hairpin. The phage ϕ80-induced RNA has a similar hairpin structure and is also cleaved by RNase P (reviewed in ref. 3). In other viral examples, the antisense RNA precursor C4 from bacteriophages P1 and P7 is cleaved by RNase P resulting in the inhibition of Ant (antirepressor) synthesis.[3] tRNA-like structures from the turnip yellow mosaic virus (RYMV) can also be cleaved by RNase P RNA from *E. coli*,[10] thus leading to research in using RNase P as an antiviral agent for animals.[11]

RNase P has also been shown to cleave the 5′ regions of some riboswitches, including the adenine riboswitch from *B. subtilis*[12] and co-enzyme B_{12} riboswitches (*btu*B) from *E. coli* and *B. subtilis*.[13] RNase P cleavage of riboswitches is not a general mechanism, as not all riboswitches are cleaved by RNase P (e.g., TPP, FMN, SAM and lysine riboswitches of *E. coli*[13] and the *xpt-pbuX* guanine riboswitch of *B. subtilis*[12] are not cleaved). The riboswitches that are cleaved do not display a tRNA-like cleavage site or even a model cleavage site.[13] The *B. subtilis btu*B riboswitch cleavage site is located in a region that has a short single-stranded region next to a putative double-stranded region but such a structural motif was not found in the *E. coli btu*B riboswitch.[13] The *pbu*E gene is controlled by the adenine riboswitch and in the absence of adenine, the riboswitch folds into an 'off' structure which includes a transcription terminator. The *pbu*E adenine riboswitch is cleaved at two sites C (−27) and C (−138), with cleavage efficiency of the latter dependent on the length of the substrate, with shorter substrates displaying decreased efficiency.[12]

BACTERIAL SMALL RNAs AND SIGMA FACTORS

In bacteria small RNAs (sRNAs) typically ~100 nt long have been identified (~100 of them in *E. coli*), mostly in intergenic regions flanked by recognisable promoters and/or Rho-independent terminators[14] (reviewed in refs. 15, 16). Studied mainly in *E. coli* and *Salmonella*, they have been associated with many metabolic pathways including stress responses to oxygen levels[17] and drugs,[18] carbohydrate uptake and metabolism,[19] iron metabolism[20] and quorum sensing,[21] and act to regulate gene expression post-transcriptionally. Some sRNAs such as Spot42 have been known since the early 1970s as an abundant RNA species,[19] whereas others are just coming to light as the roles of small RNAs in bacterial mechanisms is more widely studied. Systems biology approaches using bacterialike *E. coli* have allowed for the role of some sRNAs within the transcriptional regulatory network to be examined (reviewed within ref. 18). Furthermore, because of their small size and lack of translation, sRNAs have been hypothesised to be cost effective to the cell from a regulation point of view. Mathematical modelling of the sRNA-mRNA interaction has shown that this is a good mechanism when quick responses to external stimuli are required, consistent with experimental results.[22] A few examples of sRNA-based regulation are described below.

As a master regulator of stress response in *E. coli* the sigma factor RpoS (σ^s) responds to multiple stresses with almost 500 genes directly or indirectly under its control.[17] In an example of RNA-based regulation (Fig. 2), *Rpo*S is regulated by at least four small RNAs and requires the RNA chaperone protein Hfq.[17] Two Hfq-binding sRNAs DsrA and RprA also positively regulate *rpo*S translation by base pairing to the

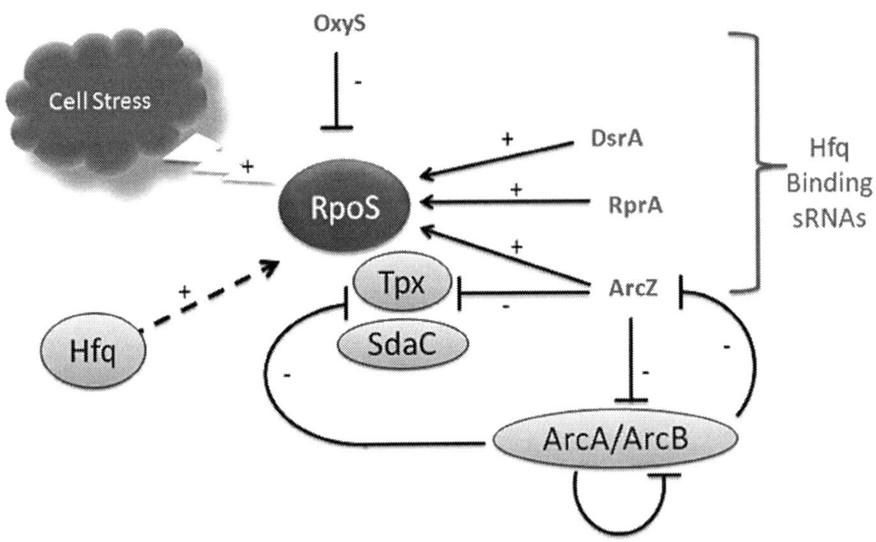

Figure 2. Example of bacterial small RNA networking in cell stress response. The stress response sigma factor RpoS has ~500 genes under its control. RpoS itself is regulated by multiple Hfq binding small RNAs (OxyS, DsrA, RprA, ArcZ). ArcZ represses and is repressed by arcB transcription, but also affects other proteins Tpx and SdaC. The ArcA/ArcB complex is known to be another key regulator complex. A color version of this figure is available at www.landesbioscience.com/curie.

upstream part of an inhibitory stem-loop in the *rpo*S mRNA leader sequence and freeing the ribosome binding site.[17] *Oxy*S, an sRNA expressed under oxidative stress, represses expression by a less understood mechanism (reviewed in ref. 17). In a manner similar to that of miRNA-target feedback loops seen in eukaryotes, one small RNA ArcZ directly represses and is repressed by *arc*B transcription, and is also a positive regulator of *Rpo*S.[17] It is suggested that each of these three positive regulators of *Rpo*S (ArcZ, DsrA and RprA) have different strengths in their ability to open the RpoS structure, and thus may regulate *Rpo*S to different levels.[17] ArcZ has other targets than RpoS including *tpx* which encodes a lipid hydroperoxide peroxidise, and *sda*C, a putative serine transporter in *Salmonella*.[23] The complexity of this small RNA network is further increased if we add the negative regulation of *tpx* and *arc*B by ArcA, and that the ArcA/ArcB system is one of the central regulators in *E. coli*.[17,24]

The Hfq protein, as well as mediating small RNA interactions with their targets also has connections to RNA degradation via a multi-enzyme complex called the degradosome.[25-27] The degradosome is made up primarily of four enzymes including the endoribnuclease RNase E, and the exoribonuclease PNPase.[27] The degradosome appears widespread in bacteria, but a similar complex, the exosome is found in archaea and eukaryotes assembling on a core that resembles PNPase.[25,28] Hfq can bind to the degradosome to promote RNase E cleavage and subsequent degradation of target mRNAs after base-pairing with small RNAs.[27,29] An interesting feature to emerge is that the interaction between Hfq and RNase E is not direct, but is likely to be mediated by another, yet to be identified RNA.[27] Hence, it is possible that the degradation of RNA may be regulated by other RNA species (for more on the Degradosome please see Chapter 15, pages 233-235).

In another example of RNA-based networking, catabolite repression prevents transcription of genes required for a less desired substrate when more than one carbon source is available for a bacterial species. Monod[30] first observed that *E. coli* and *B. subtilis*, when given two carbon sources at the same time, will often degrade one substrate first (the one giving faster growth) before the other (summarized in refs. 31, 32). In *Pseudomonas aeruginosa*, the small RNA CrcZ binds to the Crc protein, to come under the control of the CbrA/CbrB system.[32] The CbrA/CbrB system adjusts CrcZ small RNA levels in response to different carbon sources allowing a gradual mode of catabolite repression operating at the post-transcriptional level.[32] Another small RNA CyaR is regulated by the *Crp* gene (a global regulator of many sRNAs) to increase expression when cyclic AMP (cAMP) levels are high.[14]

sRNAs are heavily involved in the interaction of the bacterial cell with its environment. The outer membrane of the Gram-negative cell prevents toxic compounds from entering the cell,[33] and contains channels formed from OMP porin proteins. Some of these OMP proteins are regulated by small RNAs, including *Omp*C regulated by the MicC RNA and *Omp*A regulated by MicA RNA.[33] MicA requires Hfq both for its own intracellular stability and for annealing to the ompA mRNA.[33] The OmpX protein (bacterial adhesion and mammalian cell interation protein) is down-regulated by the small RNA CyaR (discussed above) linking the starvation response to protein which aid in moving to a new source of nutrients.[14]

Another OMP-type protein ChiP (chitoporin) is down regulated by the small RNA ChiX.[17,34] ChiX regulation of ChiP is different from the standard pathway as the sRNA does not become destablilised after base-pairing with its target (Fig. 3A) and it is thought that it could act catalytically to degrade multiple mRNAs (Fig. 3B).[17] However, there is also competitive regulation of the ChiX sRNA by an mRNA decoy where

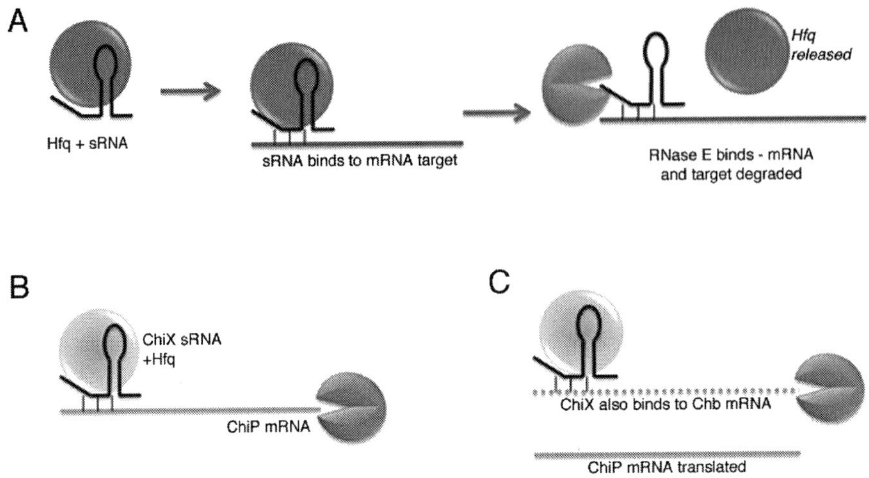

Figure 3. Examples of small RNAs and sigma factors in bacterial cellular responses. A) The typical action of an Hfq-associated sRNA. The sRNA binds to the Hfq protein and then finds its target mRNA. Once bound the Hfq detaches, permitting RNase E to bind and then both the target and sRNA are degraded. B) In response to starvation in *Salmonella*,[17] ChiX sRNA binds to its target but Hfq is not released and the target ChiP is degraded by Exonuclease I. C) Competitive regulation of ChiX is seen as the Chb mRNA acts as a decoy to soak up ChiX sRNA, permitting ChiP mRNA to be released. A color version of this figure is available at www.landesbioscience.com/curie.

base-pairing of ChiX to an intergenic region of the *chb* mRNA induces degradation of the sRNA (Fig. 3C). Upon induction by chitobiose, there is degradation of ChiX relieving the inhibition of the chip mRNA.[17]

sRNAs are important factors in the making and breaking of sugars in the cell.[19] A classic example is the regulation of the *ptsG* gene which is part of the phosophotransferase system involved in the uptake of glucose. The *ptsG* gene transports and phosphoylates glucose to G6P and while G6P is essential for glucose metabolism, accumulation within the cell causes sugar-phosphate toxicity (also called phospho-sugar stress[19]). The *Sgr*S gene codes for an Hfq-associated sRNA which has a dual function of regulating *ptsG* by base-pairing as well as coding for a small protein SgrT.[35] SgrT has been shown to also block glucose uptake without affecting levels of ptsG mRNA or *PtsG* protein, although its exact mechanism is not yet known.[35] It is interesting to note that the topological network properties associated with carbon metabolism are very different to those that control developmental processes, with short parallel transcriptional cascades for the former, and long intertwined regulatory cascades for the latter.[18] Such a mechanism again allows for a rapid change in carbon source to be maximally exploited by the bacteria.

In another example where the mechanism is more known (Fig. 4), the synthesis of glucosamine-6-phosphate (GlcN6P) is regulated by two small RNAs GlmY and GlmZ[36] (and reviewed in ref. 31). GlcN6p can be derived from amino sugars in the cell environment or by de novo synthesis with glucosamine-6-phosphate synthase (GlmS). GlmS is encoded with its partner GlmU in an operon and while GlmU is essential, GlmS is only required in the absence of amino sugars. The small RNA GlmZ is typically processed to be inactive. When GlcN6P is low, GlmY accumulates and inhibits processing of GlmZ.

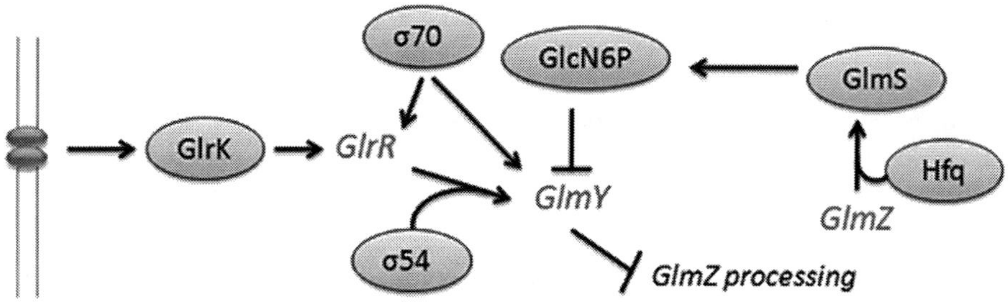

Figure 4. In phosphor-sugar stress, there is a network that activates from the cell membrane. Low levels of GlcN6P increases GlmY levels which repress GlmZ processing. GlmZ along with Hfq pairs with the glmS transcript and activates GlmS to resynthesise GlcN6P (figure based on ref. 36). A color version of this figure is available at www.landesbioscience.com/curie.

GlmZ then pairs with the *glmS* mRNA transcript activating GlmS which resynthesizes GlcN6P. The *glmS* mRNA is normally unstable due to secondary structure which buries its ribosomal binding site. However, base-pairing with sRNA GlmZ aided by the Hfq protein changes the structure and allows efficient translation of *glmS*. GlmY is also processed but how this affects the network is not yet characterised.[36] In another interesting twist, GlmY is controlled by two overlapping promoters with the same transcription start site.[36] One promoter activates transcription during the cell's exponential growth and the other operates during the cell's stationary phase. This may control the levels of GlcN6P from being high during the stationary phase when GlcN6P is not required (when cell wall and outer membrane synthesis is not required).[36]

Another interesting point about *glm*S is that the 5′-UTR contains a ribozyme (a catalytic RNA molecule that can cleave RNA), and functions as a (GlcN6P)-dependent catalyst enabling the riboswitch-like regulation of amino-sugars. In other words, the binding of GlcN6P has an effect on a gene involved in its production,[37] but without the need for the riboswitch structural elements of an aptamer and expression platform. It has been shown that GlcN6P is absolutely required for the catalytic activity,[38] demonstrating the property of RNA to be a metabolite-responsive ribozyme. Functionally, GlcN6P binding stimulates cleavage near the 5′ end of the transcript[37] and furthermore the cleaved downstream product is targeted for intracellular degradation.[39] RNase E is responsible for global mRNA decay in *E. coli*, but this enzyme is not present in a variety of prokaryotes e.g., *B. subtilis*. RNAse J1 has been shown fulfil the same role in *B. subtilis*, wherein it is responsible for the degradation of the 3′ cleavage product upon ribozyme action.[39]

These are only a few examples of identified small RNA based networks present within bacteria. With new sequencing techniques able to identify a genome's worth of RNAs in a single run, we expect there to be many others yet to be characterised. How these networks have evolved within prokaryotes is also another question wide open to be explored.

TRANSFER MESSENGER RNA

With new research the traditional roles of the main RNAs in the bacterial cell are being expanded. An example is tmRNA (transfer-messenger RNA also known as SsrA RNA or 10Sa RNA), a specialised tRNA molecule that along with a small protein (SmpB-small protein B), mediates the rescuing of stalled ribosomes. When incomplete or damaged mRNA is translated, the ribosome stalls leaving a peptidyl-tRNA on the P site of the ribosome. Alanine-charged tmRNA then enters the A-site of the stalled ribosome and accepts the peptide.[40] The translation of the sequence encoded in the tmRNA resumes the translation process to add a C-terminal peptide tag to the protein then uses an in-frame stop codon to complete the translation.[40-42] This process, known as *trans*-translation allows the tmRNA to act as both a tRNA and an mRNA as its name suggests.[43] The secondary structure of the tmRNA holds the key to how it is able to serve both functions.[41] One domain of the tmRNA has an amino acid acceptor stem chargeable with alanine and a T arm with modified nucleotides (as expected in a tRNA). However, the D arm of the tRNA does not fold-like a typical tRNA and there is no anticodon loop. Instead an mRNA-like domain (MLD) is located in a pseudoknot rich region and contains a short open reading frame with the tmRNA motif (AADENYALAA) followed by a typical stop codon. It is this domain that targets the truncated protein for degradation.[41] *Trans*-translation appears to be conserved throughout bacteria and is also present in some mitochondria and chloroplasts.[42,43] In some bacterial species the gene for the tmRNA (*ssrA*) is essential (e.g., *Neisseria gonorrhoeae*, *Mycoplasma spp.*, *Synechococcus spp.*), but in other species it is merely important for surviving challenging growth conditions.[41] As well as its typical tmRNA role tmRNA also binds in an antisense direction to the 5′ UTR of the crtMN mRNA, and thus may play an important role in the pathogenic nature of *Staphylococcus aureus*.[44]

tmRNAs are involved in sporulation in some bacterial species. Under starvation conditions *B. subtilis* enters sporulation where the cell undergoes asymmetrical division into a two chambered sporangium consisting of a forespore and a mother cell.[40] The forespore matures, aided by the mother cell until the forespore is engulfed by the mother cell, forming a free protoplast surrounded by a double membrane inside the mother cell. During sporulation different genes are expressed in each compartment which is controlled by compartment specific RNA polymerase sigma factors σ^F and σ^G in the forespore and σ^E and σ^K in the mother cell (reviewed in ref. 40). tmRNA has been shown to be necessary for the later stages of sporulation, since only 5% of *B. subtilis* cells lacking the tmRNA gene produced heat resistant spores.[40] As spore formation is regarded as a stress response to nutrient starvation, this process may also lead to increased ribosome stalling which is rescued by tmRNA. Without tmRNA, stalled ribosomes may accumulate resulting in the cell unable to translate essential spore-related proteins.[40] This may also be the case for other stress responses since tmRNA deletion also leads to hypersensitity to drugs such as aza-C (5-azacytidine) which induces DNA-protein crosslinks between cytosine methyl transferase and DNA inhibiting methylation. This cross-linking is known to lead to blocking of the translation system, and tmRNA thus relieves the blockage.[45]

CONCLUSION

In this chapter we have described in some detail RNase P, small RNAs, sigma factors and transfer messenger RNAs and how they interact in networks within the prokaryotic cell. These networks are only a few coming to light from recent studies of ncRNAs in prokaryotes. Archaeal systems are yet to be fully explored and so are many species of bacteria that are distantly related to the model species (e.g., *E. coli* and *B. subtilis*). As these studies progress, in both model and non-model prokaryotes we may then be able to begin to unravel a clearer idea of the 'general' prokaryotic RNA-infrastructure.

However, it is possible that there is such diversity amongst the bacterial and archaeal groups that such a general picture cannot be drawn. RNA studies are still in their early days in prokaryotes and there is no doubt that there is exciting work ahead.

ACKNOWLEDGEMENTS

LJC would like to thank Prof. David Penny for continued support, financial and intellectual and stimulating discussions around this topic. PJB would like to thank Prof. Peter Lockhart for his support, and is part funded by the Royal Society of New Zealand Marsden Fund (08-MAU-099).

REFERENCES

1. Esakova O, Krasilnikov AS. Of proteins and RNA: the RNase P/MRP family. RNA 2010; 16(9):1725-1747.
2. Lai LB, Vioque A, Kirsebom LA et al. Unexpected diversity of RNase P, an ancient tRNA processing enzyme: challenges and prospects. FEBS Lett 2010; 584(2):287-296.
3. Marvin MC, Engelke DR. Broadening the mission of an RNA enzyme. J Cell Biochem 2009; 108(6):1244-1251.
4. Marvin MC, Engelke DR. RNase P: increased versatility through protein complexity? RNA Biol 2009; 6(1):40-42.
5. Cho IM, Lai LB, Susanti D et al. Ribosomal protein L7Ae is a subunit of archaeal RNase P. Proc Natl Acad Sci U S A 2010; 107(33):14573-14578.
6. Jarrous N, Gopalan V. Archaeal/Eukaryal RNase P: subunits, functions and RNA diversification. Nucleic Acids Res 2010.
7. Pulukkunat DK, Gopalan V. Studies on Methanocaldococcus jannaschii RNase P reveal insights into the roles of RNA and protein cofactors in RNase P catalysis. Nucleic Acids Res 2008; 36(12):4172-4180.
8. Koutmou KS, Zahler NH, Kurz JC et al. Protein-precursor tRNA contact leads to sequence-specific recognition of 5' leaders by bacterial ribonuclease P. J Mol Biol 2010; 396(1):195-208.
9. Mohanty BK, Kushner SR. Ribonuclease P processes polycistronic tRNA transcripts in Escherichia coli independent of ribonuclease E. Nucleic Acids Res 2007; 35(22):7614-7625.
10. Mans RM, Guerrier-Takada C, Altman S et al. Interaction of RNase P from Escherichia coli with pseudoknotted structures in viral RNAs. Nucleic Acids Res 1990; 18(12):3479-3487.
11. Bai Y, Rider PJ, Liu F. Catalytic M1GS RNA as an antiviral agent in animals. Methods Mol Biol 2010; 629:339-353.
12. Seif E, Altman S. RNase P cleaves the adenine riboswitch and stabilizes pbuE mRNA in Bacillus subtilis. RNA 2008; 14(6):1237-1243.
13. Altman S, Wesolowski D, Guerrier-Takada C et al. RNase P cleaves transient structures in some riboswitches. Proc Natl Acad Sci U S A 2005; 102(32):11284-11289.
14. De Lay N, Gottesman S. The Crp-activated small noncoding regulatory RNA CyaR (RyeE) links nutritional status to group behavior. J Bacteriol 2009; 191(2):461-476.
15. Gottesman S, Storz G. Bacterial small RNA regulators: versatile roles and rapidly evolving variations. Cold Spring Harb Perspect Biol 2010; [Epub ahead of print].
16. Waters LS, Storz G. Regulatory RNAs in bacteria. Cell 2009; 136(4):615-628.
17. Mandin P, Gottesman S. Integrating anaerobic/aerobic sensing and the general stress response through the ArcZ small RNA. EMBO J 2010; 29(18):3094-3107.
18. Martinez-Antonio A, Janga SC, Thieffry D. Functional organisation of Escherichia coli transcriptional regulatory network. J Mol Biol 2008; 381(1):238-247.
19. Gorke B, Vogel J. Noncoding RNA control of the making and breaking of sugars. Genes & development 2008; 22(21):2914-2925.
20. Masse E, Salvail H, Desnoyers G et al. Small RNAs controlling iron metabolism. Curr Opin Microbiol 2007; 10(2):140-145.
21. Svenningsen SL, Tu KC, Bassler BL. Gene dosage compensation calibrates four regulatory RNAs to control Vibrio cholerae quorum sensing. EMBO J 2009; 28(4):429-439.
22. Shimoni Y, Friedlander G, Hetzroni G et al. Regulation of gene expression by small non-coding RNAs: a quantitative view. Mol Syst Biol 2007; 3:138.

23. Papenfort K, Said N, Welsink T et al. Specific and pleiotropic patterns of mRNA regulation by ArcZ, a conserved, Hfq-dependent small RNA. Mol Microbiol 2009; 74(1):139-158.
24. Soper T, Mandin P, Majdalani N et al. Positive regulation by small RNAs and the role of Hfq. Proc Natl Acad Sci U S A 2010; 107(21):9602-9607.
25. Hardwick SW, Chan VS, Broadhurst RW et al. An RNA degradosome assembly in Caulobacter crescentus. Nucleic Acids Res 2010; 39(4):1449-59.
26. Janga SC, Babu MM. Transcript stability in the protein interaction network of Escherichia coli. Mol Biosyst 2009; 5(2):154-162.
27. Worrall JA, Gorna M, Crump NT et al. Reconstitution and analysis of the multienzyme Escherichia coli RNA degradosome. J Mol Biol 2008; 382(4):870-883.
28. Evguenieva-Hackenberg E, Klug G. RNA degradation in Archaea and Gram-negative bacteria different from Escherichia coli. Prog Mol Biol Transl Sci 2009; 85:275-317.
29. Kaberdin VR, Lin-Chao S. Unraveling new roles for minor components of the E. coli RNA degradosome. RNA Biol 2009; 6(4):402-405.
30. Monod J. Recherches sur la croissance des cultures bactériennes. Paris: Hermann & cie; 1942.
31. Fujita Y. Carbon catabolite control of the metabolic network in Bacillus subtilis. Biosci Biotechnol Biochem 2009; 73(2):245-259.
32. Sonnleitner E, Abdou L, Haas D. Small RNA as global regulator of carbon catabolite repression in Pseudomonas aeruginosa. Proc Natl Acad Sci U S A 2009; 106(51):21866-21871.
33. Vogel J, Papenfort K. Small non-coding RNAs and the bacterial outer membrane. Curr Opin Microbiol 2006; 9(6):605-611.
34. Figueroa-Bossi N, Valentini M, Malleret L et al. Caught at its own game: regulatory small RNA inactivated by an inducible transcript mimicking its target. Genes & development 2009; 23(17):2004-2015.
35. Wadler CS, Vanderpool CK. Characterization of homologs of the small RNA SgrS reveals diversity in function. Nucleic Acids Res 2009; 37(16):5477-5485.
36. Reichenbach B, Gopel Y, Gorke B. Dual control by perfectly overlapping sigma 54- and sigma 70- promoters adjusts small RNA GlmY expression to different environmental signals. Mol Microbiol 2009; 74(5):1054-1070.
37. Winkler WC, Nahvi A, Roth A et al. Control of gene expression by a natural metabolite-responsive ribozyme. Nature 2004; 428(6980):281-286.
38. McCarthy TJ, Plog MA, Floy SA et al. Ligand requirements for glmS ribozyme self-cleavage. Chem Biol 2005; 12(11):1221-1226.
39. Collins JA, Irnov I, Baker S et al. Mechanism of mRNA destabilization by the glmS ribozyme. Genes Dev 2007; 21(24):3356-3368.
40. Abe T, Sakaki K, Fujihara A et al. tmRNA-dependent trans-translation is required for sporulation in Bacillus subtilis. Mol Microbiol 2008; 69(6):1491-1498.
41. Wower J, Wower IK, Zwieb C. Making the jump: new insights into the mechanism of trans-translation. J Biol 2008; 7(5):17.
42. Une M, Kurita D, Muto A et al. Trans-translation by tmRNA and SmpB. Nucleic Acids Symp Ser (Oxf) 2009; (53):305-306.
43. Hayes CS, Keiler KC. Beyond ribosome rescue: tmRNA and co-translational processes. FEBS Lett 2010; 584(2):413-419.
44. Liu Y, Wu N, Dong J et al. SsrA (tmRNA) acts as an antisense RNA to regulate Staphylococcus aureus pigment synthesis by base pairing with crtMN mRNA. FEBS Lett 2010; 584(20):4325-4329.
45. Kuo HK, Krasich R, Bhagwat AS et al. Importance of the tmRNA system for cell survival when transcription is blocked by DNA-protein cross-links. Mol Microbiol 2010; 78(3):686-700.

CHAPTER 15

LOCALIZATION OF THE BACTERIAL RNA INFRASTRUCTURE

Kenneth C. Keiler
Department of Biochemistry and Molecular Biology, Penn State University, Althouse Laboratory, University Park, Pennslyvania, USA
Email: kkeiler@psu.edu

Abstract: The bacterial RNA network includes most of the same components found in eukaryotes, and many of the interactions that underlie transcription, RNA processing and stability, translation, and protein secretion are conserved. The major difference is that all of these functions take place in a single cellular compartment. In spite of the absence of membrane-bound organelles, or in some cases because of it, key components of the RNA network are localized to specific subcellular spaces or structures to ensure proper processing and regulation. This chapter focuses on what is known about subcellular localization of the bacterial RNA network and what insights localization provides to regulation of the RNA infrastructure of the cell.

INTRODUCTION

In bacteria, as in eukaryotes, most of the RNA infrastructure is dedicated to gene expression and the majority of transcripts are rRNA, tRNA, and mRNA. Although there is little processing of mRNA in bacteria, rRNA and tRNA are produced as longer transcripts that are processed by a variety of endo- and exonucleases. Other bacterial RNAs, such as sRNA, 6S RNA, CRISPR RNA, and tmRNA, are also involved in regulating transcription or translation. Although the enzymatic processes of transcription and translation are the same in bacteria and eukaryotes, the spatial distribution of these processes is very different. In bacteria, transcription and translation both occur in the cytoplasm, and ribosomes typically initiate translation of an mRNA before it is completely transcribed. Therefore, it would not be surprising if the RNA infrastructure was more or less uniformly distributed throughout the bacterial cell. In fact, some components of

RNA Infrastructure and Networks, edited by Lesley J. Collins.

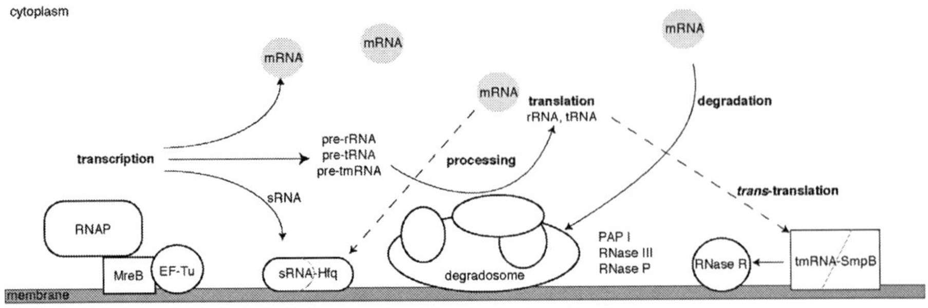

Figure 1. Location of RNA infrastructure in bacteria. A schematic cross section of the cell is shown with the membrane at the bottom. Transcription of RNAs is widely dispersed in the cytoplasm, although some RNA polymerase complexes are associated with the cytoskeletal protein MreB. Some mRNA diffuses through the cytoplasm, and some is concentrated in discrete spots. Processing of rRNA, tRNA, and tmRNA is performed by the degradosome, which is a component of the cytoskeleton, and by other nucleases which are also concentrated near the membrane. Translation occurs throughout the cytoplasm, but the sRNA-Hfq and tmRNA-SmpB ribonucleoprotein complexes that regulate translation are associated with the membrane or cytoskeletal structures. The major mRNA degrading activity is in the cytoskeletal degradosome complex, and RNase R, which degrades tmRNA, is also associated with the cytoskeleton.

the gene expression machinery, including ribosomes and RNA polymerase, are widely dispersed. However, some key factors, including the major RNA degradation machinery and translation factors, are localized in cytoskeletal structures, suggesting that the RNA infrastructure and bacterial cell architecture are intimately entwined (Fig. 1). This chapter will describe what is known about localization of the bacterial RNA infrastructure and speculate on the roles localization plays in regulating key RNA processes.

Bacteria are small compared to eukaryotic cells, and their size imposes restrictions on the extent of localization that is possible. Some eukaryotic RNA infrastructure components are localized to regions such as the nucleolus, which are approximately the same size as an entire bacterial cell. In contrast, the bacterial nucleoid occupies most of the cytoplasmic space, and most bacteria have no membrane-bound organelles. Nevertheless, bacteria do localize multicomponent structures and individual enzymes to specific subcellular addresses.[1-3] In addition, bacteria have cytoskeletal structures composed of homologs of actin, tubulin, intermediate filaments, and some unique bacterial cytoskeletal proteins.[2-4] It is not yet known how many functions the bacterial cytoskeleton shares with its eukaryotic counterpart, but it is clear that there is extensive subcellular organization in bacterial cells. RNA localization experiments have been performed predominantly in three bacterial model systems: the Gram-negative species *Escherichia coli*, the Gram-positive species *Bacillus subtilis*, and *Caulobacter crescentus*, a Gram-negative bacterium with an asymmetric cell morphology that facilitates localization studies. As described below, RNA infrastructure components in bacteria are frequently associated with the cytoskeleton or form cytoskeletal structures themselves.

LOCALIZED TRANSCRIPTION AND mRNA

Localization studies of RNA polymerase in *E. coli* and *B. subtilis* indicate that it is found throughout the nucleoid, suggesting that transcription activity is widely dispersed.[5,6] However, most transcripts that have been observed do not diffuse freely. Golding, et al., used a transcript containing repeated MS2-binding aptamer sequences and an MS2-GFP fusion protein to observe the spatial dynamics of single RNAs in live *E. coli* cells.[7] They identified three patterns of RNA localization and movement. Most of the transcripts did not move far during the time of the experiment, but moved randomly in a restricted spot, typically near the midpoint or quarter-points of the cell. These positions are significant because the mRNA was expressed from an F plasmid known to be concentrated in a similar pattern, suggesting that the mRNA is not moving far from its cognate gene.[7] A minority of transcripts diffused throughout the cell, and others moved as chains.[7] Subsequent imaging of individual mRNAs and their cognate genes in fixed *E. coli* and *Caulobacter crescentus* cells using fluorescence in situ hybridization (FISH) confirmed that the majority of each mRNA is concentrated in one spot near the site of transcription, consistent with limited diffusion.[8] It has been suggested that mRNAs may be anchored near the site of their transcription by some unknown process.[8] However, the simplest model to explain these data is that low-mobility mRNA is still tethered to the DNA by RNA polymerase, and the mRNAs diffusing throughout the cytoplasm are completed transcripts. The diffused transcripts are likely to be difficult to observe by FISH, but are evident in live cell studies. This model is attractive because many RNAs must have considerable mobility after transcription. For example, tRNAs, rRNAs, and regulatory RNAs are active at many locations in the cell so they must leave the site of their transcription. In addition, many secreted proteins are transported as they are being translated, so the mRNAs encoding these proteins must be able to move to the membrane. Nevertheless, many mRNAs may be localized near the site of their transcription for a significant portion of their lifetime.

Even though fluorescence experiments indicate that RNA polymerase can be found throughout the nucleoid, some or all of the RNA polymerase may be associated with the cytoskeleton. Experiments using *E. coli* proteins showed that RNA polymerase binds the cytoskeletal protein MreB, both in vitro and in vivo.[9] MreB is an actin homolog, which polymerizes into filaments that associate with the cytoplasmic membrane and span the length of the cell.[10-12] MreB is found in many species, but it is not yet known if the association with RNA polymerase is conserved. The interaction between MreB and RNA polymerase in *E. coli* is intriguing, because both MreB and RNA polymerase are required for correct chromosome segregation. It has been proposed that the force generated from transcription by RNA polymerase anchored to the MreB structure could push newly replicated chromosomes into different daughter cells.[9] If this model is correct, localized transcription would impact DNA segregation as well as protein synthesis, and would play a crucial role in cell biology.

SPATIAL CONTROL OF RNA DEGRADATION AND PROCESSING

Much of the RNA processing and degradation activities in bacteria are contained in a single complex, the degradosome, which is an integral part of the cytoskeleton (Fig. 2). The degradosome contains RNase E, the RhlB RNA helicase, polynucleotide phosphorylase (PNPase), and enolase.[13] This complex is responsible for initiating most of the mRNA degradation in bacteria, as well as participating in maturation of tRNA, rRNA, tmRNA, and RNase P RNA.[13,14] Electron microscopy studies initially showed that RNase E is concentrated near the membrane in *E. coli*,[15] and subsequent experiments with

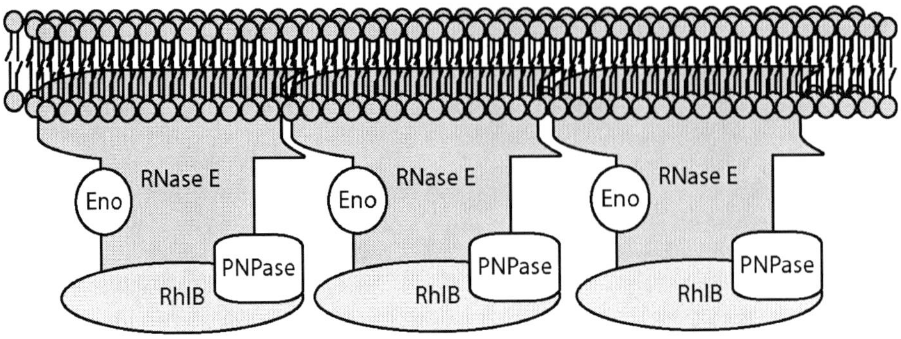

Figure 2. Localization of degradosome components. RNase E associates with the cytoplasmic membrane and polymerizes in the absence of other components. Enolase (Eno) localizes by binding to RNase E. RhlB binds RNase E, but can localize independently, suggesting that it also associates with another component of the cytoskeleton (not shown). PNPase binds RNase E and RhlB, and either interaction is sufficient to localize PNPase.

fluorescent protein fusions to RNase E and other degradosome components showed the characteristic pattern of regularly spaced spots or bands indicating a helical structure.[16,17] The degradosome filaments resemble structures formed by MreB and MinD, a cytoskeletal component in E. coli. However, studies with MinD mutants and inhibitors of MreB polymerization indicate that the degradosome structure forms independently these cytoskeletal elements.[16,17] A single domain of RNase E, distinct from the catalytic site and the sequences that bind other degradosome components, inserts in the cytoplasmic membrane.[18] Deletion of this domain from RNase E eliminates localization of the degradosome and has severe physiological consequences, discussed below.[17,19] RhlB localizes in a helical pattern independently of RNase E, suggesting that it also contains a membrane targeting sequence or that it interacts with other cytoskeletal components as well as the degradosome.[16,17] PNPase is localized through interactions with RNase E and RhlB, and enolase is localized by binding RNase E.[16,17]

The cytoskeletal structure formed by the degradosome is critical for normal cellular physiology. Cells in which degradosomes are not localized due to a deletion of the membrane targeting sequence of RNase E have severe defects in growth rate, cell division, chromosome segregation, cell morphology, and autoregulation of RNase E expression.[16,17] Although there are several possible explanations for these phenotypes, the most attractive hypothesis is that degradosome localization to the cytoskeleton is required for normal mRNA degradation and RNA processing activities, and disrupting these processes causes pleiotropic effects.[16,17,19] Cytoskeletal localization of the degradosome might provide a site for mRNA destruction to facilitate removal of particular mRNA species. In fact, over-expressed transcripts localize in a pattern very similar to that observed for the degradosome.[20] A second, nonexclusive explanation would be to sequester the nuclease activity from the site of gene expression to limit unwanted mRNA turnover. The helical structure would then provide a supply of degradosome activity that is not too distant from any point in the cell, but is nonetheless confined from free diffusion. Alternatively, the structure could be for storage of excess degradosome complexes, or could be important for RNA processing but not mRNA degradation.

Other mRNA degradation and RNA processing activities are also localized. Poly(A) polymerase, RNase III and RNase P are localized near the membrane in *E. coli*,[21,22] and RNase R is localized in a helix-like structure in *C. crescentus*.[23] The importance of localizing these enzymes is not known, but it appears that most ribonucleases are not found distributed throughout the cell and may be sequestered from nontarget RNAs.

EF-Tu IS A COMPONENT OF THE CYTOSKELETON

Translation and transcription occur simultaneously in the bacterial cytoplasm, and there have been hints that translation may be spatially organized. Fusions of fluorescent proteins to ribosomal proteins L1, S2 and L7/L12 accumulate preferentially at the cell poles and near the membrane in *E. coli* and *B. subtilis*.[5,6,24] However, it is not known what percentage of the fusion proteins are incorporated into ribosomes and how many of the ribosomes are engaged in active translation, so interpretation of these results is complicated. Cryoelectron tomography studies indicate that 70S ribosomes are distributed throughout the cytoplasm, with 15% near the membrane.[4,25] Therefore, it is likely that translation is widely dispersed in the cell.

On the other hand, EF-Tu, a key translation component and RNA-binding protein is clearly part of the bacterial cytoskeleton. EF-Tu polymerizes and colocalizes with MreB in helical filaments.[26] Several results indicate that MreB and EF-Tu form a complex that has a function distinct from protein synthesis.[26] The concentration of EF-Tu is much higher than that required for translation, and partial depletion of EF-Tu causes mislocalization of MreB, but does not affect protein synthesis. Conversely, in the absence of MreB, EF-Tu is not localized but translation is not affected. Finally, EF-Tu and MreB bind in vitro and in vivo, and EF-Tu alters the curvature of MreB filaments. Because MreB filaments are dynamic, forming and depolymerizing within the cell, but EF-Tu structures are static, it has been proposed the EF-Tu forms the tracks that control placement of MreB filaments.[26] Although the role of EF-Tu in the cytoskeleton appears to be distinct from its interactions with the ribosome and tRNAs, it does suggest an ancient connection between the RNA infrastructure and the cytoskeleton. Consistent with this idea, eEF1A, the eukaryotic homolog of EF-Tu, binds to actin in eukaryotic cells.[27,28]

LOCALIZATION OF *TRANS*-TRANSLATION COMPONENTS

The tmRNA-SmpB complex is also localized to helices in the cell. tmRNA is a specialized RNA that has properties of both a tRNA and an mRNA.[29,30] The tRNA like structure is aminoacylated by alanine-tRNA synthetase, and is bound by SmpB and EF-Tu. A distinct part of tmRNA contains an open reading frame that encodes a peptide with multiple proteolytic determinants. The tmRNA-SmpB complex can enter translating ribosomes that have stalled near the end of the mRNA and initiate a reaction known as *trans*-translation. The alanine-charged tmRNA acts like a tRNA to accept the nascent polypeptide. The mRNA is released, and the tag reading frame of tmRNA is used as a message to translate the tag peptide onto the C terminus of the nascent polypeptide. This reaction targets the newly-made protein for rapid degradation, and releases the stalled translational complex. *trans*-Translation is ubiquitous in bacteria, as genes encoding tmRNA and SmpB have been found in all sequenced bacterial genomes, and in *E. coli* approximately 1 in 250 translation initiation reactions ends in *trans*-translation.[29,30] FISH, immunofluorescence, and GFP fusion studies showed that the tmRNA-SmpB complex is localized in a helical structure in *Caulobacter crescentus* and *Shigella flexneri*.[23] This structure is not colocalized with MreB and does not require MreB filaments. It is not known if tmRNA-SmpB forms these structures independently or through interaction with another cytoskeletal component. Inactive mutants of tmRNA are still localized, suggesting that localization is not due to interaction with stalled ribosomes. In *C. crescentus*, tmRNA is specifically degraded as a function of the cell cycle. This degradation is accomplished by RNase R, a highly conserved ribonuclease. RNase R is also localized in a helical structure in *C. crescentus*, but the tmRNA-SmpB and RNase R structures do not intersect.[23] Therefore, it is possible that one function of tmRNA localization in *C. crescentus* to prevent inappropriate degradation by RNase R. Localization might also serve to sequester the tmRNA-SmpB from productive translational complexes, or to increase the concentration of tmRNA-SmpB in the vicinity of stalled translational complexes.

SPATIAL CONTROL OF GENE EXPRESSION

Small RNAs that bind the RNA chaperone Hfq are used to silence gene expression in bacteria, and localization is required for at least some of the silencing activity. sRNAs have partial complementarity to target mRNAs, and bind to the mRNA to rapidly shut off gene expression and promote degradation of the mRNA.[31] Hfq prevents degradation of sRNAs and promotes base pairing with target mRNAs. GFP fusion studies and electron microscopy have shown that Hfq is localized in close proximity to the membrane in *E. coli*.[6,32] Because sRNAs regulate the expression of many membrane proteins and the mRNAs encoding these proteins are localized to the membrane by cotranscriptional secretion, Hfq localization might be due in part to interactions with target mRNAs. However, silencing by at least one sRNA, SrgS, suggests that localization regulates sRNA-Hfq activity. SgrS binds to the 5' UTR of the *ptsG* mRNA, which encodes the glucose transporter $EIICB^{Glc}$.[33] Stress caused by excess glucose-6-phosphate induces SgrS to inhibit translation of $EIICB^{Glc}$ and promote degradation of *ptsG* mRNA by RNase E, thereby preventing further accumulation of phosphosugars.[33,34] Kawamoto, et al., showed that SgrS silencing of *ptsG* was eliminated by mutations that prevent secretion of the encoded protein.[35] SgrS activity on *ptsG* could be restored by replacing the membrane targeting sequence of $EIICB^{Glc}$ with the membrane targeting sequence from LacY, an unrelated membrane protein.[35] These observations suggest that the *ptsG* mRNA must be near the membrane to interact with SgrS, and therefore it is likely that SgrS-Hfq is localized independently of interactions with the target mRNA. It is not yet known whether sRNA-Hfq complexes are localized to regulate silencing activity, or to facilitate interactions with the degradosome.

CONCLUSION

The field of bacterial cell biology is relatively new, and many specialized RNAs have only recently been discovered, yet several interactions between the RNA infrastructure and the cytoskeleton have already been established. The best data to date suggests that most of the transcription and translation activity is not localized, although some of the machinery may interact with actin-like proteins. More compelling data indicate that systems that degrade mRNA and resolve stalled translational complexes are localized. The reasons for RNA infrastructure localization are generally not known, but both the degradosome and the tmRNA-SmpB complex might be localized to limit destructive interactions with productive RNAs. It remains to be seen if sequestration is a general rationale for localization. The localization of RNAs to regulate gene expression and the intimate connections between the RNA infrastructure and the cytoskeleton suggest that the role of localization in the RNA infrastructure is just beginning to be understood.

REFERENCES

1. Shapiro L, McAdams HH, Losick R. Why and how bacteria localize proteins. Science 2009; 326:1225-1228.
2. Thanbichler M, Shapiro L. Getting organized--how bacterial cells move proteins and DNA. Nat Rev Microbiol 2008; 6:28-40.
3. Pogliano J. The bacterial cytoskeleton. Curr Opin Cell Biol 2008; 20:19-27.
4. Briegel A, Dias DP, Li Z et al. Multiple large filament bundles observed in Caulobacter crescentus by electron cryotomography. Mol Microbiol 2006; 62:5-14.
5. Lewis PJ, Thaker SD, Errington J. Compartmentalization of transcription and translation in Bacillus subtilis. EMBO J 2000; 19:710-718.
6. Azam TA, Hiraga S, Ishihama A. Two types of localization of the DNA-binding proteins within the Escherichia coli nucleoid. Genes Cells 2000; 5:613-626.
7. Golding I, Cox EC. RNA dynamics in live Escherichia coli cells. Proc Natl Acad Sci USA 2004; 101:11310-11315.
8. Llopis PM, Jackson AF, Sliusarenko O et al. Spatial organization of the flow of genetic information in bacteria. Nature 2010; 466:77-81.
9. Kruse T, Blagoev B, Lobner-Olesen A et al. Actin homolog MreB and RNA polymerase interact and are both required for chromosome segregation in Escherichia coli. Genes Dev 2006; 20:113-124.
10. Gitai Z, Shapiro L. Bacterial cell division spirals into control. Proc Natl Acad Sci USA 2003; 100:7423-7424.
11. Kruse T, Moller-Jensen J, Lobner-Olesen A et al. Dysfunctional MreB inhibits chromosome segregation in Escherichia coli. EMBO J 2003; 22:5283-5292.
12. Jones LJ, Carballido-Lopez R, Errington J. Control of cell shape in bacteria: helical, actin-like filaments in Bacillus subtilis. Cell 2001; 104:913-922.
13. Carpousis AJ. The RNA degradosome of Escherichia coli: an mRNA-degrading machine assembled on RNase E. Annu Rev Microbiol 2007; 61:71-87.
14. Kennell D. Processing endoribonucleases and mRNA degradation in bacteria. J Bacteriol 2002; 184:4645-4657.
15. Liou GG, Jane WN, Cohen SN et al. RNA degradosomes exist in vivo in Escherichia coli as multicomponent complexes associated with the cytoplasmic membrane via the N-terminal region of ribonuclease E. Proc Natl Acad Sci USA 2001; 98:63-68.
16. Taghbalout A, Rothfield L. RNaseE and RNA helicase B play central roles in the cytoskeletal organization of the RNA degradosome. J Biol Chem 2008; 283:13850-13855.
17. Taghbalout A, Rothfield L. RNaseE and the other constituents of the RNA degradosome are components of the bacterial cytoskeleton. Proc Natl Acad Sci USA 2007; 104:1667-1672.
18. Khemici V, Poljak L, Luisi BF et al. The RNase E of Escherichia coli is a membrane-binding protein. Mol Microbiol 2008; 70:799-813.
19. Taghbalout A, Rothfield L. New insights into the cellular organization of the RNA processing and degradation machinery of Escherichia coli. Mol Microbiol 2008; 70:780-782.
20. Valencia-Burton M, Shah A, Sutin J et al. Spatiotemporal patterns and transcription kinetics of induced RNA in single bacterial cells. Proc Natl Acad Sci USA 2009; 106:16399-16404.
21. Miczak A, Srivastava RA, Apirion D. Location of the RNA-processing enzymes RNase III, RNase E and RNase P in the Escherichia coli cell. Mol Microbiol 1991; 5:1801-1810.
22. Jasiecki J, Wegrzyn G. Localization of Escherichia coli poly(A) polymerase I in cellular membrane. Biochem Biophys Res Commun 2005; 329:598-602.
23. Russell J, Keiler K. Subcellular localization of a bacterial regulatory RNA. Proc Natl Acad Sci USA 2009; 106:16405-16409.
24. Mascarenhas J, Weber MH, Graumann PL. Specific polar localization of ribosomes in Bacillus subtilis depends on active transcription. EMBO Rep 2001; 2:685-689.

25. Ortiz JO, Forster F, Kurner J et al. Mapping 70S ribosomes in intact cells by cryoelectron tomography and pattern recognition. J Struct Biol 2006; 156:334-341.
26. Defeu Soufo HJ, Reimold C et al. Bacterial translation elongation factor EF-Tu interacts and colocalizes with actin-like MreB protein. Proc Natl Acad Sci USA 2010; 107:3163-3168.
27. Gross SR, Kinzy TG. Translation elongation factor 1A is essential for regulation of the actin cytoskeleton and cell morphology. Nat Struct Mol Biol 2005; 12:772-778.
28. Liu G, Tang J, Edmonds BT et al. F-actin sequesters elongation factor 1alpha from interaction with aminoacyl-tRNA in a pH-dependent reaction. J Cell Biol 1996; 135:953-963.
29. Keiler KC. Biology of trans-Translation. Annu Rev Microbiol 2008; 62:133-151.
30. Moore SD, Sauer RT. The tmRNA system for translational surveillance and ribosome rescue. Annu Rev Biochem 2007; 76:101-124.
31. Aiba H. Mechanism of RNA silencing by Hfq-binding small RNAs. Curr Opin Microbiol 2007; 10:134-139.
32. Diestra E, Cayrol B, Arluison V et al. Cellular electron microscopy imaging reveals the localization of the Hfq protein close to the bacterial membrane. PLoS One 2009; 4:e8301.
33. Vanderpool CK, Gottesman S. Involvement of a novel transcriptional activator and small RNA in post-transcriptional regulation of the glucose phosphoenolpyruvate phosphotransferase system. Mol Microbiol 2004; 54:1076-1089.
34. Kimata K, Tanaka Y, Inada T et al. Expression of the glucose transporter gene, ptsG, is regulated at the mRNA degradation step in response to glycolytic flux in Escherichia coli. EMBO J 2001; 20:3587-3595.
35. Kawamoto H, Morita T, Shimizu A et al. Implication of membrane localization of target mRNA in the action of a small RNA: mechanism of post-transcriptional regulation of glucose transporter in Escherichia coli. Genes Dev 2005; 19:328-338.

CHAPTER 16

SMALL RNA DISCOVERY AND CHARACTERISATION IN EUKARYOTES USING HIGH-THROUGHPUT APPROACHES

Helio Pais, Simon Moxon, Tamas Dalmay and Vincent Moulton*

School of Computing Sciences, University of East Anglia, Norwich, UK

**Corresponding Author: Vincent Moulton—Email: vincent.moulton@cmp.uea.ac.uk*

Abstract: RNA silencing is a mechanism of genetic regulation that is mediated by short noncoding RNAs, or small RNAs (sRNAs). Regulatory interactions are established based on nucleotide sequence complementarity between the sRNAs and their targets. The development of new high-throughput sequencing technologies has accelerated the discovery of sRNAs in a variety of plants and animals. The use of these and other high-throughput technologies, such as microarrays, to measure RNA and protein concentrations of gene products potentially regulated by sRNAs has also been important for their functional characterisation. mRNAs targeted by sRNAs can produce new sRNAs or the protein encoded by the target mRNA can regulate other mRNAs. In either case the targeting sRNAs are parts of complex RNA networks therefore identifying and characterising sRNAs contribute to better understanding of RNA networks. In this chapter we will review RNA silencing, the different types of sRNAs that mediate it and the computational methods that have been developed to use high-throughput technologies in the study of sRNAs and their targets.

INTRODUCTION

Biological systems are regulated at multiple layers through a myriad of mechanisms. At the cellular level, normal function requires regulation of gene expression. One of the systems eukaryotic cells have in place to accomplish this task is RNA silencing, a process in which a complex formed by an RNA molecule and one or more proteins interacts either with a different RNA or DNA, causing modifications in the rates of translation

RNA Infrastructure and Networks, edited by Lesley J. Collins .

or transcription. The RNA molecules present usually contain less than 30 nucleotides and are commonly called small RNAs (sRNAs). In this chapter we will adopt this convention, although we note that there are other noncoding small RNAs that are not involved in RNA silencing (e.g., tRNAs, snoRNAs) but they will not be described here. In prokaryotic organisms there are also noncoding RNAs that regulate gene expression, but the mechanisms by which they act is completely different from eukaryotic RNA silencing and their review falls outside the scope of this chapter. For reviews on this topic see, for example, references 1-3.

One of the most important factors in the explosive growth of knowledge in the RNA silencing field has been the application of new DNA sequencing technologies. In comparison to conventional Sanger sequencing, these new technologies are characterised by producing shorter reads with very high-throughput. Researchers in the RNA silencing field have been among the first to adopt high-throughput sequencing approaches because the main limitation of the technique, i.e., the short read length, is irrelevant for the analysis of small RNA (sRNA) which are usually shorter than 30 nucleotides. High-throughput sequencing and other high-throughput methods, such as microarrays, have also been important both to profile the expression of sRNAs and transcripts potentially regulated by sRNAs (sRNA targets). The latter has been particularly useful in animal systems, where the interaction between most sRNAs and their targets is mediated by a small number of nucleotides (as few as six). Consequently, accurate computational prediction of sRNA targets is quite challenging. In this chapter we will review the field of RNA silencing, describe the small RNAs that are involved in this process and how high-throughput technologies have been used to study their biological function.

THE MECHANISMS OF RNA SILENCING

The first sRNA to be discovered, lin-4, was found in a genetic screen to study developmental defects in the worm *C. elegans*.[4,5] In plants the discovery of sRNAs was made by researchers working on viral defence.[6,7] Meanwhile, work was being done on the delivery of exogenous RNA molecules with the goal of repressing gene expression.[8] The use of double stranded RNA molecules to specifically and strongly repress gene expression led to the award of a Nobel Prize in medicine less than ten years after it was first uncovered.[9]

Since then, a multitude of different sRNA classes have been characterised, most of which are derived from longer, double stranded or fold-back RNA precursors. RNAaseIII-type enzymes called Dicers are able to recognise these precursors and process them to produce double stranded sRNAs. These are then incorporated into an Argonaute protein, where one of the strands is degraded. The other strand, called the guide strand, remains incorporated into the Argonaute, which in conjunction with other proteins forms an effector complex capable of recognizing a specific DNA or RNA target.[10] Once the effector complex has bound the target, it is able to either induce target mRNA cleavage, mRNA destabilisation without cleavage, inhibit protein translation or cause DNA and histone modifications that lead to transcriptional silencing.[11-13] The nature of the precursor, of the subsequent processing steps and of the target (DNA, mRNA or other sRNAs) determines the class a sRNA belongs to (Fig. 1).

sRNAs involved in RNA silencing have been described in a large number of eukaryotes. However, the small RNA landscapes are quite diverse among different phyla. Although some of the RNA silencing machinery is conserved between most eukaryotes, to date no

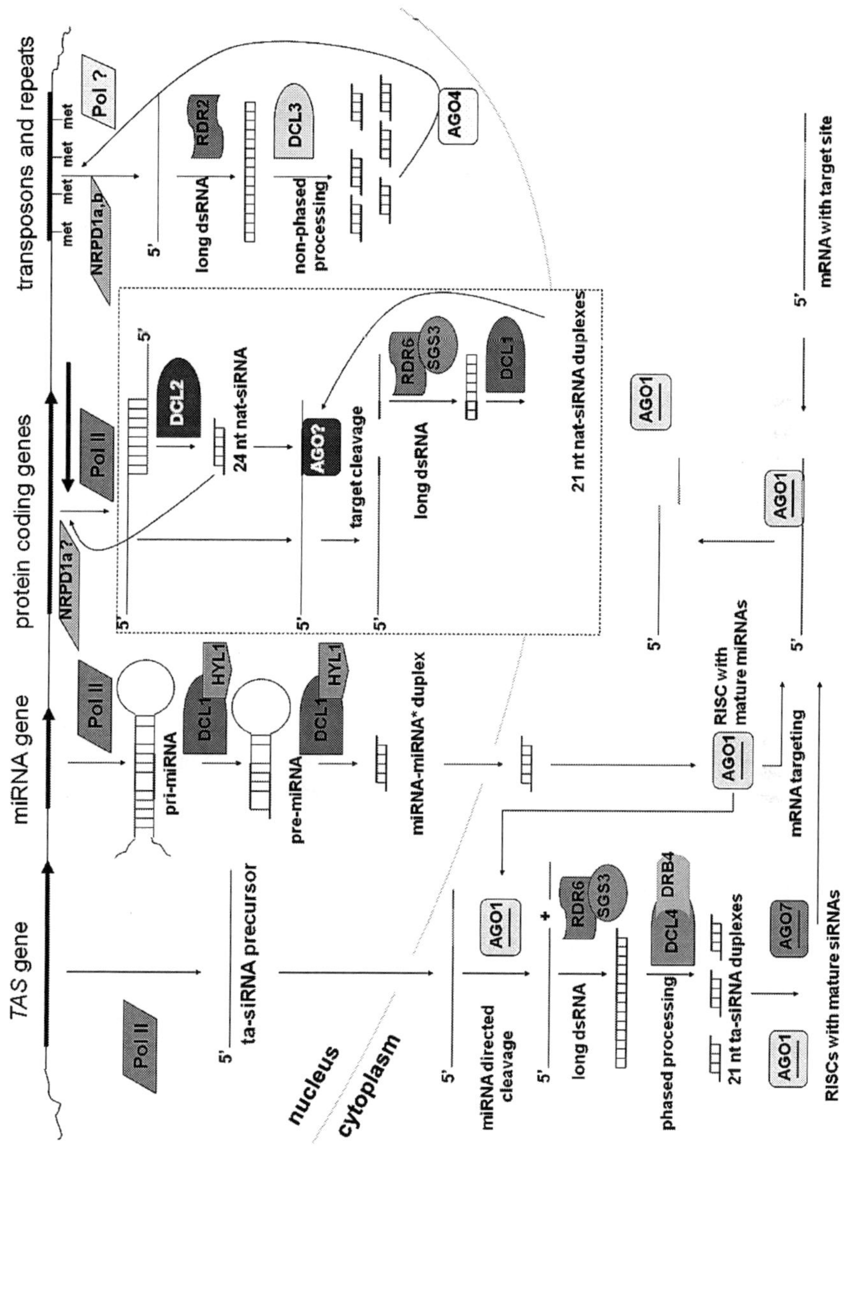

Figure 1. Small RNA production pathways. Reprinted with permission from Phillips JR, Dalmay T, Bartels D. FEBS Letters 2007; 581(19):3592-3597.

individual small RNA conserved between animals and plants has been discovered.[14,15] There are also differences in the heterogeneity of sRNAs: in plants there is a much greater diversity in sRNA producing mechanisms than in animals.[7]

miRNAs derive from precursors characterised by a secondary structure in which most nucleotides form a single stem. miRNAs are present in animals and plants but there are significant structural differences between them.[15-17] Generally, pre-miRNAs are longer and form a greater number of base pairs in plants than in animals. The biogenesis of miRNAs has been particularly well studied. This process begins with the transcription of the genomic locus where the miRNA is encoded, continues with a series of processing steps of the resulting RNA transcript and culminates with the incorporation of the mature miRNA into a silencing complex. We refer the reader to the many review articles that extensively describe miRNA biogenesis in animals,[18-21] plants[17,22] and both.[15,23]

Small interfering RNAs (siRNAs) are also present in animals and plants and again there are many differences in their biogenesis and modes of action. siRNAs originate from a longer single stranded precursor that, in plants, is turned into double stranded form by an RNA-dependent RNA polymerase (RdRp) before being processed by a Dicer enzyme.[10,14,24,25] siRNAs can target both the loci where they are produced from and other loci with high sequence homology. siRNAs can act both by DNA transcriptional silencing and RNA degradation. A sub-class of siRNAs, called natural antisense siRNAs (nat-siRNAs) is produced from partially overlapping pairs of transcripts originating from opposing strands of DNA. Production of nat-siRNAs has so far only been observed in stress response. In animals less is known about the biogenesis and identity of the targets of siRNAs. At least a subset of them are produced from overlapping transcripts and seem to be involved in transposon silencing.[26]

Piwi-interacting RNAs (piRNAs) constitute a different class of sRNAs that interact with a sub-family of the Argonaute proteins called Piwi proteins.[14,27] piRNAs are thought to be present exclusively in animals, seem to be expressed only in the germline and contribute to the stability of the cell line by silencing transposons. piRNAs are usually 24-32 nucleotides long and most of them are encoded in repetitive regions of the genome.

RNA silencing requires both an sRNA and an effector complex, formed by a set of proteins including a member of the Argonaute family. There are mainly two types of effector complex: RISCs (RNA induced silencing complexes) and RITSs (RNA-induced initiation of transcriptional silencing). RITSs, as the name indicates, are involved in transcriptional silencing, by promoting the formation of heterochromatin.[13] RISCs mediate posttranscriptional regulation and act by binding to messenger RNAs and either promoting the mRNA's degradation or influencing the rate of the mRNA's translation.[12] Each sRNA guides the RISC to a specific region of an mRNA known as the target site based on complementarity between the sRNA and its target. The rules governing the interaction between an sRNA-containing RISC and an mRNA are not currently fully understood. For example, not only it is infeasible to reliably predict an mRNA/sRNA interaction, it is also not always possible to say, *a priori*, what effect an sRNA has on a target mRNA. The four known outcomes of sRNA targeting are: cleavage of target mRNA, accelerated degradation of target mRNA, repression of translation (without mRNA degradation) and enhancement of translation.[28]

The biological processes regulated by each sRNA depend on its set of targets. For this reason a large effort has been made to extensively identify sRNA targets and in particular miRNA targets. sRNAs guide the effector complexes to their targets. The target recognition is based on nucleotide sequence complementarity. The degree of complementarity partially determines the targeting outcome; a high degree of complementarity (more common in plants) normally leads to target cleavage, a low degree of complementarity (more common in animals) leads to decreased rates of translation or increased rates of target degradation.[10] Based on experimentally validated sRNA targets sets of empirically derived rules have been created, both for high-complementarity target sites[29,30] and low-complementarity target sites.[31,32]

Given the widespread effect of sRNA guided gene expression regulation, it comes as no surprise that the expression of sRNA genes themselves is tightly regulated. As with protein coding genes, control can be exerted at many levels and with distinct levels of specificity.[33] One level of regulation takes place during the biogenesis of the sRNAs. Namely, the availability of the proteins required for the formation of a certain class of sRNAs will limit the level of those sRNAs. For example, in some cancer tissues an increased number of copies of Drosha, a gene involved in miRNA biogenesis, has been observed leading to widespread over-expression of miRNAs.[34] Another example is the control of the expression level of one of the Dicer proteins (DCL1) by a miRNA (miR-162). Because the production rate of the miRNA depends on the level of Dicer these interactions form a negative feedback loop.[35]

Another factor that influences sRNA activity is the presence of other RNA molecules that might compete with the sRNAs for one or more of these proteins. In *Arabidopsis* some noncoding RNAs whose secondary structure resembles that of pre-miRNAs seem to compete for one protein involved in plant miRNA biogenesis, HYL1.[36] An interesting consequence of the importance of the limited availability of the proteins involved in miRNA activity is that each miRNA controls the activity level of all other miRNAs, albeit passively and indirectly. For example, in zebrafish embryos, where miRNA activity is essential to promote the degradation maternal of mRNA,[37] transfection with siRNA, reduces the activity of endogenous miRNAs, presumably because the siRNA competes with the miRNAs for Argonaute.[38] There are also many mechanisms of regulating the activity of individual sRNAs. As with other types of genes, there are many factors that control the transcription rate of sRNA-containing transcripts by binding to the respective promoter region.[39] In particular, the promoter regions of miRNAs have been extensively studied[40,41] and for a large number of miRNAs some of the proteins regulating transcription have been identified.[40]

USING MICROARRAYS AND DNA SEQUENCING TO MEASURE RNA

Much of the understanding of how sRNAs are produced and what functions they perform inside the cell is derived from measurements of the sRNAs themselves, of their respective precursors and of the transcripts regulated by them. Techniques to measure the abundance of RNA molecules, such as RT-qPCR and northern blotting have been used for more than thirty years and are still used to verify results obtained through other means.[42,43] However, these methods are very time consuming and therefore can be used to study only a limited number of RNAs. Since the 1990s two high-throughput approaches that can be used to measure RNA, have been developed: microarrays and RNA sequencing (RNAseq).

Microarrays are based on the simultaneous hybridization of a large number of probes attached to a solid surface and the sample of interest.[44] The probes are designed so that each of them is complementary to a unique sequence in the sample, called the probe target. This approach has two limitations: it is possible to measure only molecules for which the sequence is known and it is possible that measurements are corrupted by cross-hybridization, that is, hybridization of a probe to a molecule other than its target. This can happen if the sample contains sequences highly similar to the target, in which case hybridization to the probe might still occur.

Another way of estimating RNA concentrations consists of sequencing a number of molecules in the sample and taking the number of times each molecule is sequenced as a measure of its abundance. The quality of the results obtained using this approach critically depends on the total number of reads that can be obtained from a single sample. The new high-throughput sequencing technologies developed over the last few years have increased this number by several orders of magnitude. The first high-throughput sequencing method to be developed is called massively parallel signature sequencing (MPSS)[45] but is now very seldom used. Currently three commercial platforms are widely used: Roche's 454/FLX system,[46] Illumina's Genome Analyzer[47] (formerly known as Solexa sequencing and succeeded by Illumina's more recent model the HiSeq 2000) and ABI's SOLiD.[48]

A known drawback of 454 sequencing is that the sequencing software uses signal intensity to determine the number of consecutive identical bases in a sequence. When multiple consecutive identical bases are encountered (especially four or more repeated bases) the software cannot reliably interpret the signal intensity (and therefore the number of bases read) which can lead to sequencing errors especially with low complexity sequences. Using the Genome Analyzer technology each nucleotide is sequenced individually, eliminating the problem that 454 technology has with homopolymeric sequences. However, the Genome Analyzer produces an increased number of errors at the 3′ end of longer sequences meaning that sequence length is a limiting factor of this technology. SOLiD sequencing is also affected by length but has an increased throughput compared to the other high-throughput platforms.

The sequencing approach overcomes the two main limitations of microarrays, with sequencing it is possible to measure both known and unknown products and in principle the measurements are independent. Furthermore, in high-throughput sequencing datasets it is possible to detect sequence variants. One problem of this method has been highlighted by two recent studies that indicate that there might be biases on the number of times a sequence is read.[49,50] That is, the expression levels, as measured by the count of a sequence, will be distorted because individual sequences have different propensities to be sequenced. Another potential disadvantage of this method is that the discovery of sRNAs is limited to ligation compatible sRNAs, in general with a 5′ mono-phosphate group and a 3′ hydroxyl group. This means that there are probably unknown sRNAs that have not been uncovered by this technology as of yet.

HIGH-THROUGHPUT APPROACHES FOR THE DISCOVERY AND FUNCTIONAL CLASSIFICATION OF sRNAs

sRNA Discovery

Before the advent of high-throughput sequencing methods some efforts were made to computationally predict miRNA genes.[30,51-55] Most of these methods try to find genomic regions that could produce RNAs with similar characteristics to those of precursor miRNAs. To find these regions secondary structure prediction algorithms are often employed.[56,57]

These algorithms return secondary structures that minimize the free energy of the RNA under a certain model. miRNA candidates display predicted secondary structures with characteristics similar to those of known miRNAs. For example, number and positions of the base pairs, number of nucleotides bridging the two arms of the stem or number and length of internal bulges in the stem.

Most methods predict many thousands of candidate miRNA sequences, indicating that such approaches suffer from a lack of specificity. To reduce the number of false positive predictions, many algorithms[30,53,54,58] employ a conservation rule, i.e., a candidate miRNA is only accepted if a homologue can be found in the genome of at least one other related species. This method of miRNA prediction and cross-species conservation checking has been successfully employed to find many novel miRNAs in both plants and animals with a high degree of accuracy. Although some miRNAs are conserved between closely related organisms, many have now been shown to be specific to individual taxonomic groups.[59,60-63] This discovery has exposed the limitations of comparative methods and has led to the need for alternative approaches to miRNA detection.

In the past, most miRNAs were formally identified using traditional Sanger sequencing after size fractionation (selecting for sequences ~20-22nt) and ligation into cloning vectors. This process was adopted in *Arabidopsis*, rice and poplar and comparison of miRNA sequences across plant families showed that the majority were conserved.[64] In animals similar studies led to the discovery of many miRNAs.[65-67] Recently, the high-throughput sequencing of sRNAs has led to the discovery of a plethora of new sRNAs, many of which are expressed at low levels and are either unique to a specific species or at least not widely conserved in related organisms. For example, this has been used to uncover miRNAs specific to the human brain,[68] QDE-2 associated RNAs (qiRNAs) in *Neurospora crassa*,[69] and endo-siRNAs in the fruit fly.[70] The use of high-throughput sequencing technologies removed the need of cloning prior to sequencing. MPSS was the first high-throughput sequencing method successfully used to discover a number of novel miRNAs in *Arabidopsis*.[71] Subsequently, 454 pyrosequencing,[59,61,63,72,73], Illumina's Genome Analyzer[74-76] and ABI's SOLiD[77,78] platform have been used to discover sRNAs.

Although high-throughput techniques have revolutionised sRNA sequencing they have led to new problems with data analysis. Previously, biologists would often manually work through small lists of sRNAs testing each for miRNA-like properties. Now, with millions of reads being produced by a single sequencing run, the need for computational techniques to process and classify sRNAs in a high-throughput manner has become apparent.

The first step of the computational analysis of this type of data consists of the identification of the genomic coordinates that could have generated each of the reads. This is done by matching the sequence reads to the genome; a procedure usually referred to as read mapping (or read aligning). A limited number of mismatches, insertions or deletions in the reads may be allowed in order to account for sequencing errors and genuine differences in relation to the reference genome, such as single nucleotide polymorphisms. Mapping millions of reads to a eukaryotic genome is a computationally intensive task and a number of software packages have been developed to perform this task optimally.[79] These new methods run in less time using less memory than other alignment tools, such as BLAST by preprocessing either the genome or the set of reads into indexes. Methods such as Bowtie and BWA[80,81] index the reference genome using a Burrows-Wheeler Transform. The PatMaN program indexes all reads into a suffix tree and uses a modified version of the Aho-Corasick algorithm to align the suffix tree to the genome.[82] Additionally, in contrast to BLAST, these methods are guaranteed to find all the matches between the set of reads and the reference genome. Usually in sRNA profiling, the reads are not assembled into contigs, but this step might be useful for the discovery of longer ncRNAs.

The genomic coordinates are subsequently used to extract regions of the genome of the same size as a typical miRNA precursor and compatible with the production of a mature miRNA in the same position where the read was mapped to. These regions are then subjected to the procedure previously described: prediction of secondary structure and comparison to secondary structures of known miRNAs. In 2008 two computational methods were released to predict miRNAs from high-throughput sequencing data: miRDeep[83] (specific to animal datasets) and miRCat[84] (at the time specific to plant datasets). Both miRDeep and miRCat show a high degree of specificity in comparison to purely computational de novo miRNA prediction algorithms. In 2009 one other software package, miRanalyser, was released.[85] This tool produces expression profiles of known miRNAs, other noncoding RNAs and predicts new miRNAs. The prediction of new miRNAs is performed using a machine learning approach and the set of known miRNAs as the test set. Because the set of predicted miRNAs does not include the known miRNAs it is difficult to assess the sensitivity of this method and to compare the quality of the predictions to those of miRCat or miRDeep. Additionally, there are two programs that specialise in profiling known miRNAs: miRprof[84] and miRExpress.[86] The existence of less abundant sequence variants of miRNAs, called isomiRs, was also uncovered using high-throughput sequencing data.[76] SeqBuster, a recently published software package[87] reports the relative abundance of the canonical miRNAs and the respective isomiRs. No doubt, by the time this chapter is in press there will be additional software packages for miRNA analysis as this is a rapidly growing field.

The recent explosion in high-throughput sequencing of sRNAs has led to a huge increase in the discovery of novel miRNAs in a wide variety of organisms. This is demonstrated by the rapid growth of the central miRNA repository miRBase[88] which has grown dramatically, from 719 entries in 2004 at the advent of high-throughput sRNA sequencing to 14,197 in the latest release (15.0). With next generation sequencing technologies becoming ever more powerful and more economical it is likely that many more miRNAs will be characterised over the coming years. The real challenge for biologists now is to try to discover functions of the thousands of miRNAs for which we currently know nothing about. The difficulties of target and functional characterisation will be covered later in this chapter.

The plant specific trans-acting siRNAs (ta-siRNAs) are another class of endogenous sRNA that can be identified using computational methods. ta-siRNAs are derived from a single-stranded RNA transcript which is targeted in two positions by a miRNA which is thought to trigger double strand formation by an RNA-dependent RNA-polymerase.[89] In double-stranded form, the precursor becomes a substrate for a Dicer enzyme (DCL4), whose progressive cleavage in 21 nt intervals leads to the "phased" pattern of mature sRNA production that is a hallmark of ta-siRNA loci. sRNA producing loci can be tested statistically in order to classify novel ta-siRNA producing regions.[90]

Other types of sRNAs are usually not so well defined by alignment patterns or sRNA sequence properties and are therefore currently difficult to classify computationally. Dicer is also known to produce sRNAs from long perfect or near-perfect double-stranded precursors in a more unpredictable pattern to that of the precisely defined miRNAs and ta-siRNAs.[91] This imprecise Dicer processing can give rise to highly complex sRNA loci, where large numbers of sRNA sequences are represented by low-abundance reads even in high-throughput data sets. In order to try to overcome problems when comparing

Table 1. Performance of sequencing technologies. Numbers obtained from manufacturer's websites on May 2010. The numbers for the high throughput platforms will change as there is very active product development on all platforms

Technology	Read Length	Nucleotides Read Per Day	Reference
Sanger sequencing (ABI3730)	Up to 900nt	Up to 3*10^6	http://www.appliedbiosystems.com
454/FLX (Titanium series)	Up to 500nt	Up to 10^9	http://454.com
Genome Analyzer (IIe)	Up to 100nt	Up to 2*10^9	http://www.illumina.com
SOLiD (version 4)	Up to 50nt	Up to 8*10^9	http://www.appliedbiosystems.com

Table 2. Bioinformatic software resources to analyse high-throughput sequencing data obtained from sRNAs. All software is freely available for noncommercial use

Name	Organism	Functionalities	Reference
miRDeep	Animals	Find miRNAs	83
miRCat	Plants and animals	Find miRNAs	84
miRanalyzer	Animals	Profile known sRNAs, find miRNAs	85
miRExpress	Animals and plants	Profile known miRNAs	80
miRProf	Animals and plants	Profile known miRNAs	84
pssRNAMiner	Plants	Prediction of ta-siRNA	121
Phasing detection tool	Plants	Prediction of ta-siRNA	84
SeqBuster	Animals	Analysis of isomiRs	87
SiLoCo	Plants	Identification of sRNA loci	84
NiBLS	Plants	Identification of sRNA loci	92

read counts between sRNAs in multiple samples it is often advisable to try to group sRNAs into transcriptional units or loci using either genome or transcript annotations. The SiLoCo method groups sRNAs such that each locus contains a minimum number of sRNAs and such that the gap between two consecutive sRNAs is below a maximum gap.[84] More recently the NiBLS algorithm, based on graph properties, has been developed.[92] This method builds a graph where its nodes are individual sRNAs and edges are created between sRNAs close to each other. The algorithm identifies subsets of nodes with high clustering coefficients. The sRNA loci are then defined by the minimum and maximum coordinates among the sequences in each subset.

Target Identification

Currently, the experimental validation of sRNA targets and hence function, is a very time-consuming and expensive process. In plants, sRNA mediated target site cleavage is widespread. The resulting cleavage products have two properties that facilitate their recognition. First, the cleavage point is very well defined: i.e., the nucleotides that are complementary to nucleotides 10 and 11 of the sRNA. Second, the 3′ cleavage product is generally stable. A consequence of the first property is that it is possible to distinguish between cleavage products and other mRNA degradation products. The implication of the second property is that it is possible to clone and sequence these cleavage products. Additionally, the 3′ cleavage product contains a 5′ mono-phosphate group which helps to ligate an RNA oligo, making it possible to identify the exact cleavage position.[93] Traditionally this has been done using a process called 5′ Rapid Amplification of cDNA Ends (5′ RACE). In essence this process allows the sequencing of cleavage products from the transcript predicted to be targeted by a given sRNA. The sequences can then be aligned to the full length mRNA and, if the mRNA is regulated, then cleavage products should begin at the position after the miRNA mediated cleavage was predicted to occur.

Recently a new high-throughput approach has been described which allows researchers to carry out a high-throughput target validation analysis.[94,95] This degradome sequencing approach captures all cleaved mRNA fragments in the transcriptome of the input sample and using suitable bioinformatics tools such as CleaveLand allows the prediction and validation of all plant miRNA targets in a single experiment.[96]

For targets that are not cleaved there are two main approaches used for experimental validation.[97] The first one requires the ability to manipulate the concentration or level of activity of the sRNA. For real targets the respective protein concentration levels should change in response to different sRNA activity levels. The second approach consists of copying the sequence of the target site to a reporter gene and measuring its activity. The target site should contain not only the region complementary to the sRNA but also the corresponding flanking regions. For a real target, introducing point mutations to the copied sequence should result in decreased activity of the reporter gene.

The limitations of experimental methods have led to the development of computational methods to predict sRNA targets. In most of these methods a combination of three features is taken into account: sequence complementarity between miRNAs and target sites, other thermodynamic factors and conservation of putative target sites across species. Different methods exist to predict the target sites with high degree of complementarity more typical of plants and target sites with low degree of complementarity, more usual in animals. For the former the most used method is a rule based algorithm presented in a paper by Schwab and colleagues.[29] For the latter TargetScan[98] and PicTar[99] have been used to produce predictions of conserved miRNA targets. Two other methods, RNA22 and Pita are widely used to predict nonconserved targets.[100,101] The predictions generated by these methods include many false positives and false negatives, with these rates much higher in the low complementarity predictions.

Alternative hybrid target prediction methods using microarrays or high-throughput sequencing have also been developed. One set of approaches relies on measuring the concentrations of either mRNAs or proteins under different activity levels of a single sRNA. Targeting relationships are inferred for the genes that show concentration changes and contain potential target sites. mRNA concentrations can be estimated using microarrays and more recently, high-throughput sequencing. Large scale measurement of protein concentrations is also possible, although with the current technology only a fraction of proteins can be measured.[102,103]

In the first experiment that used mRNA measurements obtained after miRNA activity manipulation, the abundances of miR-1 and miR-124 were increased by transfecting cells with siRNAs with identical sequences of those two miRNAs.[104] Evidence for widespread targeting at the mRNA level was found by analysing the sequences of down-regulated genes. In particular the sequence complementary to the first eight nucleotides of the miRNAs, thought to be important for target recognition, was found to be highly over represented among these genes. In other experiments the activity of individual miRNAs was decreased, either by suppressing activity or deleting the miRNA gene.[105-108] In both types of experiments the reduced miRNA activity is expected to lead to a de-repression of the targeted mRNAs and a consequent increase in mRNA concentrations. Finally, in a subsequent experiment the effects of over-expressing and suppressing the same miRNA were jointly assessed.[109] The combination of the results from these two experiments increased the specificity of the set of target candidates. This method of miRNA target prediction, looking for differences in mRNA concentrations under different miRNA activities, has two limitations: first, it identifies only the fraction of targets regulated at the mRNA level; second, some of the predicted targets might be false positives, since the observed mRNA concentration change can be caused by factors other than miRNA targeting. However, it usually generates a lower number of false positives than purely computational methods.

Another approach using high-throughput measurement of mRNAs has also been developed in recent years. This approach relies on the immuno-precipitation of RISCs bound to targeted mRNAs and subsequent profiling of these mRNAs using microarrays.[110-113] The profiling must be followed by computational analysis of the sequences of the profiled mRNAs to identify the putative target sites and sRNAs that target each mRNA. More recently, a study reported the sequencing of a sample where RISC-mRNA and RISC-sRNA complexes were immuno-precipitated in the same sample and both mRNAs and sRNAs were subjected to high-throughput sequencing. This combined data allowed for a more accurate determination of sRNA/mRNA regulatory interactions.[114]

Further sRNA Characterisation

As we have seen, sRNA classes are partially defined by their biogenesis pathways. High-throughput sequencing of small RNAs in cells that lack one or more than one of the proteins involved in these pathways has been used to classify sRNAs and in particular to distinguish miRNAs from other types of sRNAs.[72,115] This might be useful to predict the targeting pathways in which each sRNAs is involved.

High-throughput technologies have also been used to study the transcriptional regulation of sRNAs. This can be done by immuno-precipitating transcription factors and the DNA they are bound to, followed by DNA profiling, resorting either to microarrays or high-throughput sequencing. This method has been used to determine transcription start sites and promoter regions[116] and to establish transcriptional regulatory interactions between transcription factors and sRNAs.[39] These types of studies have allowed the construction of regulatory networks involving both sRNAs and regulatory proteins.[117,118]

CONCLUSION

High-throughput methods have been extremely important for the characterization of sRNAs, namely for the discovery and classification of the genes that encode sRNAs, the identification of the genes regulated by sRNAs and the genes that regulate the expression of sRNAs. In the near future the incorporation of new technological developments is going to have a great influence in the progress of the research. A new method to prepare libraries for high-throughput sequencing that requires much smaller quantities of RNA has also been recently published,[119] making it possible to generate much more accurate measurements of RNA concentrations at a single-cell level. These new protocols will allow easier sequencing of bacterial and other non polyA-tailed RNAs and it is likely that there will be a rapid increase in research in these areas. New sequencing technologies capable of single-molecule resolution have been recently described and are expected to be widely available in the next few years.[120] The expected throughput increase and cost decrease of sequencing will likely cause a shift in the usage of these technologies: in the future sequencing will be used less to catalogue sRNAs and more to understand their function, for instance by simultaneously profiling sRNAs and their targets across multiple tissues and developmental stages. These datasets will allow us not only to increase the amount and quality of information on regulatory interactions involving sRNAs, that is, to uncover the topology of the regulatory networks involving sRNAs, but also to improve our understanding of the dynamics of these networks and of the regulation of biological processes by sRNAs.

ACKNOWLEDGEMENTS

The authors thank Dr. Frank Schwach for helpful discussions during the preparation of this manuscript. H.P. is a student Instituto Gulbenkian de Ciência's PhD Programme in Computational Biology (sponsored by Fundação para a Ciência e a Tecnologia [FCT], Fundação Calouste Gulbenkian and Siemens SA Portugal) and was supported by FCT fellowship SFRH/BD/33204/2007. Work supported by Biotechnology and Biological Sciences Research Council (BB/E004091/1).

REFERENCES

1. Gottesman S. The small RNA regulators of Escherichia coli: roles and mechanisms*. Annual review of microbiology 2004; 58:303-328.
2. Wagner EG, Simons RW. Antisense RNA control in bacteria, phages and plasmids. Annual review of microbiology 1994; 48:713-742.
3. Wassarman KM. Small RNAs in bacteria: diverse regulators of gene expression in response to environmental changes. Cell 2002; 109(2):141-144.
4. Wightman B, Ha I, Ruvkun G. Posttranscriptional regulation of the heterochronic gene lin-14 by lin-4 mediates temporal pattern formation in C. elegans. Cell 1993; 75(5):855-862.
5. Lee RC, Feinbaum RL, Ambros V. The C. elegans heterochronic gene lin-4 encodes small RNAs with antisense complementarity to lin-14. Cell 1993; 75(5):843-854.
6. Mueller E, Gilbert J, Davenport G et al. Homology-dependent resistance: transgenic virus resistance in plants related to homology-dependent gene silencing. Plant J. 1995; 7(6):1001-1013.
7. Voinnet O. Systemic Spread of Sequence-Specific Transgene RNA Degradation in Plants Is Initiated by Localized Introduction of Ectopic Promoterless DNA. Cell 1998; 95(2):177-187.
8. Nellen W, Lichtenstein C. What makes an mRNA anti-sensitive? Trends Biochem Sci 1993; 18(11):419-423.

9. Fire A, Xu S, Montgomery MK et al. Potent and specific genetic interference by double-stranded RNA in Caenorhabditis elegans. Nature 1998; 391(6669):806-811.
10. Carthew RW, Sontheimer EJ. Origins and Mechanisms of miRNAs and siRNAs. Cell 2009; 136(4):642-655.
11. Eulalio A, Huntzinger E, Izaurralde E. Getting to the root of miRNA-mediated gene silencing. Cell 2008; 132(1):9-14.
12. Filipowicz W, Jaskiewicz L, Kolb FA et al. Post-transcriptional gene silencing by siRNAs and miRNAs. Curr Opin Struct Biol 2005; 15(3):331-341.
13. Verdel A, Jia S, Gerber S et al. RNAi-mediated targeting of heterochromatin by the RITS complex. Science 2004; 303(5658):672-676.
14. Ghildiyal M, Zamore PD. Small silencing RNAs: an expanding universe. Nat Rev Genet 2009; 10(2):94-108.
15. Bartel DP. MicroRNAs: genomics, biogenesis, mechanism and function. Cell 2004; 116(2):281-297.
16. Bushati N, Cohen SM. microRNA Functions. Annu Rev Cell Dev Biol 2007; 23(1):175-205.
17. Jones-Rhoades MW, Bartel DP, Bartel B. MicroRNAs and Their Regulatory Roles in Plants. Annu Rev Plant Biol 2006; 57:19-53.
18. Cullen BR. Transcription and processing of human microRNA precursors. Mol Cell2004; 16(6):861-865.
19. Du T, Zamore PD. microPrimer: the biogenesis and function of microRNA. Development 2005; 132(21):4645-4652.
20. Faller M, Guo F. MicroRNA biogenesis: there's more than one way to skin a cat. Biochim Biophys Acta 2008; 1779(11):663-667.
21. Kim VN, Han J, Siomi MC. Biogenesis of small RNAs in animals. Nat Rev Mol Cell Biol 2009; 10(2):126-139.
22. Voinnet O. Origin, Biogenesis and Activity of Plant MicroRNAs. Cell 2009; 136(4):669-687.
23. Kim VN. MicroRNA biogenesis: coordinated cropping and dicing. Nat Rev Mol Cell Biol 2005; 6(5):376-385.
24. Chen X. Small RNAs and Their Roles in Plant Development. Annu Rev Cell Dev Biol 2009; 25(1):21-44.
25. Schwach F, Moxon S, Moulton V et al. Deciphering the diversity of small RNAs in plants: the long and short of it. Brief Funct Genomic Proteomic 2009; 8(6):472-481.
26. Okamura K, Lai EC. Endogenous small interfering RNAs in animals. Nat Rev Mol Cell Biol 2008; 9(9):673-678.
27. Thomson T, Lin H. The biogenesis and function of PIWI proteins and piRNAs: progress and prospect. Annu Rev Cell Dev Biol 2009; 25(1):355-376.
28. Filipowicz W, Bhattacharyya SN, Sonenberg N. Mechanisms of posttranscriptional regulation by microRNAs: are the answers in sight? Nat Rev Genet 2008; 9(2):102-114.
29. Schwab R, Palatnik JF, Riester M et al. Specific Effects of MicroRNAs on the Plant Transcriptome. Dev Cell 2005; 8(4):517-527.
30. Jones-Rhoades MW, Bartel DP. Computational identification of plant microRNAs and their targets, including a stress-induced miRNA. Mol Cell 2004; 14(6):787-799.
31. Rajewsky N. microRNA target predictions in animals. Nat Genet 2006; 38 Suppl 1(6s):S8-S13.
32. Bartel DP. MicroRNAs: target recognition and regulatory functions. Cell 2009; 136(2):215-233.
33. Ding XC, Weiler J, Grosshans H. Regulating the regulators: mechanisms controlling the maturation of microRNAs. Trends in biotechnology 2009; 27(1):27-36.
34. Muralidhar B, Goldstein L, Ng G et al. Global microRNA profiles in cervical squamous cell carcinoma depend on Drosha expression levels. The Journal of Pathology 2007; 212(4):368-377.
35. Xie Z, Kasschau KD, Carrington JC. Negative feedback regulation of dicer-like1 in arabidopsis by microRNA-guided mRNA degradation. Current Biology 2003; 13(9):784-789.
36. Piriyapongsa J, Jordan IK. Dual coding of siRNAs and miRNAs by plant transposable elements. RNA 2008; 14(5):814-821.
37. Giraldez AJ, Mishima Y, Rihel J et al. Zebrafish MiR-430 promotes deadenylation and clearance of maternal mRNAs. Science 2006; 312(5770):75-79.
38. Zhao X, Fjose A, Larsen N et al. Treatment with small interfering RNA affects the microRNA pathway and causes unspecific defects in zebrafish embryos. FEBS Journal 2008; 275(9):2177-2184.
39. Marson A, Levine SS, Cole MF et al. Connecting microRNA genes to the core transcriptional regulatory circuitry of embryonic stem cells. Cell 2008; 134(3):521-533.
40. Zhou X, Ruan J, Wang G et al. Characterization and identification of microRNA core promoters in four model species. PLoS Comput Biol 2007; 3(3).
41. Megraw M, Baev V, Rusinov V et al. MicroRNA promoter element discovery in Arabidopsis. RNA 2006; 12(9):1612-1619.
42. Alwine JC, Kemp DJ, Stark GR. Method for detection of specific RNAs in agarose gels by transfer to diazobenzyloxymethyl-paper and hybridization with DNA probes. Proc Natl Acad Sci USA 1977; 74(12):5350-5354.
43. Becker-André M, Hahlbrock K. Absolute mRNA quantification using the polymerase chain reaction (PCR). A novel approach by a PCR aided transcript titration assay (PATTY). Nucleic Acids Res 1989; 17(22):9437-9446.

44. Brown PO, Botstein D. Exploring the new world of the genome with DNA microarrays. Nat Genet 1999; 21(1 Suppl):33-37.
45. Reinartz J, Bruyns E, Lin J et al. Massively parallel signature sequencing (MPSS) as a tool for in-depth quantitative gene expression profiling in all organisms. Brief Funct Genomic Proteomic 2002; 1(1):95-104.
46. Margulies M, Egholm M, Altman WE et al. Genome sequencing in microfabricated high-density picolitre reactors. Nature 2005; 437(7057):376-380.
47. Bennett S. Solexa Ltd. Pharmacogenomics 2004; 5(4):433-438.
48. Shendure J, Porreca GJ, Reppas NB et al. Accurate multiplex polony sequencing of an evolved bacterial genome. Science 2005; 309(5741):1728-1732.
49. Willenbrock H, Salomon J, Søkilde R et al. Quantitative miRNA expression analysis: comparing microarrays with next-generation sequencing. RNA (New York, N.Y.) 2009; 15(11):2028-2034.
50. Linsen SE, de Wit E, Janssens G et al. Limitations and possibilities of small RNA digital gene expression profiling. Nature methods 2009; 6(7):474-476.
51. Lai E, Tomancak P, Williams R et al. Computational identification of Drosophila microRNA genes. Genome Biol 2003; 4(7).
52. Lim LP, Glasner ME, Yekta S et al. Vertebrate microRNA genes. Science 2003; 299(5612).
53. Bonnet E, Wuyts J, Rouzé P et al. Detection of 91 potential conserved plant microRNAs in Arabidopsis thaliana and Oryza sativa identifies important target genes. Proc Natl Acad Sci USA 2004; 101(31):11511-11516.
54. Wang X, Zhang J, Li F et al. MicroRNA identification based on sequence and structure alignment. Bioinformatics 2005; 21(18):3610-3614.
55. Lim LP, Lau NC, Weinstein EG et al. The microRNAs of caenorhabditis elegans. Genes Dev 2003; 17(8):991-1008.
56. Hofacker IL. Vienna RNA secondary structure server. Nucleic Acids Res 2003; 31(13):3429-3431.
57. Zuker M. Mfold web server for nucleic acid folding and hybridization prediction. Nucleic Acids Res2003; 31(13):3406-3415.
58. Adai A, Johnson C, Mlotshwa S et al. Computational prediction of miRNAs in Arabidopsis thaliana. Genome Res 2005; 15(1):78-91.
59. Barakat A, Wall K, Leebens-Mack J et al. Large-scale identification of microRNAs from a basal eudicot (Eschscholzia californica) and conservation in flowering plants. Plant J. : for cell and molecular biology 2007; 51(6):991-1003.
60. Bentwich I, Avniel A, Karov Y et al. Identification of hundreds of conserved and nonconserved human microRNAs. Nat Genet 2005; 37(7):766-770.
61. Fahlgren N, Howell MD, Kasschau KD et al. High-throughput sequencing of Arabidopsis microRNAs: evidence for frequent birth and death of MIRNA genes. PloS one 2007; 2(2):e219.
62. Szittya G, Moxon S, Santos DM et al. High-throughput sequencing of Medicago truncatula short RNAs identifies eight new miRNA families. BMC genomics 2008; 9:593.
63. Yao Y, Guo G, Ni Z et al. Cloning and characterization of microRNAs from wheat (Triticum aestivum L.). Genome Biol 2007; 8(6):R96.
64. Axtell MJ, Bartel DP. Antiquity of microRNAs and their targets in land plants. Plant Cell 2005; 17(6):1658-1673.
65. Lagos-Quintana M, Rauhut R, Lendeckel W et al. Identification of novel genes coding for small expressed RNAs. Science (New York, N.Y.) 2001; 294(5543):853-858.
66. Lee RC, Ambros V. An extensive class of small RNAs in Caenorhabditis elegans. Science (New York, N.Y.) 2001; 294(5543):862-864.
67. Lau NC, Lim LP, Weinstein EG et al. An abundant class of tiny RNAs with probable regulatory roles in caenorhabditis elegans. Science 2001; 294(5543):858-862.
68. Berezikov E, Thuemmler F, Laake LW et al. Diversity of microRNAs in human and chimpanzee brain. Nat Genet 2006; 38(12):1375-1377.
69. Lee H, Chang S, Choudhary S et al. qiRNA is a new type of small interfering RNA induced by DNA damage. Nature 2009; 459(7244):274-277.
70. Ghildiyal M, Seitz H, Horwich MD et al. Endogenous siRNAs derived from transposons and mRNAs in drosophila somatic cells. Science 2008; 320(5879):1077-1081.
71. Wang XJ, Reyes J, Chua NH et al. Prediction and identification of Arabidopsis thaliana microRNAs and their mRNA targets. Genome Biol 2004; 5(9).
72. Lu C, Kulkarni K, Souret FF et al. MicroRNAs and other small RNAs enriched in the Arabidopsis RNA-dependent RNA polymerase-2 mutant. Genome Res 2006; 16(10):1276-1288.
73. Moxon S, Jing R, Szittya G et al. Deep sequencing of tomato short RNAs identifies microRNAs targeting genes involved in fruit ripening. Genome Res 2008; 18(10):1602-1609.
74. Glazov EA, Cottee PA, Barris WC et al. A microRNA catalog of the developing chicken embryo identified by a deep sequencing approach. Genome Res 2008; 18(6):957-964.

75. Kuchenbauer F, Morin RD, Argiropoulos B et al. In-depth characterization of the microRNA transcriptome in a leukemia progression model. Genome Res 2008; 18(11):1787-1797.
76. Morin RD, O'Connor MD, Griffith M et al. Application of massively parallel sequencing to microRNA profiling and discovery in human embryonic stem cells. Genome Res 2008; 18(4):610-621.
77. Goff LA, Davila J, Swerdel MR et al. Ago2 immunoprecipitation identifies predicted microRNAs in human embryonic stem cells and neural precursors. PLoS ONE 2009; 4(9).
78. Cai Y, Yu X, Zhou Q et al. Novel microRNAs in silkworm (Bombyx mori). Funct Integr Genomics 2010; 10(3):405-15.
79. Trapnell C, Salzberg SL. How to map billions of short reads onto genomes. Nat Biotechnol 2009; 27(5):455-457.
80. Li H, Durbin R. Fast and accurate short read alignment with Burrows-Wheeler transform. Bioinformatics (Oxford, England) 2009; 25(14):1754-1760.
81. Langmead B, Trapnell C, Pop M et al. Ultrafast and memory-efficient alignment of short DNA sequences to the human genome. Genome Biol 2009; 10(3).
82. Prüfer K, Stenzel U, Dannemann M et al. PatMaN: rapid alignment of short sequences to large databases. Bioinformatics 2008; 24(13):1530-1531.
83. Friedländer MR, Chen W, Adamidi C et al. Discovering microRNAs from deep sequencing data using miRDeep. Nat Biotech 2008; 26(4):407-415.
84. Moxon S, Schwach F, Dalmay T et al. A toolkit for analysing large-scale plant small RNA datasets. Bioinformatics 2008; 24(19):2252-2253.
85. Hackenberg M, Sturm M, Langenberger D et al. miRanalyzer: a microRNA detection and analysis tool for next-generation sequencing experiments. Nucl Acids Res 2009; 37(suppl_2):W68-W76.
86. Wang WC, Lin FM, Chang WC et al. miRExpress: Analyzing high-throughput sequencing data for profiling microRNA expression. BMC Bioinformatics 2009; 10(1).
87. Pantano L, Estivill X, Martí E. SeqBuster, a bioinformatic tool for the processing and analysis of small RNAs datasets, reveals ubiquitous miRNA modifications in human embryonic cells. Nucleic Acids Res2010; 38(5):e34.
88. Griffiths-Jones S, Saini HK, van Dongen S et al. miRBase: tools for microRNA genomics. Nucleic Acids Res2008; 36(suppl_1):D154-D158.
89. Axtell MJ, Jan C, Rajagopalan R et al. A two-hit trigger for siRNA biogenesis in plants. Cell 2006; 127(3):565-577.
90. Chen H, Li Y, Wu S. Bioinformatic prediction and experimental validation of a microRNA-directed tandem trans-acting siRNA cascade in Arabidopsis. Proc Natl Acad Sci USA 2007; 104(9):3318-3323.
91. Elbashir SM, Lendeckel W, Tuschl T. RNA interference is mediated by 21- and 22-nucleotide RNAs. Genes Dev 2001; 15(2):188-200.
92. MacLean D, Moulton V, Studholme DJ. Finding sRNA generative locales from high-throughput sequencing data with NiBLS. BMC Bioinformatics 2010; 11:93.
93. Llave C, Kasschau KD, Rector MA et al. Endogenous and silencing-associated small RNAs in plants. The Plant cell 2002; 14(7):1605-1619.
94. German MA, Pillay M, Jeong DH et al. Global identification of microRNA–target RNA pairs by parallel analysis of RNA ends. Nat Biotechnol 2008; 26(8):941-946.
95. Addo-Quaye C, Eshoo TW, Bartel DP et al. Endogenous siRNA and miRNA targets identified by sequencing of the Arabidopsis degradome. Current Biology 2008; 18(10):758-762.
96. Addo-Quaye C, Miller W, Axtell MJ. CleaveLand: a pipeline for using degradome data to find cleaved small RNA targets. Bioinformatics (Oxford, England) 2009; 25(1):130-131.
97. Kuhn DE, Martin MM, Feldman DS et al. Experimental validation of miRNA targets. Methods (San Diego, Calif.) 2008; 44(1):47-54.
98. Lewis BP, Burge CB, Bartel DP. Conserved seed pairing, often flanked by adenosines, indicates that thousands of human genes are microRNA targets. Cell 2005; 120(1):15-20.
99. Krek A, Grün D, Poy MN et al. Combinatorial microRNA target predictions. Nat Genet 2005; 37(5):495-500.
100. Kertesz M, Iovino N, Unnerstall U et al. The role of site accessibility in microRNA target recognition. Nat Genet 2007; 39(10):1278-1284.
101. Miranda KC, Huynh T, Tay Y et al. A pattern-based method for the identification of MicroRNA binding sites and their corresponding heteroduplexes. Cell 2006; 126(6):1203-1217.
102. Baek D, Villen J, Shin C et al. The impact of microRNAs on protein output. Nature 2008; 455(7209):64-71.
103. Selbach M, Schwanhäusser B, Thierfelder N et al. Widespread changes in protein synthesis induced by microRNAs. Nature 2008; 455(7209):58-63.
104. Lim LP, Lau NC, Garrett-Engele P et al. Microarray analysis shows that some microRNAs downregulate large numbers of target mRNAs. Nature 2005; 443(7027):769-773.
105. Krützfeldt J, Rajewsky N, Braich R et al. Silencing of microRNAs in vivo with `antagomirs'. Nature 2005; 438(7068):685-689.

106. Elmen J, Lindow M, Silahtaroglu A et al. Antagonism of microRNA-122 in mice by systemically administered LNA-antimiR leads to up-regulation of a large set of predicted target mRNAs in the liver. Nucleic Acids Res2007; 36(4):1153-1162.
107. Elmén J, Lindow M, Schütz S et al. LNA-mediated microRNA silencing in nonhuman primates. Nature 2008; 452(7189):896-899.
108. Zhao Y, Ransom JF, Li A et al. Dysregulation of cardiogenesis, cardiac conduction and cell cycle in mice lacking miRNA-1-2. Cell 2007; 129(2):303-317.
109. Nicolas FE, Pais H, Schwach F et al. Experimental identification of microRNA-140 targets by silencing and overexpressing miR-140. RNA 2008; 14(12):2513-2520.
110. Zhang L, Hammell M, Kudlow BA et al. Systematic analysis of dynamic miRNA-target interactions during C. elegans development. Development 2009; 136(18):3043-3055.
111. Karginov FV, Conaco C, Xuan Z et al. A biochemical approach to identifying microRNA targets. Proc Natl Acad Sci USA 2007; 104(49):19291-19296.
112. Easow G, Teleman AA, Cohen SM. Isolation of microRNA targets by miRNP immunopurification. RNA 2007; 13(8):1198-1204.
113. Hong X, Hammell M, Ambros V et al. Immunopurification of Ago1 miRNPs selects for a distinct class of microRNA targets. Proc Natl Acad Sci USA 2009; 106(35):15085-15090.
114. Chi SW, Zang JB, Mele A et al. Argonaute HITS-CLIP decodes microRNA–mRNA interaction maps. Nature 2009; 460(7254):479-486.
115. Babiarz JE, Ruby JG, Wang Y et al. Mouse ES cells express endogenous shRNAs, siRNAs and other Microprocessor-independent, Dicer-dependent small RNAs. Genes Dev 2008; 22(20):2773-2785.
116. Corcoran DL, Pandit KV, Gordon B et al. Features of Mammalian microRNA Promoters Emerge from Polymerase II Chromatin Immunoprecipitation Data. PLoS ONE 2009; 4(4).
117. Martinez NJ, Ow MC, Barrasa MI et al. A C. elegans genome-scale microRNA network contains composite feedback motifs with high flux capacity. Genes Dev 2008; 22(18):2535-2549.
118. Shalgi R, Lieber D, Oren M et al. Global and local architecture of the mammalian microRNA-transcription factor regulatory network. PLoS Comput Biol 2007; 3(7):e131.
119. White R, Blainey P, Fan HC et al. Digital PCR provides sensitive and absolute calibration for high throughput sequencing. BMC Genomics 2009; 10:116.
120. Ansorge WJ. Next-generation DNA sequencing techniques. N Biotechnol 2009; 25(4):195-203.
121. Dai X, Zhao PX. pssRNAMiner: a plant short small RNA regulatory cascade analysis server. Nucleic Acids Res2008; 36(Web Server issue):W114-W118.

CHAPTER 17

HOW OLD ARE RNA NETWORKS?

Toni Daly,[1,3] X. Sylvia Chen[2] and David Penny*,[3]

[1]Allan Wilson Centre of Molecular Ecology and Evolution, Massey University, Palmerston North, New Zealand; [2]Department of Biochemistry, University of Otago, Dunedin, New Zealand; [3]Institute of Molecular BioSciences, Massey University, Palmerston North, New Zealand

**Corresponding Author: David Penny—Email: d.penny@massey.ac.nz*

Abstract: Some major classes of RNAs (such as mRNA, rRNA, tRNA and RNase P) are ubiquitous in all living systems so are inferred to have arisen early during the origin of life. However, the situation is not so clear for the system of RNA regulatory networks that continue to be uncovered, especially in eukaryotes. It is increasingly being recognised that networks of small RNAs are important for regulation in all cells, but it is not certain whether the origin of these networks are as old as rRNAs and tRNA. Another group of ncRNAs, including snoRNAs, occurs mainly in archaea and eukaryotes and their ultimate origin is less certain, although perhaps the simplest hypothesis is that they were present in earlier stages of life and were lost from bacteria. Some RNA networks may trace back to an early stage when there was just RNA and proteins, the RNP-world; before DNA.

INTRODUCTION

With new classes of RNA continuing to be discovered, it appears as if many classes of RNA are likely to occur in eukaryotes, bacteria and archaea—but what about the actual RNA-protein networks in which they are involved. For example, small RNA regulation of gene expression has been seen in both eukaryotes and prokaryotes and even viruses use small RNAs. Although the basic mechanism of target recognition and cleavage is similar in all these groups, the proteins and interactions differ. This chapter considers some ideas for the time of origin of some key classes of RNA and their associated networks. We readily accept the idea that some RNA-protein interactions are very old but when we get down to the elaborate pathways of RNA-based regulation there was an early assumption that such regulation only evolved when organisms became more complex (e.g., multicellular).

RNA Infrastructure and Networks, edited by Lesley J. Collins.

Now we see that the networks of RNA-protein interaction are more general; but can we infer the presence of particular RNA functions in **Ida** (Initial Darwinian Ancestor), or in the later **Luca** (the Last Universal Common Ancestor), or in the even later **Fred** (Fairly Remote Eukaryotic Daddy).[1,2]

When it comes to understanding the main roles of RNA in modern cell biology there is a potential problem of the 'alphabet soup' formed from so many classes of small RNAs. Defining these subgroups is important for identifying subclasses that are reasonably closely related—they have well-defined homologies and clearly related functions as collated and updated in the RNA database Rfam.[3] However, our interest here is at the other end of the spectrum—once we have these many classes and networks of small RNAs, can we put them into higher level evolutionary groups with more general functions. As an example, Boria et al[4] identified sbRNAs (stem bulge RNAs) in nematodes as evolutionary homologues of the more widespread Y-RNAs. Y-RNAs are involved with the protein R_0 in a network of interactions that both assist misfolded RNAs but are also involved with DNA transcription. This equivalence of R_0 and sbRNAs reinforces the message that RNA networks evolve, and as we understand more about how RNA and its associated proteins evolve, we can begin to surmise how such regulation could have evolved much earlier in the beginning of life.

REGULATORY NETWORKS OF SMALL RNAs

Networks involving small RNAs can regulate translation with some proteins also used in defence networks. Small (~20-30 nt) RNAs have many functions including, fine-tuning expression of temporal and tissue specific mRNA, destruction of aberrant mRNAs via cleavage, repression, up-regulation and also translational control via DNA methylation and maintenance of histone architecture.

Three basic regulatory classes of ncRNA networks are recognized in eukaryotes, determined by their biogenesis and mode of operation[5] (siRNA-based, miRNA-based and piRNA-based). miRNA sequences are usually found in intergenic and intronic regions of eukaryotes and in some viruses. They have their own promoters and regulatory units. RNA Pol II transcribes long unstable primary RNAs that form hairpin loops (pri-miRNA). Some miRNA genes are found in the UTR regions of coding sequences (indicating that the transcript can be processed as either a miRNA or an mRNA) and some are even found in exons. A complex cellular milieu of miRNAs allows tuning of thousands of genes through combinatorial interactions within 3′ untranslated regions and could account for cell specificity and/or temporal control.[6] miRNAs were once thought to be cell-specific but recent work has shown that in plants at least, miRNAs move extensively between cells and can control protein levels in remote tissue.[7] In a viral example, the herpes virus (dsDNA) produces a range of pri-miRNA transcripts which are processed to pre-miRNAs in the host nucleus. They utilise the endogenous miRNA network using the host RNAse III endonuclease (Drosha) and a dsRNA binding protein (Pasha). These are then exported via the same pathway as endogenous miRNAs. These viral miRNAs can extend the lytic phase of the cell, by suppression of apoptosis. Other DNA viruses seem to produce just one or two miRNAs.[8] miRNAs have an obligate nuclear processing phase, so viruses that enter the nucleus are more likely to encode them. HIV is an RNA retrovirus that encodes at least three miRNAs.[9,10] Similarly with bacteria, the success of the plant bacterium *Agrobacterium tumefaciens* is enhanced by knocking out the siRNA pathway, but still requires an intact plant miRNA pathway.[11]

In animals, the network of miRNA-based reactions involves exportin-5 mediating nuclear export of pre-miRNAs, a cytoplasmic RNAse III endonuclease Dicer, together with the dsRBP Loqs, (*Drosophila*) and trans-activator binding protein (TRBP) (in humans), transforms the pre-miRNA into the mature transcript. Plants lack a Drosha and Pasha homologue and instead use a nuclear Dicer-like endonuclease (DCL1), which makes similar cuts, but they also require a dsRBP (HYL1) to accurately process the miRNA precursors.[12] Such changes over tine are what we expect from an evolutionary process and this makes it likely that the plant/animal ancestor had a comparable system.

Animal miRNAs often have limited complementarity, either as a duplex (miRNA:miRNA*) or with the target mRNA. Usually the strand with the lower 5′ end stability enters the Argonaute domain of the RISC complex (RNA induced silencing complex). Argonautes are an ancient protein family known as 'slicers'. Plant and animal miRNAs usually load into Ago1 although miRNAs have been found in all four human Agos. It was believed that the miRNA* strand is degraded, but recent work in *Drosophila* indicates that the miRNA* strand can be modified by 2′-O-methylation at the 3′ end then loaded into Ago2 in the RISC complex, the domain usually occupied by siRNA.[13]

In animals, introns can be linearized by the lariat debranching enzyme and the resulting RNA folds to a pre-miRNA structure. These spliced lariats are processed by Dicer and have been termed mirtrons. They have the same action as a miRNA, but do not require the cleavage capabilities of Drosha. Initially discovered in *D. melanogaster* and *C. elegans*, mirtrons have now been found in mammals, including primate specific sets.[14] Additionally the HIV virus encodes a miRNA; hiv1-miR-TAR that uses the cellular mirtron pathway.[9] It has been suggested[14] that this pathway could predate Drosha mediated cleavage. An alternate would be that they have evolved independently three times in different lineages and that viruses have taken advantage of both pathways. Most animal Agos now lack cleavage ability[15] and suppress mRNA translation by physical impedance, or by suppression of transcription via heterochromatin and DNA methylation.[16]

Animal miRNAs usually suppress translation by binding to the target mRNA, but the exact mechanism remains under debate. Three models[17] for preventing initiation are; miRNA:RISC complexes compete for ribosome binding, compete for cap binding, or prevent circularization of the mRNA by inducing de-adenylation (circularization enhances mRNA translation). The latter two strategies could also promote the degradation of mRNA unprotected by cap or polyA tail. Another suggestion is that even if translation has begun, the miRNA:RISC complex could cause ribosomes to drop off prematurely. However, miRNAs can also up-regulate translation (particularly when the cell is under stress).[18] Binding position is relevant—an example being if the interaction of mammalian miR10a:RISC is within the 3′UTR of ribosomal proteins then miR10a suppresses translation, but in the 5′UTR it could activate it.[19] Thus there is a wide diversity of miRNA processing and the existence of small RNA networks appears universal in eukaryotes.

Recently small RNAs derived from tRNA were discovered.[20] These load preferentially into Ago3 and Ago4 where one of the classes of tRNA derived small RNAs (tsRNA) has a moderate capability to down-regulate mRNA. Ago3 and Ago4 may act as a buffer by loading unstructured small RNAs preventing Ago2 from becoming overloaded.[21] Intriguingly they are confined to the cytoplasm, though their processing signature indicates that they are processed in the nucleus. Cellular localisation also plays a part in the standard miRNA-processing network. The miRNA:RISC complex in the cytoplasm often relocates to P bodies enriched in mRNA degradation proteins.[22] Smaller complexes composed of just siRNA or miRNA and Ago2 can also be found in the nucleus. Possibly, a small proportion of the cytoplasmic Ago2:RISC complex is stripped of Dicer and TRBP prior to, or during translocation to the nucleus. In the case of siRNA or perfectly complementary miRNAs, the target is cleaved and the Ago2 is released and exported to the cytoplasm. In *C. elegans*,[23] NRDE3 (an Ago) together with siRNA is necessary and sufficient for location to the nucleus and silencing of nuclear mRNAs. Although there are 27 Agos in nematodes, there appear to be little redundancy in that mutants lacking this protein (*nrde3*⁻ nematodes) are defective for all nuclear mRNA silencing.

miRNA-mediated cleavage appears similar to the RNA interference mechanism mediated by siRNAs (see later). However, there are subtle differences in their requirement for complementarity; siRNA and plant miRNAs have perfect complementarity to their target, but the animal miRNA:miRNA* duplex contains mismatches bulges and GU wobble pairs. Their biogenesis is more strikingly different; most siRNAs silence their encoded targets and are processed from complementary dsRNA transcripts from invasive nucleic acid during infection (exo-siRNA) and so processing is often in the cytoplasm. In contrast, miRNA sequences are encoded in the organism's genome and so the first steps in the processing pathway occur in the nucleus and requires export. In plants and worms, each siRNA precursor gives rise to many siRNA duplexes, but only one miRNA:miRNA* is generated from each pre-miRNA. Additionally miRNAs are nearly always conserved in related organisms but exo-siRNAs are rarely conserved.[24]

Endogenous-siRNA (endo-siRNA) is involved in a pathway that could evolutionarily link the nuclear and cytoplasmic processing. This process offers protection from genomically integrated parasitic DNA so its transcripts are derived from the genome and processed in the nucleus. The aim is the suppression of parasitic DNA either by guiding histone modification and DNA methylation, or cleavage of mRNA. The endo-siRNA pathway shares some of the miRNA proteins. For instance, *Drosophila* uses the double stranded RNA binding protein (dsRBP) Loquacious (Loqs) and Dicer 2 for the endo-siRNA pathway and another isoform of Loqs is used for the miRNA pathway in partnership with Dicer 1.[25] The commonality is the endogenous root of the RNA, but the need to quell exogenous transcripts would likely have arisen before silencing one's own.

Piwi interacting RNAs (piRNAs) were originally detected in *Drosophila* germline cells and they are slightly larger (~24-31 nt) RNAs with the 2′ O-methyl 3′ ending reminiscent of siRNAs. They are particularly known for the suppression of transposons and repeat sequences (see later), but flies and vertebrates have an additional type of piRNA expressed in germline cells during the pachytene and prepachytene stages of meiosis (known as primary piRNA). They are derived from the 3′UTR of cellular transcripts involved in diverse cellular processes. They do not exhibit the 'ping-pong' type amplification where secondary piRNAs are produced via an Aub/Ago3 dependent loop as seen in the suppression of transposons (see later), and they are depleted for repeat elements. The mechanism of primary piRNA production is poorly understood but it is thought that they are generated at the same time as the mRNA is being translated in the cytoplasm[26] and may be derived directly from mRNAs selected for piRNA production. There is some crossover of pathways here because in *Drosophila* at least, primary piRNAs that are involved with the suppression of transposable elements have been found in the somatic cells surrounding germ cells. These cells lack Aub and Ago3 and appear to come from one cluster (flamenco) and target one class of transposon (gypsy elements). Gypsy elements mobilise initially by being copied into RNA, but are then processed into virus-like particles in the cytoplasm.[27,28] It is intriguing that the siRNA, pathway which specialises in dealing with viral RNA, appears to have neglected gypsy elements and that in flies at least, a variant of the piRNA pathway has developed to deal with them. The important conclusion is again that there are many variations in details within eukaryotes, but the same basic mechanism seems to re-occur.

RNA REGULATION AND DEFENCE AGAINST THE DARK ARTS

All cells and even viruses, are subject to de novo invasion by both exogenous and endogenous nucleic acids including transposons, viruses, pseudogenes—essentially what has been called 'selfish DNA'. We refer to this potential invasion of DNA as the 'Dark Arts' and (with acknowledgement to JK Rowling) 'Defence against the Dark Arts' heavily involves RNA networks.[29] Exogenous nucleic acid can occasionally be advantageous for an organism, but in most cases it will be deleterious and possibly lethal.

We first discuss the small RNAs involved in RNA interference (RNAi) pathways in eukaryotes and viruses and the following section the clustered, regularly interspaced, short palindromic repeats (CRISPR arrays) found in bacteria and archaea. We expect that at all stages during the origin of life, the same basic biological principles apply,[30] and that parasites would always have been present.[31] Viruses have also evolved small-RNA based mechanisms which use the conserved RNAi machinery of the host.[8] The co-evolution of these mechanisms is important for understanding molecular epidemiology 'in action'.

We consider two forms of siRNA networks as described earlier; exogenous siRNAs (exo-siRNA) from invasive transcripts and short RNAs produced from endogenous transcripts (endo-siRNA). Exo-siRNAs are processed from long double stranded RNAs (dsRNA), often transcribed from viral or plasmid sources during infection in animals, fungi, protists and plants. It forms a basis for antiviral defence and requires at least three key proteins, the RNaseIII endonuclease—Dicer, an RNA dependant RNA polymerase (RdRP) and a member of the Argonaute (Ago) family.

There are 3 clades of Argonaute; 'Argonaute-like', 'Piwi-like' (both found in prokaryotes) and 'group 3 Argonautes' found in *C. elegans*.[32] The Ago proteins are found in all three domains, but are quite diverse in sequence. They possibly originated as a structural support for catalytic RNA and eventually took on the catalytic role leaving the RNA as a guide? There is large variation between eukaryotes in their complement of these proteins (see Table 1) but our main question here is whether there are basic mechanisms across all eukaryotes. To be more precise, did the last common ancestor of eukaryotes (Fred) have this defence mechanism, which has subsequently been modified in the different eukaryote lineages? It has been unclear whether mammals use the exo-siR system given the complex immune system available to them. There is some evidence that mammalian viruses encode RNAi suppressors and that implies that they could use exo-siR,[33] and more recently virally encoded siRNAs have been found in mammalian cells. For example, the abundance of virally encoded siRNAs increases in cells defective for the interferon pathway (IFN α/β) (initially thought to preclude the possibility of exo-siRNA in mammals).[34]

Exo-siRNA deals with invading RNA, but eukaryote genomes are largely comprised of selfish DNA that has accumulated over time and a variant of the siRNA system (endo-siRNA) deals with this. In plants RDR2 copies transcripts from silent loci, principally transposons, thought to be produced by RNA pol IV. This produces long dsRNA which is

Table 1. Some of the key enzymes in the RNAi pathways and variation between species

Species	RNase III Endonuclease	RNA Dependent RNA Polymerase	Argonaute-Like	Piwi-Like
S. pombe	DCR1	RDRP1	Ago1	
A. thaliensis	DCL1-4	RDR1, RDR2, RDR6	Ago1-10	
D. melongaster	DCR1-2, Drosha	A subunit of RNA pol II has RdRP capability	Ago1-2	Piwi, Aub, Ago3
C. elegans	DCR1, Drosha	Rrf-1, Rrf-3	Alg1-2, T22B3.2, T23D8.7, ZK757.3	PRG1-2, ERGO1 Plus at least 18 group 3 Argonautes
H. sapiens	DCR1, Drosha	RdRP composed of hTERT in complex with *RMRP*[37]	Ago1-4	HILI, HIWI, HIWI2 PIWIL3

cleaved to make sets of cis-acting siRNA (casiRNA) that affect the transcripts of the gene loci that produced them. RDR6 can copy miRNAs from the *TAS* locus, which then enters the RNAi pathway and is trans-acting (tasiRNA); i.e., produced from a discrete locus. Plants also have a third method of producing siRNA via RDR6 by copying overlapping transcripts, one produced constitutively and one produced in times of biotic stress. These are natural antisense transcript-derived siRNAs (natsiRNA).[35] This continued evolution of RNA-protein networks makes it harder to recognise specific networks, but does not obscure the general mechanisms of defensive RNA networks.

Classical RdRP homologs have not been found in mammals, yet endo-siRNAs appear ubiquitous in eukaryotes. This defence is important in flies, and an elongation subunit of RNA pol II has been found to have RdRP capability and is involved in the RNAi pathway.[36] An alternate RdRP is also available to mammals; an RdRP composed of hTERT in complex with *RMRP* has been shown to produce siRNAs in humans.[37] Endogenous dsRNA can also derive from DNA transcripts that would result in RNA prone to forming duplexes (Fig. 1A),[38] e.g., bi-directional transcription of a single gene, trans-interaction between transcripts from a founder gene and pseudogene, or overlapping transcription across two genes (convergent transcription). Long single stranded RNA transcripts that fold back on themselves could form from long transcripts of transposons in inverted orientation, or from duplicated inverted pseudogenes, or from structured loci. All of these could result in hairpin duplex structures that can be processed using the siRNA pathway with the addition of a double stranded RNA binding protein (dsRBP) (Fig. 1B).

In relation to the Dark Arts, exo-siRNA defends against exogenous viral and plasmid nucleic acid infection and endo-siRNAs against RNA from endogenous repeating units, transposons, integrated viral sequences and pseudogenes. This defence can occur in any cell, but is especially important in organisms (such as animals) where there is a division into somatic and germ-line cells. The matched endogenous repeat sequences were originally termed repeat associated small interfering RNAs (rasiRNA). This term has been retained for plant endo-siRNAs that target repeat sequences. However, piRNAs aren't simply longer siRNAs—their biogenesis is different, because they are independent of Drosher and Dicer endonucleases and they interact with a specific clade of Ago proteins

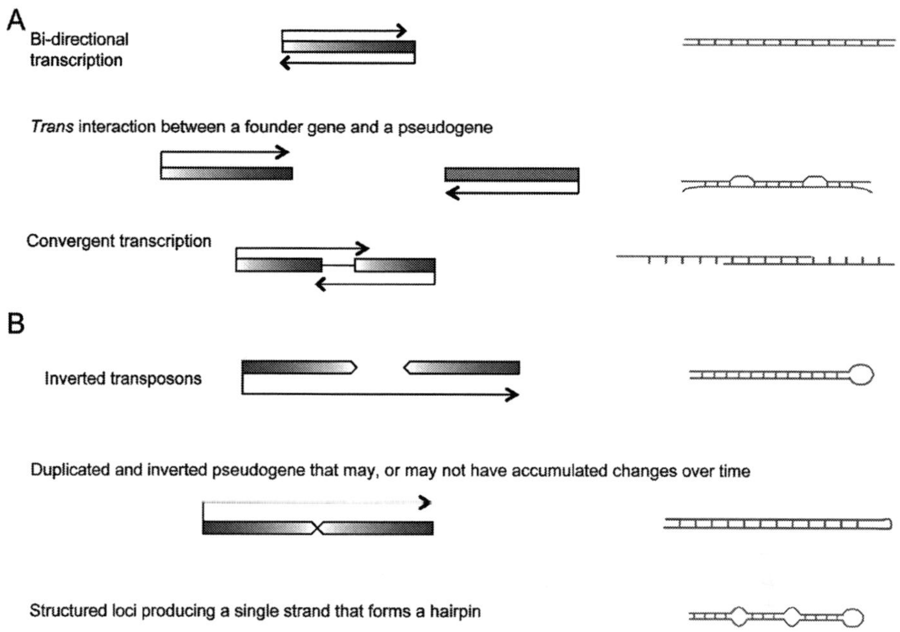

Figure 1. A) Transcription of endogenous DNA that gives rise to dsRNA. Bidirectional transcription resulting in dsRNA can arise when sense and antisense promoters are found in the 5′UTR. Parent and pseudogene transcripts can form endo-siRNAs that regulate the parent gene. Pseudogenes are essentially unsuccesful duplications, 'faulty' in some way, yet many are retained and don't appear to accumulate as many mutations as would be expected. siRNAs can also derive from overlapping transcripts, this occurs naturally in plants giving rise to natsiRNAs produced at times of stress. Convergent transcription in the geminivirus gives rise to a very short overlap but is sufficient to produce the dsRNA needed for exo-siRNA. B) Transcripts prone to forming hairpin loops form from the 'read through' of transposons or pseudogenes that have duplicated and inverted. A color version of this figure is available at www.landesbioscience.com/curie.

known as Piwi, AUB (Aubergine) and Ago3 in *Drosophila* (MILI1, MIWI and MIWI2 in mice, HILI, HIWI1, HIWI2 and HIWI3 in humans). The evolution of RNA networks continues; it is not a 'once and for all time' setup.

Transposons make up the bulk of most eukaryote DNA and are classed as either Type I transposons (that have an RNA intermediate and can be thought of as 'copy and paste'), or Type II elements (where the segment of DNA can move, 'cut and paste'). piRNAs are processed from long endogenous predominantly antisense transposon or repeat sequence transcripts often found concentrated in hotspots on the genome.[39] Their biogenesis is different from the other pathways made possible because the Piwi clade of Argonautes have retained their catalytic capability. Although precursor piRNAs are mostly antisense, a small number of sense piRNAs are produced. These load into Ago3 and cleave antisense retrotransposon transcripts or the antisense precursor piRNA leaving a 5′ antisense product or secondary piRNA. These load preferentially into Piwi or Aub and bind and cleave transposon mRNA forming a 5′ sense piRNA that loads into Ago3 and guides cleavage of an antisense mRNA or the antisense precursor. This is known as a 'forward feed' or 'ping pong' amplification loop and can generate large numbers of piRNAs without the need for

Dicer processing. Plants and worms have also amplification steps in their siRNA pathways, generating secondary siRNAs though the mechanisms are quite different. Silencing is achieved principally by DNA methylation via the RITS (RNA-induced Initiation of Transcriptional Gene Silencing) complex containing Piwi and AUB. In mammals this epigenetic regulatory role is heritable through maternal transmission.[40] A comparison of some features of RNAi defence networks is shown in Figure 2.

Some genomic regions give rise to both endo-siRNA and piRNA and target the same transcripts, indicating cooperation/interaction between networks. Indeed, integration across the entire RNAi networks seems necessary, otherwise a viral infection might overwhelm the machinery required to suppress transposons to the mutual benefit of virus and transposon. Alternatively, there could be some redundancy among the molecular machinery that keeps all pathways working. For instance, it is known that the plant

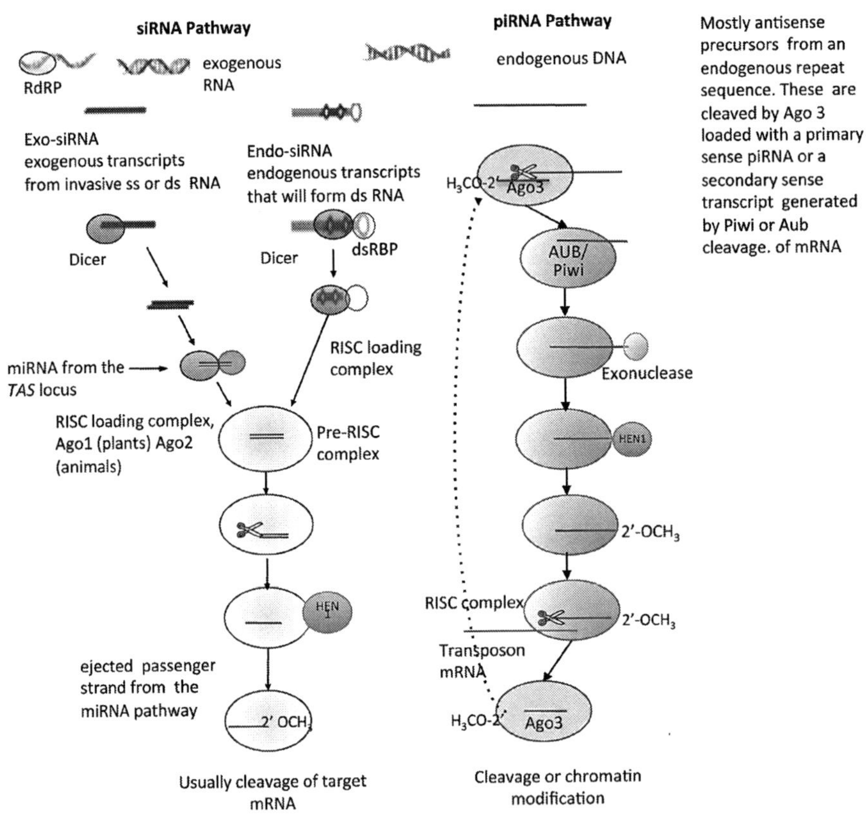

Figure 2. A comparison of RNAi networks involved with the defence of the Dark Arts. Current invasions can be quashed by exo-siRNAs utilising the RNA derived from the invader, either conveniently provided, for example by the geminivirus, or formed into double stranded RNA by RdRP. Endogenous invasive DNA is dealt with by the endo-siRNA pathway, sharing many proteins with the exo-siRNA and also the miRNA networks. The piRNA pathway is also adapted to prevent expression of endogenous invasive DNA, usually by modification of DNA architecture and specifically in the germline. Although the piRNA pathway is independent of Dicer it utilises a variant of the Argonaut protein confined to germline cells (see Table 1). A color version of this figure is available at www.landesbioscience.com/curie.

DCL2 can take over the functions of DCL4. Is this redundancy driven by fortuitous gene duplication, or in response to the plant dependence on the RNAi networks?

Viral particles evolve rapidly and even miRNA genes within related viruses may have little homology although they maintain sequence similarity with their mRNA targets for host mRNAs and with some cellular miRNAs. Viruses can suppress host miRNA and siRNA pathways via RNA silencing suppressors (RSS) at many steps. Some DNA viruses score 'own goals' by producing dsRNA, an example being the cassava mosaic virus, a geminivirus that produces dsRNA via convergent transcription from opposite promoters. The plant cytoplasmic DCL4 slices these into siRNAs that suppress the viral transcripts. However, the Dark Arts are never idle; viruses have evolved RSS to turn off the host defence, by inactivating DCL4. Nevertheless, the plant can retaliate and use DCL2 as a substitute for DCL4.[41] The geminiviruses that give rise to convergent transcripts can inhibit the RdRP needed for the miRNA and siRNA pathways in plants,[42] and prevent RNA-directed DNA methylation (Fig. 3). Plants have counter-evolved a method of methylating and silencing geminivirus minichromosomes using RNA-directed methylation via Ago4.[43] Geminivirus counteracts by suppressing methylation globally. A similar arms race occurs wherever viral nucleic acid is found, that is, everywhere. From

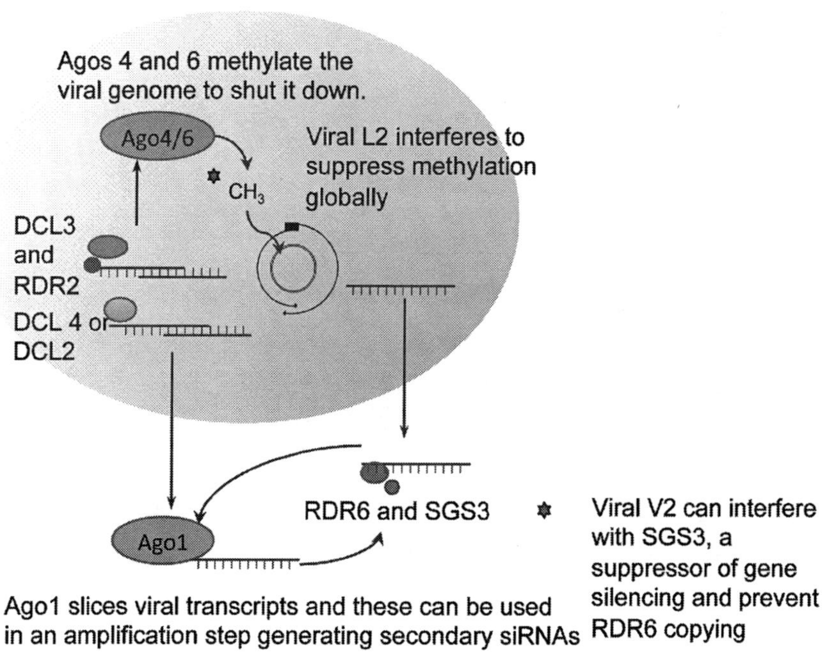

Figure 3. Geminiviruses are ssDNA viruses that replicate via rolling replication forming dsDNA without a dsRNA stage. Transcription is bidirectional and the transcript overlaps resultin a dsRNA of a small segment of the *AC1* gene, which is indispensable for replication. DCL4 preferentially dices this into small (22nt) virus-derived RNAs (exo-siRNA) which load into Ago1 and slice mRNA transcripts arising from the *AC1* gene. The cleaved mRNA can be converted by plant RDR6 into secondary siRNAs in an amplification step, which also converts other viral ss transcripts into siRNAs. An alternate fate of the viral overlapping transcript is to be extended by RDR2 and diced by DCL3 in the nucleus, where it enters Agos 4 or 6 and shuts down viral replication. However, the viruses can interfere with this shutdown at different stages. A color version of this figure is available at www.landesbioscience.com/curie.

an evolutionary perspective the RNAi networks must be very old. Not only because RNAi is ubiquitous, but because it makes use of ancient proteins and is involved with the maintenance of other very old proteins such as the highly conserved histones which serve to enhance copying fidelity.

So what of bacteria and archaea, how does their arsenal stack up against parasitic nucleic acids? "Clustered, regularly interspaced, short palindromic repeats" (CRISPRs)[44] occur extensively in bacteria and archaea as a defence against phages, plasmids and transposons, using small RNAs to interfere with invasion. CRISPRS appear ubiquitous in archaea and the apparent lower incidence (~40%) in bacteria may be an artefact of lab maintained strains. However, an alternate explanation is that some pathogenic bacteria, reliant on plasmids for antibiotic resistance or virulence factors, have mechanisms to avoid the acquisition of CRISPR systems. Some *E. coli* strains have CRISPR arrays containing spacer sequences derived from *cas* genes termed 'antiCRISPRs' that can destroy the DNA of invading CRISPR/*cas* systems.[45] These species lack endogenous *cas* genes yet the antiCRISPR is selectively retained presumably to permit horizontal gene transfer, or for an as yet unidentified function. The CRISPR system is dynamic and evolutionary and an outline is given in Figure 4 (for more detail on the CRISPR system, please see Chapter 13, pages 213-218).

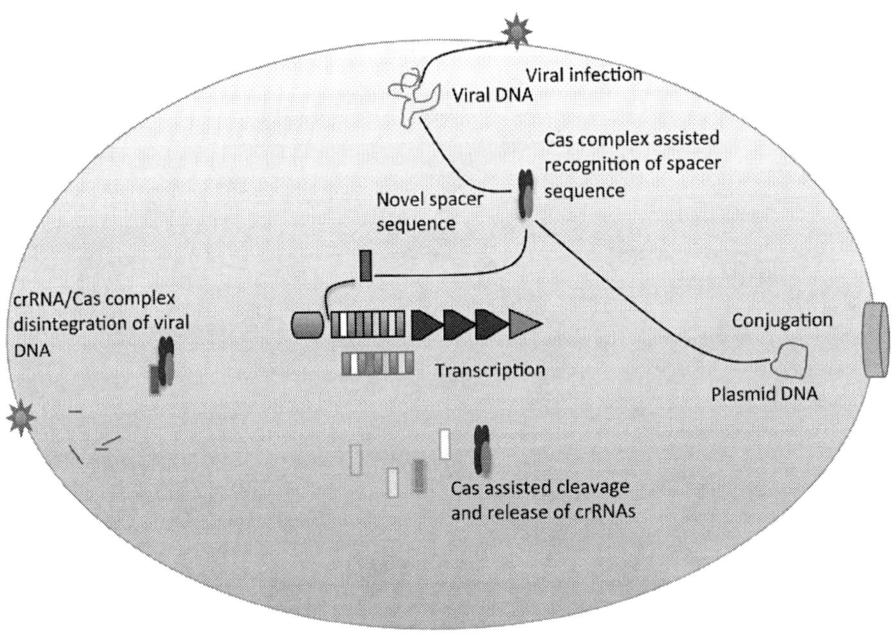

Figure 4. The CRISPR system is a combined exo and endo-siRNA that results in a dynamic, heritable immune system. crRNAs are produced from an endogenous transcript akin to endo-siRNA and diced using Cas proteins, these are analogous but not homologous to the Dicing proteins of the siRNA system. The spacer sequence is acquired from a current infection (top and right), though it isn't clear how rapidly this can be utilised, it would seem that it would be slower than eukaryote exo-siRNA due to the integration step. A major difference between the prokaryote and eukaryote defence against the Dark Arts is that evidence thus far would indicate that cleavage is directed against invading DNA (left), rather than mRNA. A color version of this figure is available at www.landesbioscience.com/curie.

Spacers between the CRISPR repeats are usually from sources external to the organism (e.g., from the phage) and possession of a spacer sequence matching a challenging phage gives resistance to that phage.[46] A suite of conserved proteins involved in DNA transactions was identified as typically occurring close to the CRISPR locus and called *cas* genes—(CRISPR associated). They have functional domains typical of nucleases, polymerases, helicases and nucleotide binding proteins[47] and are required for the acquisition of the spacer sequences. Although the mechanism is speculative, studies have shown that the core *cas* genes Cas1 and Cas2 occur at all CRISPR loci and are involved with acquisition of the spacer sequences. Disruption to these genes results in loss of ability to acquire new spacer sequences, but does not prevent the function of existing ones. Entire CRISPR/*cas* systems can also be found in plasmids,[48] and the same subset can be found in distantly related organisms. This indicates that they can be acquired from plasmids—by the very mechanism that they normally prevent.

Constitutive transcription of the entire repeat/spacer array gives long precursors which are processed into short crRNAs. Each crRNA corresponds to one spacer sequence flanked by two partial repeats.[47] The significance of the flanking sequences is that the resulting crRNAs can recognise 'self' DNA because the ends of the crRNAs will be able to attach by Watson-Crick pairing along their entire length, whereas predatory DNA will result in the crRNA being free at both ends.[49] Disruption of invading target DNA with a self-splicing intron that restores the mRNA, results in loss of resistance in *Staphylococcus epidermidis*, despite having the correct spacer sequence. This indicates that mRNA is not the target.[50] Bidirectional processing was found in the archaeon *Sulfolobus*, suggesting the possibility that RNA duplexes could form and given that archaea contain Argonautes, could this possibly represent a form of siRNA pathway akin to eukaryotes.

RNA cleavage has been demonstrated in vitro in another archaeon *Pyrococcus* via an RNP complex comprised of crRNA and Cas proteins (Cmr1-Cmr6) encoded from a Cas associated module known as RAMP (repeat associated mysterious protein, mainly found in thermophilic archaea).[51] RAMP proteins can be located distally from the CRISPR and do not appear to move with the CRISPR array. DNA cleavage wasn't detected in *P. furiosus*, this may mean that the *Pyrococcus* CRISPR system is directed at mRNA, or directed at the genome of an RNA phage. The spacer sequences in *P. furiosus* have not been identified. If they are aimed at RNA phage, this is not surprising given that so few RNA phage have been sequenced and spacer sequences matching any RNA phage have not (yet) been found. It is possible that species with RAMP encoded Cas proteins have an alternate mechanism of silencing invasive nucleic acid.

For prokaryotes, with both a single center for DNA replication (Ori) and a short life span, it is vital that the genome is small so that replication can be carried out quickly.[52] It would be costly to allow CRISPR arrays to grow unchecked and there is evidence of spacer deletion at the 3′ end of the array.[47] This would allow historic invasions to be forgotten if the cost of maintaining the surveillance was high. There are also higher levels of expression of mature crRNA from the leader end of the transcript (from more recently acquired spacers) and lower levels of distal sequences.[47] Such a strategy ensures protection against current threats, whilst keeping a low level of surveillance for the re-emergence of an old foe. The Dark Arts cannot be so easily subdued. Mutations in a 3-5 proto-spacer adjacent motif (PAM) can prevent proto-spacer sequence recognition, or target sequence mutation can evade crRNA target recognition, phage from *Leptospirillum* can reshuffle 25 nt blocks[53] confounding the slightly larger spacer sequences of the crRNA. And so it goes on. For the convenience of molecular evolutionists, CRISPRs give a historical insight into past host and pathogen encounters!

Thus it seems that all living cells can use some type of small RNA to defend themselves. Furthermore, we expect that there always would have been viruses, so the simplest hypothesis at present is that small RNAs have always been used in Defence Against the Dark Arts. Eukaryotes, bacteria and archaea have all evolved slightly different methods of dealing with parasitic DNA, but the use of RNA is a common thread. The diversity in mechanisms shows evolution has continued, but reinforces that the whole network system is very old.

OTHER REGULATORY RNAs

Our knowledge of noncoding RNA regulated pathways has expanded markedly in the past decade.[54-56] The roles of networks of regulatory and catalytic ribonucleoprotein particles (RNPs) in eukaryotes have been well studied, including classical examples such as the small-nuclear RNPs (snRNPs) in mRNA splicing,[57,58] small-nucleolar RNPs (snoRNPs) in rRNA processing—either the 2′ hydroxyl of ribose is to be methylated, or a uracil converted to pseudouracil,[59] and the RNase P in tRNA processing.[60] Genome wide expression studies of eukaryotic model organisms have shown that over 90% of the genome is transcribed, which suggests that the richness of noncoding regulatory elements extends well beyond what is currently known.[61]

The conserved infrastructural network of RNP-mediated networks in eukaryotic cells indicates a pre-eukaryotic origin for many RNP functions.[62] Both experimental and computational genome-wide mining of noncoding RNAs have revealed that many regulatory RNAs have highly flexible expression patterns.[63,64] An example of such RNAs are snoRNAs that have been found in all eukaryotes and form one of the largest families of noncoding RNAs.[3] They are not restricted to rRNA biogenesis, snoRNAs in the Cajal body ("small Cajal body RNAs"—scaRNAs)[65] modify snRNAs that function in the spliceosome. They have been identified as precursor for smaller RNAs;[66] a human HBII-52 snoRNA (SNORD 115) is processed into smaller RNAs, which regulate alternative splicing of the serotonin-receptor gene,[67] and alternative splicing itself appears very ancient in eukaryotes.[68] Similarly, a C/D box snoRNA from the Epstein-Barr virus is processed into smaller RNAs.[69] Many snoRNAs in modern genomes appear to arise by duplications.[70,71] Large numbers of new snoRNAs are continuously being identified through experimental and bioinformatic screens.[72] All these findings, together with the existence of many orphan snoRNAs without known targets[73] suggest additional network interactions. Identification of snoRNA-derived small RNAs from such a wide range of organisms also implies that some alternative regulatory roles of snoRNAs have an ancient origin. In some eukaryotic lineages with smaller genomes (such as diplomonads and microsporidia) and in archaea, there are fewer annotated snoRNAs, but those that are there appear functional.[74]

Cis-regulatory RNAs have been identified in both eukaryotes and prokaryotes,[3] and classic examples include riboswitches in bacteria and plants,[75] iron-response elements (IREs),[76] and the eukaryotic histone 3′UTR stem-loop.[77] Cis-regulatory RNAs are usually located in the untranslated regions (UTRs) in mRNAs, though some are in coding regions.[78] In association with RNA-binding proteins they regulate translation, splicing, stability and localization. For proteins, there are hundreds of identified RNA-binding protein domains,[79] but the identification of complementary protein-binding motifs in RNA is still at an early stage. An important role of cis-regulatory RNA elements is mRNA localization in eukaryotes. Expression of many eukaryotic RNAs involves packaging of mRNA-protein complexes into transport particles, trafficking of the particles within the cytoplasm and finally translated at their target destination. A few examples are localization of budding yeast ASH1 mRNA to the bud tip,[80] localization of β-actin mRNA to the leading edge of migrating fibroblasts,[81] localization of Nonos mRNA to the posterior end of the *Drosophila* embryo,[82] and localization of human Vimentin mRNA to the perinuclear region of the cytoplasm.[3,83] The noncoding motifs on the 3′-UTRs of the above mRNAs function to bind specific proteins for the transport of the mRNAs. Existence of these RNAs in all three kingdoms of life also hints at this network arising prior to the universal common ancestor (Luca).

HOW OLD ARE THE DIFFERENT INTERACTIONS OF RNA?

In general a good case can be made for many of the general classes of RNA networks to arise relatively early in evolution,[84,85] but we need to outline the options carefully. Current approaches attribute importance to several overlapping earlier stages[86] during the origin of life (see Fig. 5). We start with the familiar world of DNA, proteins and RNA

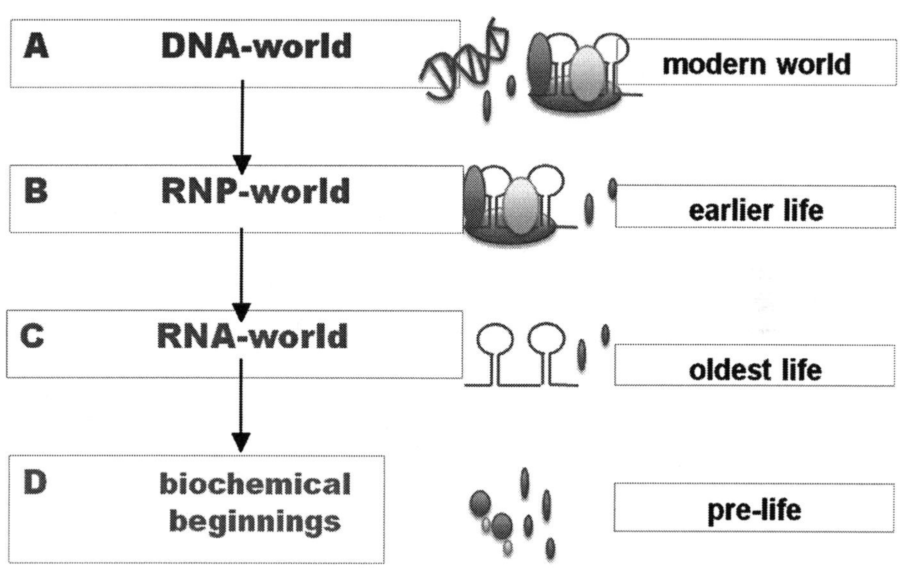

Figure 5. Working backwards through four stages for the origin of life. (A) Our modern world has DNA as the main information storage molecule and its double-stranded structure allows a higher replication accuracy. The standard model suggests an earlier RNA-protein (RNP) world (B) where RNA was the main molecule for genetic information and proteins had the main catalytic role. Many regulatory networks may date back to this time. At a still earlier stage (C), the standard model is an RNA-word where both catalysis and information storage was carried out by RNA—presumably aided by short (noncoded) peptides. The error rate is assumed to be relatively high and thus only relatively short RNAs could be coded. Earlier still there must have been a chemical stage (D) where perhaps autocatalytic cycles were important. It is not specified when membranes (allowing protocells) first arose. A color version of this figure is available at www.landesbioscience.com/curie.

and work backwards to earlier stages. The biological world that we are familiar with has DNA as the main coding molecule and this allows several steps of error checking (against the complementary strand as just one example). The replication accuracy of DNA-based system is much higher than for RNA-coding systems and as far as we can tell, the size of even the larger eukaryote genomes is no longer limited by the accuracy of replication of DNA.

Our modern world (Fig. 5A) has the three familiar groups of macromolecules—proteins, RNA and DNA. On any evolutionary scenario there will have been earlier and simpler worlds and under the standard theory it is predicted that our modern world was preceded (Fig. 5B) by a ribonuclear protein world (RNP) with RNA having the coding roles and proteins doing most of the catalysis. Part of the evidence that DNA evolves last comes from the necessity for protein enzymes (ribonucleotide reductases) that use free radical mechanisms to reduce an OH of ribose (in a ribonucleotide RNA precursor) to a deoxy-ribonucleotide.[87] As far as we know, only proteins (not RNA) can catalyse such complex free radical reactions. In the proposed ribonucleoprotein-world (RNP of Fig. 5B) protein is doing most catalysis (especially of small molecules), leaving RNA with the main coding function and RNA-protein networks would probably have had many regulatory roles.

From our knowledge of modern biochemistry, protein enzymes copying single-stranded RNA are much less accurate and have higher error rates than copying the double stranded DNA molecule. This lower accuracy means that RNA genomes would be much shorter than DNA-based organisms. Indeed, we see today that RNA viruses have much shorter genomes than double-stranded DNA viruses.[88] Effectively there is a positive feedback cycle of increasing fidelity of replication allowing longer coding sequences, which allows a further increase in replication fidelity—we call this the Darwin-Eigen cycle.[89] But the important point here is that from this stage we will have many RNA-protein interactions and so, in principle, some of the RNA-protein networks could have been established at this stage. Or were they lost in Luca and then later re-established—but only in eukaryotes? Losing them and later re-establishing them, seems a less likely hypothesis.

At this point (ribonucleic acids and proteins) we are still on reasonably firm ground; but what comes earlier? Currently the best hypothesis for the earlier stage (Fig. 4C) is that RNA is the first polymer that had the ability to reproduce itself (either directly, or indirectly through a cycle and with or without assistance of short noncoded peptides). It has long been considered[90] that enzyme cofactors with a ribodinucleotide structure (such as NAD, FAD, NADP) were remnants of an early RNA-world. The evidence for this hypothesis of the early role of RNA is increasing.[91] In this proposed RNA-world, RNA would have had coding, catalytic and regulatory roles. An interesting point is that some RNAs and their interactions, appear to date from this proposed RNA-world. RNA is not as an effective catalyst as proteins[85] (probably because proteins form more specific and stable 3D structures) and so this leaves space for improved protein catalysts in going from the second to the third stage of Figure 5. There had to be earlier stages before organised polymers formed (whether RNA, protein, or DNA) and we summarize these stages as 'biochemical beginnings' (Fig. 5D).[1] Basically by definition, we do not expect any macromolecules to have persisted from this stage.

Some classes of functional and regulatory RNA are so widespread that they are accepted as 'universal' and are inferred to have been present in the last common ancestor, Luca. These include rRNA, tRNA, mRNA, RNase P, SRP RNA. A striking discovery was that the catalytic core of the ribosome (forming the amino-acyl bonds linking amino acids into proteins) was composed of RNA. In other words, the ribosome was a ribozyme.[92] Thus the three core aspects of protein synthesis—the messenger, transfer RNA and ribosome—are all RNA molecules that have a network of interactions. Evolutionarily this makes sense if an RNA-world preceded an RNP-world. However there are few convincing theories for the origin of protein synthesis because we need these three classes of RNA before proteins can be synthesized. In an evolutionary context, we cannot evolve something 'because it will be useful in the distant future'—it must have a function 'here and now'. Certainly rRNA, tRNA and mRNA must date back to an RNA-world (before proteins). Sometimes we can define useful subquestions—such as whether the intron/exon structure, with its associated RNAs involved in splicing, is ancestral to all modern eukaryotes.[93] Some RNAs and their interactions may yield progress relatively easily; others will be more difficult for now.

A recent discovery is that tRNAs are part of the proof-reading that increases accuracy,[94] the tRNA(Leu) has catalytic activity in removing a mischarged tRNA(Leu), even though protein does help stabilise some of the intermediates along the catalytic pathway. This means that we need to consider that tRNAs are not passive in amino acid recognition; they are also part of the regulatory network. Although not so well known as the RNAs involved in protein synthesis, the signal recognition particle, 7S in eukaryotes,[95] also appears to occur in all living systems, so this RNA-protein network also appears ancient.

Were these RNA interactions all present in Luca? We will not follow this question here, but have written on it[89] under the names introns 'first', 'early' or 'late'. Because the position of the root of the eukaryote tree is not certain,[96] our strategy[93] has been to search for RNAs that occur in all of the five (or six) deepest lineages of eukaryotes. For example, the intron/exon structure of eukaryote genomes, with its requirement for splicing, appears universal in all modern eukaryotes, so it is expected to have been in the last common eukaryote ancestor (Fred). But that only puts the problem back further. For example, did snRNAs arise before, or after, another milestone—the endosymbiotic event that led to the origin of mitochondria? Details on the relative ages of classes will depend on the models of genome reduction.[97] There are plenty of interesting and fundamental questions for the next decade.

CONCLUSION

Every eukaryote group appears to have networks that handle small RNAs a little differently. However, this is likely a normal evolutionary process creating variants arising from an early ancestral system in Fred. Perhaps it is still premature to decide how the ancestral eukaryote functioned in this respect, but the discovery of different forms of RNA networks in so many groups of eukaryotes has turned the focus back to earlier stages.

Parasitic RNA would have been present in the RNA and RNP-worlds; that is the nature of biology. Thus we expect that at least from the time of Luca there would have been mechanisms to protect the early genome from parasites of various kinds—whether viral or transposon-like. We expect that the Defence Against the Dark Arts idea would have always been a problem for cells since they all require some type of defence mechanism. Although the machinery differs, prokaryotes and eukaryotes have networks with roughly the same function using small RNAs to keep the Dark Arts at bay. It seems likely that the siRNA system was present in Luca as the Argonautes are in all three domains and are almost universally conserved, indicating their age and importance. The endo and piRNA pathways may have evolved to suppress parasitic nucleic acid that escaped or overrun the exo-siRNA pathway.

Increasingly, more RNAs seem to be universal. Bacteria and viruses have many cis-regulatory elements which bind proteins or metabolites (riboswitches), plants and fungi have riboswitches and now it looks like that cis-regulatory elements (in 5′- and 3′-UTRs) are very common in eukaryotes too. Large scale transcription studies in human and mouse at least have uncovered many UTR-derived RNAs, many of which are likely to contain cis-regulatory elements and they may work in a similar way as those in bacteria and virus (binding to specific proteins and having regulatory roles in transcription).

At present we must keep an open mind about how old the RNA and proteins systems are that form the RNA networks. At present, we favour the simplest hypothesis that the RNA networks are, in their basic form, very old. But we must keep testing this hypothesis, looking for small RNA networks involved in regulation in as wide a range of groups as possible. It is a stimulating time for research on RNA networks.

ACKNOWLEDGEMENTS

We thank Lesley Collins for important editing and comments for this chapter.

REFERENCES

1. Yarus M. Getting past the RNA world: The initial Darwinian ancestor. Cold Spring Harbor Perspectives in Biology 2010. doi: 10.1101/cshperspect.a003590.
2. Penny D, Collins LJ. Evolutionary genomics leads the way. In: Caetano-Anolles, ed. Evolutionary Genomics and Systems Biology. Hoboken:Wiley-Blackwell, 2010:3-16.
3. Gardner PP, Daub J, Tate JG et al. Rfam: updates to the RNA families database. Nucl Acids Res 2009; 37:D136-D140.
4. Boria I, Gruber AR, Tanzer A et al. Nematode sbRNAs: homologs of vertebrate Y RNAs. J Mol Evol 2010; 70:346-358.
5. Ghildiyal M, Zamore PD. Small silencing RNAs: an expanding universe. Nat Genet 2009; 10:102-108.
6. Bartel DP, Chen CZ. Micromanagers of gene expression: the potentially widespread influence of metazoan microRNAs. Nat Rev 2004; 5:396-400.
7. Carlsbecker A, Lee JY, Roberts CJ et al. Cell signalling by microRNA165/6 directs gene dose-dependent root cell fate. Nature 2010; 465:316-321.
8. Cullen BR. Viral and cellular messenger RNA targets of viral microRNAs. Nature 2009; 457:421-425.
9. Ouellet DL, Plante I, Landry P et al. Identification of functional microRNAs released through asymmetrical processing of HIV-1 TAR element. Nucl Acids Res 2008; 36:2353-2365.
10. Lin J, Cullen BR. Analysis of the interaction of primate retroviruses with the human RNA interference machinery. J Virol 2007; 81:12218-12226.
11. Dunoyer P, Himber C, Voinnet O. Induction, suppression and requirement of RNA silencing pathways in virulent agrobacterium tumefaciens infections. Nat Genet 2006; 38:258-263.
12. Dong Z, Han MS, Federoff N. The RNA-binding proteins HYL1 and SE promote accurate in vitro processing of pri-miRNA by DCL1. PNAS 2008; 105:9970-9975.
13. Okamura K, Liu N, Lai EC. Distinct mechanisms for microRNA strand selection by *Drosophila* Argonautes. Mol Cell 2009; 36:431-444.
14. Berezikov E, Chung W, Willis J et al. Mammalian mirtron genes. Mol Cell 2007; 28:328-333.
15. Zamore PD, Haley B. Ribo-gnome: the big world of small RNAs. Science 2005; 309:1519-1524.
16. Moazed D. Small RNAs in transcriptional gene silencing and genome defence. Nature 2009; 457:413-420.
17. Carthew RW, Sonheimer EJ. Origins and mechanisms of miRNAs and siRNAs. Cell 2009; 136:642-655.
18. Leung AKL, Sharp PA. MicroRNAs: A safeguard against turmoil? Cell 2007; 130:581-585.

19. Orom UA, Nielsen FC, Lund AH. MicroRNA-10a binds the 5′UTR of ribosomal protein mRNAs and enhances their translation. Mol Cell 2008; 30:460-471.
20. Lee YS, Shibata Y, Malhotra A et al. A novel class of small RNAs: tRNA-derived RNA fragments (tRFs). Genes and Dev 2009; 23:2639-2649.
21. Haussecker D, Huang Y, Lau A et al. Human tRNA-derived small RNAs in the global regulation of RNA silencing. RNA 2010; 16:673-695.
22. Kulkarni M, Ozgur S, Stoecklin G. On track with P-bodies. Biochem Soc Trans 2010; 38:242.
23. Guang S, Bochner AF, Pavelec DM et al. An Argonaute transports siRNAs from the cytoplasm to the nucleus. Science 2008; 321:537-541.
24. Bartel B. MicroRNAs directing siRNA biogenesis. Nat Struct Mol Biol 2005; 12:569-571.
25. Miyoshi K, Miyoshi T, Hartig JV et al. Molecular mechanisms that funnel RNA precursors into endogenous small-interfering RNA and microRNA biogenesis pathways in *Drosophila*. RNA. 2010; 16:506-515.
26. Robine N, Lau NC, Balla S et al. A broadly conserved pathway generates 3′UTR-directed primary piRNAs. Curr Biol 2009; 19:2066-2076.
27. Malone CD, Hannon GJ. Small RNAs as guardians of the genome. Cell 2009; 136:656-668.
28. Li C, Vagin VV, Lee S et al. Collapse of germline piRNAs in the absence of Argonaute3 reveals somatic piRNAs in flies. Cell 2009; 137:509-521.
29. Gottesman S. Micros for microbes: noncoding regulatory RNAs in bacteria. Trends Genet. 2005; 21:399-404.
30. de Nooijer S, Holland BR, Penny D. Eukaryote origins: there was no Garden of Eden? PLoS One 2009; 4:e5507.
31. Boerlijst MC, Hogeweg P. Spiral wave structure in prebiotic evolution—hypercycles stable against parasites. Physica D 1991; 48:17-28.
32. Yigit E, Batista PJ, Bei Y et al. Analysis of the C. elegans Argonaute family reveals that distinct Argonautes act sequentially during RNAi. Cell 2006; 127:747-757.
33. de Vries W, Berkhout B. RNAi suppressors encoded by pathogenic human viruses. Int J Biochem Cell Biol 2008; 40:2007-2012.
34. Parameswaran P, Sklan E, Wilkins C et al. Six RNA viruses and forty-one hosts: viral small RNAs and modulation of small RNA repertoires in vertebrate and invertebrate systems. PLoS Pathogens 2010; 6:e1000764.
35. Ruiz-Ferrer V, Voinnet O. Roles of plant small RNAs in biotic stress responses. Annu Rev Plant Biol 2009; 60:485-510.
36. Lipardi C, Paterson BM. Identification of an RNA-dependent RNA polymerase in *Drosophila* involved in RNAi and transposon suppression. PNAS 2009; 106:15645-15650.
37. Maida Y, Yasukawa M, Furuuchi M et al. An RNA dependent RNA polymerase formed by hTERT and the RNase MRP RNA. Nature 2009; 461:230-235.
38. Watanabe T, Totoki Y, Toyoda A et al. Endogenous siRNAs from naturally formed dsRNAs regulate transcripts in mouse oocytes. Nature 2008; 453:539-543.
39. Hartig JV, Tomari Y, Forstemann K. piRNAs—the ancient hunters of genome invaders. Genes Dev 2007; 21:1707-1713.
40. Brennecke J, Malone CD, Aravin AA et al. An epigenetic role for maternally inherited piRNAs in transposon silencing. Science 2008; 322:1387-1392.
41. Ding S, Voinnet O. Antiviral immunity directed by small RNAs. Cell 2007; 130:413-426.
42. Wang H, Buckley KJ, Yang X et al. Adenosine kinase inhibition and suppression of RNA silencing by geminivirus AL2 and L2 proteins. J Virol 2005; 79:7410-7418.
43. Raja P, Sanville BC, Buchmann RC et al. Viral genome methylation as an epigenetic defence against geminiviruses. J Virol 2008; 82:8997-9007.
44. Jansen R, Embden JD, Gaastra W et al. Identification of genes that are associated with DNA repeats in prokaryotes. Mol Microbiol 2002; 43:1565-1575.
45. Touchan M, Rocha EPC. The small, slow and specialized CRISPR and anti-CRISPR of Escherichia and Salmonella. PloS One 2010; 5(6):e11126.
46. Bolotin A, Quinquis B, Sorokin A et al. Clustered regularly interspaced short palindrome repeats (CRISPRs) have spacers of extrachromosomal origin. Microbiology 2005; 151:2551-2561.
47. Wiedenheft B, Zhou K, Jinek M et al. Structural basis for DNase activity of a conserved protein implicated in CRISPR- mediated genome defense. Structure 2009; 17:904-912.
48. Andersson AF, Banfield JF. Virus population dynamics and acquired virus resistance in natural microbial communities. Science 2008; 320:1047-1050.
49. Marraffini LA, Sontheimer EJ. Self versus nonself discrimination during CRISPR RNA-directed immunity. Nature 2010; 463:568-571.
50. Marraffini LA, Sontheimer EJ. CRISPR interference limits horizontal gene transfer in staphylococci by targeting DNA. Science 2008; 322:1843-1845.

51. Hale C, Kleppe K, Terns RM et al. Prokaryotic silencing (psi)RNAs in Pyrococcus furiosus. RNA 2008; 14:2572-2579.
52. Penny D, Poole AM. Lateral gene transfer: some theoretical aspects. NZ BioScience 2003; 12:32-35.
53. Andersson AF, Banfield JF. Virus population dynamics and acquired virus resistance in natural microbial communities. Science 2008; 320:1047-1050.
54. Licatalosi DD, Darnell RB. RNA processing and its regulation: global insights into biological networks. Nat Rev Genet 2010; 11:75-87.
55. Lioliou E, Romilly C, Romby P et al. RNA-mediated regulation in bacteria: from natural to artificial systems. New Biotech 2010; 27:222-235.
56. Sashital DG, Doudna JA. Structural insights into RNA interference. Curr Opin Struct Biol 2010; 20:90-97.
57. Newman AJ, Nagai K. Structural studies of the spliceosome: blind men and an elephant. Curr Opin Struct Biol 2010; 20:82-89.
58. Staley JP, Woolford JL Jr. Assembly of ribosomes and spliceosomes: complex ribonucleoprotein machines. Curr Opin Cell Biol 2009; 21:109-118.
59. Bachellerie JP, Cavaille J, Huttenhofer A. The expanding snoRNA world. Biochimie 2002; 84:775-790.
60. Lai LB, Vioque A, Kirsebom et al. Unexpected diversity of RNase P, an ancient tRNA processing enzyme: challenges and prospects. FEBS Lett 2010; 584:287-296.
61. Carninci P. RNA dust: where are the genes? DNA Res 2010; 17:51-59.
62. Collins LJ, Penny D. The RNA infrastructure: dark matter of the eukaryotic cell? Trends Genet 2009; 25:120-128.
63. Dieci G, Preti M, Montanini B. Eukaryotic snoRNAs: a paradigm for gene expression flexibility. Genomics 2009; 94:83-88.
64. Slezak-Prochazka I, Durmus S, Kroesen BJ et al. MicroRNAs, macrocontrol: regulation of miRNA processing. RNA 2010; 16:1087-1095.
65. Kiss AM, Jady BE, Darzacq X et al. A Cajal body-specific pseudouridylation guide RNA is composed of two box H/ACA snoRNA-like domains. Nucl Acids Res 2002; 30:4643-4649.
66. Taft RJ, Glazov EA, Lassmann T et al. Small RNAs derived from snoRNAs. RNA 2009; 15:1233-1240.
67. Kishore S, Khanna A, Zhang et al. The snoRNA MBII-52 (SNORD 115) is processed into smaller RNAs and regulates alternative splicing. Hum Mol Genet 2010; 19:1153-1164.
68. Irimia M, Rukov JL, Penny D et al. Functional and evolutionary analysis of alternatively spliced genes is consistent with an early eukaryotic origin of alternative splicing. BMC Evol Biol 2007; 7:188.
69. Hutzinger R, Feederle R, Mrazek J et al. Expression and processing of a small nucleolar RNA from the Epstein-Barr virus genome. PLoS Pathog 2009; 5:e1000547.
70. Chen CL, Liang D, Zhou et al. The high diversity of snoRNAs in plants: identification and comparative study of 120 snoRNA genes from Oryza sativa. Nucl Acids Res 2003; 31:2601-2613.
71. Zemann A, op de Bekke A, Kiefmann M et al. Evolution of small nucleolar RNAs in nematodes. Nucl Acids Res 2006; 34:2676-2685.
72. Raabe CA, Sanchez CP, Randau et al. A global view of the nonprotein-coding transcriptome in Plasmodium falciparum. Nucl Acids Res 2010; 38:608-617.
73. Hertel J, Hofacker IL, Stadler PF. SnoReport: computational identification of snoRNAs with unknown targets. Bioinformatics 2008; 24:158-164.
74. Gardner PP, Bateman A, Poole AM. SnoPatrol: how many snoRNA genes are there? J Biol 2010; 9:4.
75. Henkin TM. Riboswitch RNAs: using RNA to sense cellular metabolism. Genes Dev 2008; 22:3383-3390.
76. Pantopoulos K. Iron metabolism and the IRE/IRP regulatory system: an update. Ann N Y Acad Sci 2004; 1012:1-13.
77. Dominski Z, Marzluff WF. Formation of the 3′ end of histone mRNA: getting closer to the end. Gene 2007; 396:373-390.
78. Nguyen MQ, Zhou Z, Marks CA et al. Prominent roles for odorant receptor coding sequences in allelic exclusion. Cell 2007; 131:1009-1017.
79. Finn RD, Mistry J, Tate J et al. The Pfam protein families database. Nucl Acids Res 2010; 38:D211-D222.
80. Gonsalvez GB, Urbinati CR, Long RM. RNA localization in yeast: moving towards a mechanism. Biol Cell 2005; 97:75-86.
81. Condeelis J, Singer RH. How and why does beta-actin mRNA target? Biol Cell 2005; 97:97-110.
82. Kugler JM, Lasko P. Localization, anchoring and translational control of oskar, gurken, bicoid and nanos mRNA during *Drosophila* oogenesis. Fly 2009; 3:15-28.
83. Bermano G, Shepherd RK, Zehner ZE et al. Perinuclear mRNA localisation by vimentin 3′-untranslated region requires a 100 nucleotide sequence and intermediate filaments. FEBS Lett 2001; 497:77-81.
84. Collins LJ, Chen XS. Ancestral RNA: the RNA biology of the eukaryotic ancestor. RNA Biol 2009; 6:1-8.
85. Jeffares DC, Poole AM, Penny D. Relics from the RNA world. J Mol Evol 1998; 46:18-36.
86. Penny D. An interpretive review of the origin of life research. Biol Philos 2005; 20:633-671.

87. Poole A, Penny D, Sjöberg BM. Confounded Cytosine! Tinkering and the evolution of DNA. Nat Rev Mol Cell Biol 2001; 12:147-151.
88. Duffy S, Shackelton LA, Holmes EC. Rates of evolutionary change in viruses: patterns and determinants. Nat Rev Genet 2008; 9:267-276.
89. Penny D, Hoeppner MP, Poole AM et al. An overview of the introns-first theory. J Mol Evol 2009; 69:527-540.
90. White HB. Coenzymes as fossils of an earlier metabolic state. J Mol Evol 1976; 7:101-104.
91. Yarus M. Life from an RNA World: The Ancestor Within. Cambridge:Harvard University Press, 2010.
92. Steitz TA, Moore PB. RNA, the first macromolecular catalyst: the ribosome is a ribozyme. Trends Biochem Sci 2003; 28:411-418.
93. Collins LJ, Penny D. Complex spliceosomal organization ancestral to extant eukaryotes. Mol Biol Evol 2005; 22:1053-1066.
94. Hagiwara Y, Field MJ, Nureki O et al. Editing mechanism of aminoacyl-tRNA synthetases operates by a hybrid ribozyme/protein catalyst. J Am Chem Soc 2010; 132:2751-2758.
95. Rosenblad MA, Gorodkin J, Knudsen B et al. SRPDB: signal recognition particle database. Nucl Acids Res 2003; 31:363-364.
96. Keeling PJ, Burger G et al. The tree of eukaryotes. Trends Ecol Evol 2005; 20:670-676.
97. Poole AM, Penny D. Lateral gene transfer, some theoretical aspects. NZ BioScience 2003:32-35.

INDEX

Symbols

A

B

C

D

E

G

H

I

L

M

N

P

R

S

T

U

V

X

Y

Z